Bioreactors for Stem Cell Expansion and Differentiation

Gene and Cell Therapy Series

Series Editors
Anthony Atala & Graça Almeida-Porada

Published Titles

Placenta: The Tree of Life
edited by Ornella Parolini

Cellular Therapy for Neurological Injury
edited by Charles S. Cox, Jr

Regenerative Medicine Technology: On-a-Chip Applications for Disease
Modeling, Drug Discovery and Personalized Medicine
edited by Sean V. Murphy and Anthony Atala

Therapeutic Applications of Adenoviruses
edited by Philip Ng and Nicola Brunetti-Pierri

Gene and Cell Delivery for Invertebral Disk Degeneration
edited by Raquel Madeira Gonçalves and Mario Adolfo Barbosa

Bioreactors for Stem Cell Expansion and Differentiation
edited by Joaquim M. S. Cabral and Cláudia Lobato da Silva

For more information about this series, please visit:
https://www.crcpress.com/Gene-and-Cell-Therapy/book-series/
CRCGENCELTHE

Bioreactors for Stem Cell Expansion and Differentiation

Edited by

Joaquim M. S. Cabral
Instituto Superior Técnico, Universidade de
Lisboa, Lisboa, Portugal

Cláudia Lobato da Silva
Instituto Superior Técnico, Universidade de
Lisboa, Lisboa, Portugal

CRC Press
Taylor & Francis Group
Boca Raton London New York

CRC Press is an imprint of the
Taylor & Francis Group, an **informa** business

CRC Press
Taylor & Francis Group
6000 Broken Sound Parkway NW, Suite 300
Boca Raton, FL 33487-2742

First issued in paperback 2020

© 2019 by Taylor & Francis Group, LLC
CRC Press is an imprint of Taylor & Francis Group, an Informa business

No claim to original U.S. Government works

ISBN-13: 978-1-4987-9590-6 (hbk)
ISBN-13: 978-0-367-73293-6 (pbk)

Visit the Taylor & Francis Web site at
http://www.taylorandfrancis.com

and the CRC Press Web site at
http://www.crcpress.com

Contents

Series Preface

Gene and cell therapies have evolved in the past several decades from a conceptual promise to a new paradigm of therapeutics, able to provide effective treatments for a broad range of diseases and disorders that previously had no possibility of cure.

The fast pace of advances in the cutting-edge science of gene and cell therapy, and supporting disciplines ranging from basic research discoveries to clinical applications, requires an in-depth coverage of information in a timely fashion. Each book in this series is designed to provide the reader with the latest scientific developments in the specialized fields of gene and cell therapy, delivered directly from experts who are pushing forward the boundaries of science.

In this volume of the Gene and Cell Therapy book series, *Bioreactors for Stem Cell Expansion and Differentiation*, the editors have assembled a remarkable team of outstanding investigators and clinicians, who are experts in stem cell biology, 2D and 3D stem cell culture, large-scale expansion, designer bioreactors and microcarriers, sensing, monitoring, and modeling. The chapters provide cutting-edge information on next-generation systems used for expansion and/or differentiation of stem cells from different sources, at a production scale able to meet the needs of the fields of regenerative medicine and high-throughput screening. In addition, the authors offer novel insights into bioreactor-based culture systems specific for tissue engineering, and discuss sophisticated and cost-effective manufacturing strategies that are able to comply with good manufacturing practices, geared to overcoming technological shortcomings that currently preclude advancing towards product commercialization.

This book brings together a team of outstanding investigators presenting thought-provoking articles in the aforementioned fields, and provides a breadth of knowledge across disciplines, giving the reader a full understanding of the state-of-art and a sneak preview into the future of these fields.

We would like to thank the volume editors, Cláudia Lobato da Silva and Joaquim M. S. Cabral, and the authors, all of whom are remarkable experts, for their valuable contributions. We would also like to thank our senior acquisitions editor, C.R. Crumly, and the CRC Press staff for all their efforts and dedication to the Gene and Cell Therapy book series.

Anthony Atala
Wake Forest Institute for Regenerative Medicine

and

Graça Almeida-Porada
Wake Forest Institute for Regenerative Medicine

Preface

Stem cells have the capacity to self-renew (e.g., make copies of themselves) and to differentiate into multiple specialized cell types, being an attractive cell source for cellular therapies, tissue engineering, and toxicological studies. The successful implementation of stem cell-based therapies and tissue engineering approaches will require the reproducible production of well-characterized and functional cells or tissues under highly controlled culture conditions.

Stem cell cultivation has been performed, at a research level, in traditional culture flasks under static conditions, which are limited in terms of cell productivity, their non-homogeneous nature resulting in concentration gradients (pH, oxygen, nutrients, metabolites), difficulty of monitoring and control, and extensive handling is required for feeding/harvesting procedures.

Bioreactors have been successfully employed in the traditional pharmaceutical industry for decades, allowing a robust and cost-effective production of a wide range of drugs, and can represent a suitable alternative to overcome the limitations of static systems for stem cell cultivation. In addition, the knowledge in bioprocessing science gathered from these fully characterized pharmaceutical processes has the potential to be translated into the challenging specifications of stem cell culture. Indeed, the design of bioreactors targeting the expansion (i.e., proliferation under an undifferentiated state) of stem cells or their differentiation into a specific cell/tissue must take into consideration the challenging specificities of stem cell cultivation, namely: (1) the complexity of non-homogeneous stem cell populations; (2) the occurrence of multiple cell-to-cell and matrix-to-cell interactions; (3) the suspension/adherent nature of the cells; and (4) the market approach (autologous (patient-specific) *versus* allogeneic (donor-based)), and the ability to comply to Good Manufacturing Practices (GMP), featuring an integrated quality assurance system in order to obtain the authorization from regulatory agencies towards the production of a safe and efficient medicinal product.

Different types of bioreactors have been developed for stem cell culture, which will be depicted in this volume of the Cell and Gene Therapy book series entitled *Bioreactors for Stem Cell Expansion and Differentiation*.

Chapter 1 by Elena F. Jacobson and Emmanuel S. Tzanakakis, covers the aspects relevant to the cultivation of human pluripotent stem cells (PSCs) as aggregates, depicting the requirements of human PSC pluripotency and genomic stability, various bioreactor designs for three-dimensional (3D) stem cell cultivation, different culture media, and feeding strategies. Carlos A.V. Rodrigues and co-workers provide, in Chapter 2, a review on the developments in human PSC culture in bioreactor-based systems, particularly focusing on 3D cell culture and bioreactor configurations and the critical parameters for a successful and efficient expansion and differentiation of human PSC into cells of the three germ layers, including the generation of organoids. Chapter 3, written by Mark C. Allenby and colleagues, summarizes the *state of the art* of *ex vivo* methodologies and culture systems for the manufacturing of erythroid cells from human PSC (including embryonic stem cells (ESCs) and induced pluripotent stem cells (iPSCs)), discussing some of the issues to be addressed to facilitate translation into a clinical-grade product. Chapter 4, by Maria João Sebastião, Bernardo Abecasis and collaborators, provides a review on the methodologies currently being applied for the isolation, culture, and characterization of cardiac stem/progenitor cells (CPCs), highlighting the use of bioreactor systems for the expansion of CPCs in 3D culture strategies and the available analytical tools for cell-based product and process characterization. Arindom Sen and colleagues, in Chapter 5, focus on the development of robust protocols for the large-scale expansion and differentiation of human neural progenitor cell (NPC) populations for the treatment of neurodegenerative disorders. It describes the specificities of human NPC handling and characterization methods, the large-scale expansion of human NPC populations in suspension bioreactors, and animal model studies where differentiated cells, derived from bioreactor-generated human NPCs, were transplanted to assess their therapeutic potential for Huntington's disease and spinal cord pain. Chapter 6, written by Aletta Schnitzler and co-workers, focuses on the bioprocessing of human stem cells for therapeutic use through single-use bioreactors combined with microcarrier technology. Particularly for the manufacturing of human mesenchymal stem/stromal cells (MSCs), the authors also describe strategies for the downstream processing (such as, cell harvest from the microcarriers) and include a case study describing the expansion of human MSCs in a 50-L closed, single-use stirred-tank bioreactor, finalizing with some regulatory perspectives for the manufacturing of cell-based products. Chapter 7, by Marta H.G. Costa and co-workers, provides a comprehensive review focused on bioreactor systems for the cultivation of hematopoietic stem and progenitor cells (HSPC), describing several culture parameters and tailored bioreactor configurations that could contribute towards the establishment of a robust and scalable platform for the manufacture of HSPC. Chapter 8, by Sunghoon Jung and colleagues, reviews the bioprocessing challenges of human MSC manufacturing for large-scale allogeneic products, specifically focusing on the upstream process and including different bioreactor configurations. Also, in the context of cell manufacturing, Ioannis Papantoniou and collaborators in Chapter 9 present several sensing and monitoring tools available for bioreactor implementation. In addition, a discussion on the use of time-series data from several sensors, which could allow predictive model-based control and, thus, increasing the robustness of bioreactor operation, is included.

Besides the pivotal role of bioreactors in stem cell manufacturing, bioreactor technology is also a powerful tool for the successful implementation of tissue engineering approaches. Overall, besides allowing a controlled system for the cultivation of different cell types, bioreactors can be used to provide a set of tissue-specific physiological stimuli *in vitro* for proper tissue maturation. In this context, Ana Gonçalves and co-workers, in Chapter 10, advocate the use of bioreactor systems as attractive tools to provide biomechanical signaling to cell/tissue constructs, under closely monitored and tightly controlled environments. Specifically, this chapter provides a comprehensive review on the field of tendon tissue engineering, discussing the role of biomechanical stimulation in tendons and the most frequently used bioreactor systems. Also, in the context of tissue engineering, Chapter 11 written by Sara Morini and colleagues, describes the most used 2D and 3D liver models employed in liver tissue engineering, while highlighting the importance of bioreactor systems in the field and concluding with a focus on the regulatory and technological challenges related to liver bioengineering.

In this book, the major advances and challenges focusing on the use of bioreactors in cell therapy manufacturing and tissue engineering settings are presented. Research efforts must continue towards the development of bioreactors for stem cell expansion and differentiation, with a translational focus and under a GMP-guided research, to overcome the foreseen technological barriers when these therapies reach the commercialization stage.

About the Editors

Joaquim M. S. Cabral is Professor of Biological Engineering at Instituto Superior Técnico (IST), Universidade de Lisboa, Portugal. He is the Head of Department of Bioengineering and Director of the iBB—Institute for Bioengineering and Biosciences at IST. He obtained his diploma in Chemical Engineering from Instituto Superior Técnico in 1976. His PhD and "Habilitation" degrees in Biochemical Engineering were obtained from IST in 1982 and 1988, respectively. After a post-doctoral study at Massachusetts Institute of Technology in 1983–1984, He joined the IST faculty as an Assistant Professor and became a full Professor in 1992.

Joaquim Cabral's research aims at contributing to the development of biochemical engineering science through novel developments in bioprocess engineering and stem cell bioengineering, through the integration of molecular and cell biology in engineering. He has published over 400 research papers, and co-authored or edited seven books. His research interests at the Stem Cell Engineering Research Group (SCERG) of iBB—Institute for Bioengineering and Biosciences, IST are focused on stem cells research for tissue engineering and regenerative medicine, stem cell bioprocessing and manufacturing, and the development of novel stem cell bioreactors and advanced bioseparation and purification processes. He serves on several editorial boards of scientific journals in the areas of biotechnology and stem cells and tissue engineering.

Cláudia Lobato da Silva, born in Lisboa, Portugal, received a diploma in Chemical Engineering (Biotechnology), Instituto Superior Técnico (IST), Technical University of Lisboa (Universidade Técnica de Lisboa, UTL), Portugal, in 2001, and got a PhD in Biotechnology in 2006 at IST-UTL in collaboration with the University of Nevada, Reno, USA. She is Assistant Professor (with Tenure) at the Department of Bioengineering, IST, Universidade de Lisboa, Lisboa.

Cláudia Lobato da Silva's research interests at the Stem Cell Engineering Research Group (SCERG) of iBB—Institute for Bioengineering and Biosciences, IST, include expansion of human stem cells, namely hematopoietic and mesenchymal stem/stromal cells, cellular therapies with human adult stem cells, isolation and purification of stem cells, and bioreactors for stem cell culture. She is co-author of more than 65 scientific articles published in international peer-reviewed journals and 12 book chapters. Presently, she is Member of the Pedagogical Council, Commission for the Quality of Curricular Units, at IST, and serves on the editorial board of *BMC Biotechnology* as associate editor. Since 2015, She is the Vice-President of the Portuguese Society of Stem Cells and Cellular Therapies (Sociedade Portuguesa de Células Estaminais e Terapias Celulares (SPCE-TC)).

Contributors

Bernardo Abecasis
Instituto de Tecnologia Química e
  Biológica António Xavier
Universidade Nova de Lisboa
and
iBET
Instituto de Biologia Experimental e
  Tecnológica
Oeiras, Portugal

Jean-Marie Aerts
Prometheus, Division of Skeletal
  Tissue Engineering
KU Leuven
and
M3-BIORES
KU Leuven
Leuven, Belgium

Mark C. Allenby
Biological Systems Engineering
  Laboratory
Department of Chemical
  Engineering
London, United Kingdom

Helen Almeida
Instituto de Investigación Sanitária de
  Aragón (IIS Aragón)
Zaragoza, Spain

Joana I. Almeida
Instituto de Investigación Sanitária de
  Aragón (IIS Aragón)
Zaragoza, Spain

Manuel Almeida
Instituto de Investigación Sanitária de
  Aragón (IIS Aragón)
Zaragoza, Spain

Paula M. Alves
Instituto de Tecnologia Química e
  Biológica António Xavier
Universidade Nova de Lisboa
and
iBET
Instituto de Biologia Experimental e
  Tecnológica
Oeiras, Portugal

Janmeet Anant
EMD Millipore
Bedford, Massachusetts

Pilar Sainz Arnal
Instituto de Investigación Sanitária de
  Aragón (IIS Aragón)
and
Instituto Aragonés de Ciencias de la
  Salud (IACS)
Zaragoza, Spain

Manjula Aysola
EMD Millipore
Bedford, Massachusetts

Behnam A. Baghbaderani
Cell Therapy Development
Emerging Technologies
Lonza Walkersville, Inc.
Walkersville, Maryland

Pedro M. Baptista
Instituto de Investigación Sanitária
    de Aragón (IIS Aragón)
and
Center for Biomedical Research
    Network Liver and Digestive
    Diseases (CIBERehd)
Zaragoza, Spain
and
Instituto de Investigación Sanitaria
    de la Fundación Jiménez Díaz
and
Biomedical and Aerospace
    Engineering Department
Universidad Carlos III de Madrid
Madrid, Spain
and
Fundacion ARAID
Zaragoza, Spain

Leo A. Behie
Canada Research Chair in
    Biomedical Engineering
    (Emeritus)
Schulich School of Engineering
University of Calgary
Calgary, Alberta, Canada

Dominika Berdecka
3B's Research Group—Biomaterials,
    Biodegradables and Biomimetics
University of Minho
Headquarters of the European
    Institute of Excellence on Tissue
    Engineering and Regenerative
    Medicine
and
ICVS/3B's—PT Government
    Associate Laboratory
Guimarães, Portugal

Mariana Branco
3B's Research Group—Biomaterials,
    Biodegradables and Biomimetics
University of Minho
Headquarters of the European Institute
    of Excellence on Tissue Engineering
    and Regenerative Medicine
Guimarães, Portugal

Joaquim M. S. Cabral
Department of Bioengineering and
    iBB – Institute of Bioengineering
    and Biosciences
Instituto Superior Técnico
Universidade de Lisboa
and
The Discoveries Centre for
    Regenerative and Precision Medicine
Lisbon Campus
Instituto Superior Técnico
Universidade de Lisboa
Lisboa, Portugal

Manuel J.T. Carrondo
iBET
Instituto de Biologia Experimental e
    Tecnológica
Oeiras, Portugal
and
Departamento de Química
Faculdade de Ciências e Tecnologia
Universidade Nova de Lisboa
Monte da Caparica, Portugal

Marta H.G. Costa
Department of Bioengineering and
    iBB – Institute of Bioengineering
    and Biosciences
Instituto Superior Técnico
Universidade de Lisboa
Lisboa, Portugal

Cláudia Lobato da Silva
Department of Bioengineering and
  iBB – Institute of Bioengineering
  and Biosciences
Instituto Superior Técnico
Universidade de Lisboa
and
The Discoveries Centre for
  Regenerative and Precision Medicine
Lisbon Campus
Instituto Superior Técnico
Universidade de Lisboa
Lisboa, Portugal

Pablo Royo Dachary
Liver Transplant Unit
Surgery Department
Lozano Blesa University Hospital
Zaragoza, Spain

Sébastien de Bournonville
Prometheus, Division of Skeletal Tissue
  Engineering
KU Leuven
Leuven, Belgium
and
Biomechanics Research Unit
GIGA-R In Silico Medicine
Université de Liège
Liege, Belgium

Maria Margarida Diogo
Department of Bioengineering and
  iBB – Institute of Bioengineering
  and Biosciences
Instituto Superior Técnico
Universidade de Lisboa
and
The Discoveries Centre for
  Regenerative and Precision Medicine
Lisbon Campus
Instituto Superior Técnico
Universidade de Lisboa
Lisboa, Portugal

Frederico Castelo Ferreira
Department of Bioengineering and
  iBB – Institute of Bioengineering
  and Biosciences
Instituto Superior Técnico
Universidade de Lisboa
and
The Discoveries Centre for
  Regenerative and Precision Medicine
Lisbon Campus
Instituto Superior Técnico
Universidade de Lisboa
Lisboa, Portugal

Agustín García Gil
Instituto de Investigación Sanitária de
  Aragón (IIS Aragón)
and
Liver Transplant Unit
Surgery Department
Lozano Blesa University Hospital
Zaragoza, Spain

Liesbet Geris
Prometheus, Division of Skeletal Tissue
  Engineering
KU Leuven
and
Skeletal Biology and Engineering
  Research Center
KU Leuven
and
M3-BIORES
KU Leuven
Leuven, Belgium
and
Biomechanics Research Unit
GIGA-R In Silico Medicine
Université de Liège
Liege, Belgium
and
Biomechanics Section
KU Leuven
Leuven, Belgium

Ana Rita Gomes
3B's Research Group—Biomaterials,
    Biodegradables and Biomimetics
University of Minho
Headquarters of the European Institute
    of Excellence on Tissue Engineering
    and Regenerative Medicine
Guimarães, Portugal

Manuela E. Gomes
3B's Research Group—Biomaterials,
    Biodegradables and Biomimetics
University of Minho
Headquarters of the European Institute
    of Excellence on Tissue Engineering
    and Regenerative Medicine
and
ICVS/3B's—PT Government Associate
    Laboratory
and
The Discoveries Centre for
    Regenerative and Precision Medicine
Headquarters at University of Minho
Guimarães, Portugal

Patricia Gomes-Alves
Instituto de Tecnologia Química e
    Biológica António Xavier
Universidade Nova de Lisboa
and
iBET
Instituto de Biologia Experimental e
    Tecnológica
Oeiras, Portugal

Ana I. Gonçalves
3B's Research Group—Biomaterials,
    Biodegradables and Biomimetics
University of Minho
Headquarters of the European Institute
    of Excellence on Tissue Engineering
    and Regenerative Medicine
and
ICVS/3B's—PT Government Associate
    Laboratory
Guimarães, Portugal

Priyanka Gupta
Prometheus, Division of Skeletal
    Tissue Engineering
KU Leuven
and
Skeletal Biology and Engineering
    Research Center
KU Leuven
Leuven, Belgium

Elena F. Jacobson
Department of Chemical and
    Biological Engineering
Tufts University
Medford, Massachusetts

Sunghoon Jung
Bioprocess Research &
    Development
PBS Biotech, Inc.
Camarillo, California

Michael S. Kallos
Department of Chemical and
    Petroleum Engineering
Schulich School of Engineering
University of Calgary
and
Biomedical Engineering Graduate
    Program
University of Calgary
Calgary, Alberta, Canada

Mark Lalli
EMD Millipore
Bedford, Massachusetts

Toon Lambrechts
Prometheus, Division of Skeletal
    Tissue Engineering
KU Leuven
and
M3-BIORES
KU Leuven
Leuven, Belgium

Brian Lee
PBS Biotech, Inc.
Camarillo, California

Sara Llorente
Instituto de Investigación Sanitária de
    Aragón (IIS Aragón)
and
Liver Transplant Unit, Gastroenterology
    Department
Lozano Blesa University Hospital
Zaragoza, Spain

Niki Loverdou
Prometheus, Division of Skeletal Tissue
    Engineering
KU Leuven
Leuven, Belgium
and
Biomechanics Research Unit
GIGA-R In Silico Medicine
Université de Liège
Liege, Belgium

Alberto Lue
Instituto de Investigación Sanitária de
    Aragón (IIS Aragón)
and
Liver Transplant Unit, Gastroenterology
    Department
Lozano Blesa University Hospital
Zaragoza, Spain

Athanasios Mantalaris
Biological Systems Engineering
    Laboratory
Department of Chemical Engineering
London, United Kingdom

Ivar Mendez
Department of Surgery and Royal
    University Hospital
University of Saskatchewan and
    Saskatoon Health Region
Saskatoon, Saskatchewan, Canada

Sara Morini
Instituto de Investigación Sanitária
    de Aragón (IIS Aragón)
Zaragoza, Spain
and
Department of Bioengineering
    and iBB-Institute for
    Bioengineering and
    Biosciences
Instituto Superior Técnico
Universidade de Lisboa
Lisboa, Portugal

Julie Murrell
EMD Millipore
Bedford, Massachusetts

Diogo E.S. Nogueira
3B's Research Group—Biomaterials,
    Biodegradables and Biomimetics
University of Minho
Headquarters of the European
    Institute of Excellence on
    Tissue Engineering and
    Regenerative Medicine
Guimarães, Portugal

Iris Plá Palacín
Instituto de Investigación
    Sanitária de Aragón (IIS
    Aragón)
Zaragoza, Spain

Krishna M. Panchalingam
Cell Therapy Process
    Development
Lonza Walkersville, Inc.
Walkersville, Maryland

Nicki Panoskaltsis
Department of Hematology
Imperial College London
London, United Kingdom

Ioannis Papantoniou
Prometheus, Division of Skeletal Tissue
    Engineering
KU Leuven
and
Skeletal Biology and Engineering
    Research Center
KU Leuven
Leuven, Belgium

Rui L. Reis
3B's Research Group—Biomaterials,
    Biodegradables and Biomimetics
University of Minho
Headquarters of the European Institute
    of Excellence on Tissue Engineering
    and Regenerative Medicine
and
ICVS/3B's—PT Government Associate
    Laboratory
and
The Discoveries Centre for
    Regenerative and Precision Medicine
Headquarters at University of Minho
Guimarães, Portugal

Carlos A.V. Rodrigues
Department of Bioengineering and
    iBB – Institute of Bioengineering
    and Biosciences
Instituto Superior Técnico
Universidade de Lisboa
and
The Discoveries Centre for
    Regenerative and Precision Medicine
Lisbon Campus
Instituto Superior Técnico
Universidade de Lisboa
Lisboa, Portugal

Márcia T. Rodrigues
3B's Research Group—Biomaterials,
    Biodegradables and Biomimetics
University of Minho
Headquarters of the European
    Institute of Excellence on
    Tissue Engineering and
    Regenerative Medicine
and
ICVS/3B's—PT Government
    Associate Laboratory
Guimarães, Portugal

Natalia Sánchez-Romero
Instituto de Investigación Sanitária
    de Aragón (IIS Aragón)
Zaragoza, Spain

Susana Brito dos Santos
Biological Systems Engineering
    Laboratory
Department of Chemical
    Engineering
London, United Kingdom

Aletta C. Schnitzler
EMD Millipore
Bedford, Massachusetts

Maria João Sebastião
Instituto de Tecnologia Química e
    Biológica António Xavier
Universidade Nova de Lisboa
and
iBET
Instituto de Biologia Experimental e
    Tecnológica
Oeiras, Portugal

Arindom Sen
Department of Chemical and Petroleum
  Engineering
Schulich School of Engineering
University of Calgary
and
Biomedical Engineering Graduate
  Program
University of Calgary
Calgary, Alberta, Canada

Margarida Serra
Instituto de Tecnologia Química e
  Biológica António Xavier
Universidade Nova de Lisboa
and
iBET
Instituto de Biologia Experimental e
  Tecnológica
Oeiras, Portugal

Trinidad Serrano-Aulló
Instituto de Investigación Sanitária de
  Aragón (IIS Aragón)
and
Liver Transplant Unit
Gastroenterology Department
Lozano Blesa University Hospital
Zaragoza, Spain

Teresa P. Silva
3B's Research Group—Biomaterials,
  Biodegradables and Biomimetics
University of Minho
Headquarters of the European Institute
  of Excellence on Tissue Engineering
  and Regenerative Medicine
Guimarães, Portugal

Emmanuel S. Tzanakakis
Department of Chemical and Biological
  Engineering
Tufts University
Medford, Massachusetts
and
Tufts Clinical and Translational Science
  Institute
Tufts Medical Center
Boston, Massachusetts

Laurens Verscheijden
Instituto de Investigación Sanitária de
  Aragón (IIS Aragón)
Zaragoza, Spain

1 Large-Scale Culture of 3D Aggregates of Human Pluripotent Stem Cells

Elena F. Jacobson and Emmanuel S. Tzanakakis

CONTENTS

1.1 INTRODUCTION

As more therapies involving cells derived from human pluripotent stem cells (PSCs) are pursued, the need to produce clinically relevant quantities of stem cells in a scalable and economical manner becomes more pressing. Large-scale adherent, two-dimensional (2D) culture systems, including roller bottles, Nunc™ Cell Factory™ Systems, Pall Xpansion® Multiplate Bioreactor System, Corning® CellSTACK® Cell Culture Chambers, HYPERStack®, and CellCube® Module (Merten 2015) have been utilized, but stirred-suspension bioreactors (SSBs) remain ubiquitous in the biopharmaceutical industry for the cultivation of diverse cell types. Static culture modalities have several limitations, including difficulties in monitoring and controlling culture parameters (e.g., pH, temperature, dissolved O_2 [DO]), being labor intensive and

1

inherently lower surface-to-culture volume ratios compared to SSBs. In addition, concentration gradients are more pronounced in 2D cultures, contributing to significant batch-to-batch variability and adversely affecting cell growth and other specific cell characteristics.

Importantly, platforms for human PSC cultivation developed around the SSBs since they would be easier to translate from laboratory to commercial production than entirely novel designs. Several efforts have been reported for the expansion and differentiation of human PSCs in SSBs, owing to the advantages of these bioreactors, including their scalability and the continuous control of the culture environment. Human PSCs, which include human induced pluripotent stem cells (iPSCs) and embryonic stem cells (ESCs), can be cultured in SSBs as aggregates, after encapsulation (e.g., alginate beads or biomatrices) or adherent on microcarriers. In this chapter, we will focus on the large-scale culture of human PSC aggregates, including requirements for large-scale stem cell culture, cell aggregate culture systems, primary culture variables/parameters, types of culture media, and feeding strategies.

1.2 REQUIREMENTS FOR LARGE-SCALE STEM CELL EXPANSION AND DIFFERENTIATION

The bioprocesses envisioned to produce human PSC-based therapeutics broadly involve two stages: (1) expansion of human PSCs while preserving their pluripotency; and, once a prescribed cell quantity is reached, the cells are subjected to (2) differentiation toward a desired cell or tissue type. From a bioprocess economics viewpoint, integration of these two steps is desirable. Large-scale expansion of human PSCs requires several rounds of mitosis in an artificial milieu. This increases the risk for aberrant loss of stem cell pluripotency during expansion, and the emergence and accumulation of genomic irregularities (Garitaonandia et al. 2015, Prakash Bangalore et al. 2017). Thus, it is essential to have close surveillance of genotype and phenotype of cultured human PSCs.

1.2.1 Testing for Pluripotency

Various methods are available for checking the pluripotency of human PSCs, including immunostaining, quantitative polymerase chain reaction (qPCR), microarrays, western blotting, and flow cytometry. Most often, cultured human PSCs are tested for the expression of markers, including NANOG, POU5F1 (also known as Oct4), SOX2, TRA-1-60, TRA-1-81, SSEA3, and SSEA4, characteristic of the pluripotent state. Typically, the marker levels are compared to those of human PSCs in a reference condition (e.g., dish cultures). These assays are relatively straightforward to run, although the results can be affected by antibody or oligonucleotide primer quality, the cell sample, and cell/DNA/RNA/protein preparation and processing. Also, some pluripotent markers are more persistent during the onset and progression of stem cell differentiation, necessitating caution when interpreting the results. For example, the detection of the Oct4A, but not the Oct4B isoform, is proper for assessing stemness, given that the latter isoform is expressed in non-PSCs, including peripheral blood mononuclear cells (Kotoula et al. 2008).

While the detection of pluripotency markers is simple, functional assays are preferable. The latter are performed based on the cells' ability to differentiate into all three germ layers: endoderm, ectoderm, and mesoderm. This is typically done via subcutaneous injection of the cells into immunocompromised mice. If the cells are pluripotent, they form tumors, termed teratomas, comprising diverse cell types. Alternatively, this assay can be carried out by transplanting cells onto the chorioallantoic membrane of avian/chicken embryos. Teratoma formation is considered the gold standard for demonstrating the pluripotent state of human PSCs, but teratocarcinoma cell lines and genetically abnormal human ESCs and iPSCs can express pertinent teratoma markers (TRA-1-60, DNMT3B, and REX1) at comparable levels to normal human PSCs (Chan et al. 2009). Thus, teratoma formation indicates the ability of human PSCs for tri-lineage differentiation but does not preclude the presence of genomic anomalies (see 1.2.2). The lengthy nature of the protocol (a few weeks between cell injection and teratoma formation) also makes streamlining of the assay difficult, particularly for large-scale human PSC cultivation. Furthermore, the lack of assay standardization between studies, including the number of injected cells (200–5 million), the site of injection, the type of immunodeficiency and strain of the mouse, the passage number of injected cells, cell harvest method, and histomorphological analysis, makes inter-study comparison difficult (Muller et al. 2010).

An alternative method for testing the ability of cells to differentiate into all three germ layers is the formation of embryoid bodies (EBs) *in vitro* followed by the detection of genes indicating multilineage commitment. While PSC specification within EBs can be biased by culture manipulations and the reagents used (e.g., serum), this is a faster and less expensive way to determine if human PSCs retain their ability to give rise to various progeny *in vitro*. In contrast to EB differentiation, which is largely random, directed differentiation of human PSCs into mesoderm, endoderm, and ectoderm and the creation of functional cells (e.g., beating cardiomyocytes [Lian et al. 2013]) can also be used to verify the potential of human PSCs for multi-lineage specification.

Another piece of evidence regarding the pluripotent state of human PSCs can be provided through the analysis of their epigenetic state. Epigenetics are heritable gene expression modifications that do not involve DNA sequence changes. Some epigenetic changes include DNA methylation, which is the addition of a methyl group to the DNA to turn the gene off, and modifications of the proteins (histones) that package and order the DNA. Modifications in any of the five histone proteins (H1–H5) can cause transcriptional activation or repression. Chromatin is made of DNA and a histone and in pluripotent cells chromatin is very dynamic and rearranges, which changes the accessibility of the DNA for replication and transcription. On the other hand, differentiation of stem cells leads to a more structured, condensed, and heterochromatic genome. Generally, in PSCs there is a change from acetylated histone H3 and H4 to increased global levels of trimethylated lysine-9 H3, which leads to gene inactivation when stem cells start to differentiate. A specific example is that the active regions of the NANOG and Oct3/4 promoters are enriched for acetylation of H4 and trimethylated lysine-4 of H3 in mouse ESCs while these modifications are absent in the trophectoderm, and they instead enrich methylated lysine-9 of H3 (Kimura et al. 2004, Lee et al. 2004, Atkinson and Armstrong 2008).

Overall, teratoma formation, epigenetic footprint analysis, EB culture, directed *in vitro* differentiation, and pluripotency marker expression represent a battery of assays used to demonstrate the pluripotent nature of cultured human PSCs. However, such assays do not prove that cells can indeed give rise to different tissues and organs in a developing embryo. This can be achieved by injecting the PSCs into murine embryos and analyzing chimeric litters. Chimera formation, which cannot be demonstrated for human PSC due to obvious ethical reasons, could confirm that PSCs are capable of maturing into germline cells and contributing to the normal development of whole embryos (Buta et al. 2013).

1.2.2 TESTING GENOMIC INSTABILITY, KARYOTYPE, EPIGENOMIC CHANGES

Genomic abnormalities of cultured human PSCs can be assessed by molecular profiling microarrays (Ivanova et al. 2002, Ramalho-Santos et al. 2002, Wang et al. 2011), next-generation sequencing (NGS) for epigenomics and glycomics (Barski et al. 2007, Wang, Schones, and Zhao 2009), and mass spectrometry (Phanstiel et al. 2008, Bhanu et al. 2016). Karyotyping is also very important method for reporting on several chromosomal abnormalities, including tumorigenic mutations, in human ESCs and iPSCs maintained in culture (Amps et al. 2011, Taapken et al. 2011).

Conventional karyotyping methods are relatively straightforward, rapid and inexpensive to implement. A cell's karyogram can be exposed by chromosomal staining with Giemsa (commonly termed G-banding), which reveals gross chromosomal abnormalities (e.g., trisomy) (Mahdieh and Rabbani 2013). However, conventional karyotyping only allows the detection of deletions or duplications >5 Mb, even with high resolution G-banding. In addition, the analysis is typically limited to a small number of cells (e.g., <50) rather than the whole cell population (Hulten et al. 2003).

To determine more specific chromosomal changes, NGS and/or fluorescence *in situ* hybridization (FISH), in particular comparative genomic hybridization (CGH), can be used (Mahdieh and Rabbani 2013). FISH is fast and has a higher resolution than other karyotyping techniques, but only reveals abnormalities at a specific locus. In addition, CGH can only detect 5–10 Mb blocks of over- or under-represented chromosomal DNA, while balanced rearrangements, such as inversions or translocations, are more challenging to uncover (Squire et al. 2002).

Quantitative fluorescent polymerase chain reaction (qf-PCR) can be used to detect common autosomal aneuploidies (trisomies 21, 18, and 13), sex chromosome aneuploidies, and triploidy. While qf-PCR is rapid and substantially cheaper than full karyotyping, it cannot be used to detect all chromosomal rearrangements and may lead to false positives (Chitty et al. 2006).

Since large-scale stem cell production is ultimately sought, high-throughput karyotyping will be necessary. Bacterial artificial chromosome (BAC)-based assays, featuring DNA immobilized on polystyrene beads (e.g., KaryoLite BoBs® from PerkinElmer [Waltham, MA]), are fast, cost effective, sensitive for some mosaicisms and can be run with a small sample quantity. Moreover, the data analysis is easy, and the results are representative of the whole cell population. However, balanced translocations and small-scale changes generally cannot be detected by BAC-based assays. Additionally, the proportion of karyotypical abnormal cells needs to be ≥30% for reliable detection of irregularities (Lund et al. 2012).

The methods mentioned earlier provide a wide variety of approaches for testing genomic instability, karyotype and epigenomic changes. Each have their own pros and cons. Depending on the application, these methods can be used individually or in combination to obtain a clearer picture of the state of human PSCs in culture. Ensuring genomic, chromosomal, and epigenomic stability will be critical for the large-scale culture of human PSCs.

1.3 CELL AGGREGATE CULTURE SYSTEMS

Several systems have been reported for the culture of cell aggregates, including rotating wall vessels, wave-induced bioreactors, shake flasks, spinner flasks, and instrumented SSBs (Figure 1.1).

1.3.1 ROTATING WALL VESSELS

Rotating wall vessels (Figure 1.1a), which were originally developed for use by the National Aeronautics and Space Administration (NASA) in the United States, include slow-turning lateral vessels (STLVs) and high aspect rotating vessels (HARVs). Their sizes are generally limited to 5–500 mL and their operation is characterized by low shear stress (generally between 0.002 and 0.006 dyne/cm²) and efficient gas transfer, resulting in a well-mixed culture environment. Of note,

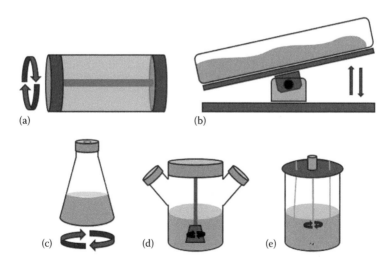

FIGURE 1.1 Schematics of culture systems: (a) Rotating wall vessel, (b) Wave-induced bioreactor, (c) Shake flask, (d) Spinner flask, and (e) Instrumented SSB.

mechanical shear stress levels from 3 to 10 dyne/cm^2 reduced cell viability of human embryonic kidney cells (Stathopoulos and Hellums 1985). In addition, 0.92 dyne/cm^2 adversely affected proliferation, morphology, and function of baby hamster kidney fibroblast cells (BHK-21) (Goodwin et al. 1993). Thus, the low shear stress environment of STLVs and HARVs appears to be beneficial for cultured cells, but the actual threshold above which shear induces damage to stem cells is yet unknown. Additionally, a certain level of shear is still necessary to reduce excessive cell clustering and the formation of oversized aggregates, which cause diffusion limitations for nutrients and O_2. To this end, aggregates with necrotic cores have been observed in HARV cultures of human ESCs (Gerecht-Nir et al. 2004). A perfusion system with a dialysis chamber has been described for STLVs, allowing continuous renewal of culture medium and removal of waste products. This is beneficial for cell proliferation and minimization of aberrant differentiation (Come et al. 2008). While scale-up for rotating wall vessels has not been reported, it remains unclear whether these are amenable to automation for monitoring and adjusting the bioreactor milieu. Despite the advantage of a low shear for cell viability and proliferation, monitoring and control of culture parameters would be necessary for large-scale cultivation, and the low shear may cause excessive cell clustering leading to oversized aggregates with considerable hindrances in O_2 and nutrient transport (Azarin and Palecek 2010).

1.3.2 Wave-Induced Bioreactors

Wave-induced bioreactors (Figure 1.1b), which are often used for hematopoietic cell expansion and other cell therapy applications, have a fairly simple design consisting of a plastic culture bag on a rocking platform to produce waves at the air-liquid interface (Kumar and Starly 2015). They are easy to scale-up with volumes as large as 580 L being used and have the benefit of operating with single-use pre-sterilized bags, which reduces sterilization and cleaning costs associated with multi-use vessels. Parameters, such as pH, temperature, and dissolved CO_2 and O_2, can be adjusted in the bioreactor, and the bag weight can be measured to determine cell mass directly (GE Healthcare Life Sciences Wave Bioreactor^{TM} System 2008). Conversely, the use of a rocking platform increases the space requirements for the system. More importantly, the low shear in wave-induced bioreactors can lead to the formation of large aggregates, wave bioreactor mixing is complicated to model and the heuristics for scale-up are not as well developed as for more common bioreactor types. Despite the possibility of aggregate formation and difficulties in bioreactor flow modelling, wave-induced bioreactors have been utilized in large-scale mammalian culture, and they could potentially be used for human PSC culture depending on the application.

1.3.3 Shake Flasks

Shake culture flasks (Figure 1.1c) require minimal initial capital investment, are simple to set up, and have controlled temperature and partial CO_2 pressure (pCO_2) because they operate in incubators. The flasks are kept on platforms, which provide shaking or orbital motion. Also, shake flasks can be single-use and pre-sterilized, and the culture is generally well-mixed. However, aggregate agglomeration may occur due to the shaking motion, leading to O_2 and nutrient diffusion limitations within the aggregates. Also, cluster dispersion throughout the culture medium is generally poor, potentially causing concentration gradients near the aggregates. In addition, monitoring of dissolved gases and pH is challenging and sampling requires removal of the flask from the incubator and transfer to an aseptic environment (Rodrigues et al. 2011). Overall, shake flasks are convenient for simple small-scale suspension culture since their size generally ranges from 25 mL to 6 L, but the potential aggregate agglomeration and lack of cluster dispersion may lead to aberrant differentiation and batch-to-batch variability (Klockner and Buchs 2012).

1.3.4 Spinner Flasks

Spinner flasks (Figure 1.1d) serve as inexpensive autoclavable or disposable, single-use surrogate of instrumented bioreactors, and their operation is straightforward. Because of the low working volumes, spinner flasks can be used to test multiple conditions relatively rapidly and at a reasonable cost. The equivalent agitation rates between spinner flasks and larger bioreactors can be determined via experiments or based on computational models (Werner et al. 2014, Ponnuru et al. 2014, Berry et al. 2016). Stirring not only creates a homogenous environment for the cells, but also induces shear allowing, to some extent, for the control of the size of aggregates, thereby reducing transport limitations of O_2 and nutrients. However, spinner flasks generally lack monitoring and the control of culture parameters (besides pCO_2 and temperature), and sampling can be labor intensive given that spinner flasks are kept in incubators (Rodrigues et al. 2011). These limitations along with the size of spinner flasks, generally 125 mL to a few liters, hinder their use for large-scale cell culture.

1.3.5 Instrumented Stirred-Suspension Bioreactors

SSBs (Figures 1.1e through 1.3) achieve a relatively homogenous culture environment, are amenable to the addition of CO_2 to the culture media for buffering, and can be easily integrated with systems for surveillance and adjustment of dissolved O_2 (DO), pH, and temperature (Figure 1.2). These features are essential to the reduction of batch-to-batch variability and for the maintenance of preferred culture parameters. Auxiliary systems are also available for retention of cells when operating the SSB in a perfusion mode. An optimal agitation rate lies between high values, which induce damaging hydrodynamic shear, and very slow stirring, which causes excessive agglomeration with adverse mass transfer limitations of gases, nutrients, cytokines, and metabolic by-products (Figure 1.3). These mass transfer limitations can lead to necrotic aggregate cores and aberrant differentiation. Consequently, a range of optimal agitation rates must be established for each cell line and SSB setup to minimize cell damage and uncontrolled differentiation while still maintaining the size of aggregates below a threshold which induces diffusion limitations.

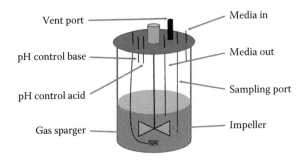

FIGURE 1.2   Schematic of instrumented SSB with configuration details.

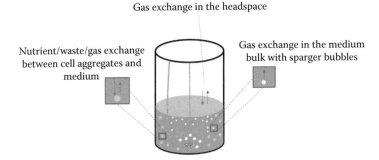

FIGURE 1.3   Representative exchanges in a bioreactor. Note that cell aggregates actually are dispersed throughout culture medium.

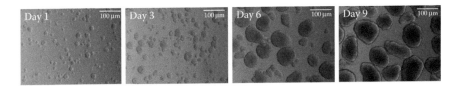

FIGURE 1.4   Images of H9 human embryonic stem cell aggregates from bioreactor culture. Conditions: 80 mL culture volume, batch mode, 37°C, pH 7–7.4.

Images of human ESCs aggregates cultured in an automated bioreactor are shown in Figure 1.4 along with cell concentrations and viability throughout culture (Figure 1.5). Scale-up of SSBs is relatively common, with reactors from 100 mL to 500 m³. These systems are utilized for industrial production of therapeutics. However, the impact of long-term culture on human PSCs in SSBs has yet to be determined. Furthermore, the cost of bioreactors limits their use outside of biopharma industry, gas sparging can shear cells, and instrumentation may disrupt flow patterns in the bioreactor (Rodrigues et al. 2011). Yet, the considerable amount of available knowledge and experience around these systems as a result of their use with mammalian cell culture make them an appealing culture modality for the culture of human PSC aggregates and its translation to a commercial setting.

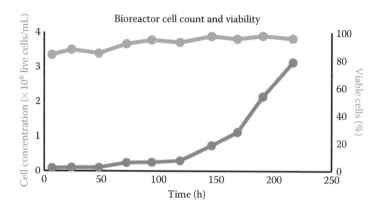

FIGURE 1.5 Cell concentration and viability assessed by Trypan Blue dye exclusion throughout the bioreactor culture.

1.4 STEM CELL CULTURE MEDIA

Multiple media are made commercially for pluripotent, feeder-free stem cell culture with proprietary compositions, while other media have defined composition (Table 1.1). Most media are xeno-free, but some contain components, such as non-recombinant serum albumin (e.g., mTeSR1™) which are derived from animals. The use of a small number of defined medium components (e.g., E8™ has only eight ingredients) reduces lot-to-lot variability and allows for medium customization based on cell line and culture use. Consequently, a smaller number of defined components is preferable but the exact compounds and their concentrations in various human PSC culture media remain proprietary, hindering efforts for cell line and system culture standardization and optimization.

TABLE 1.1
Human Stem Cell Pluripotency Media and Main Components

Medium	Main Components	Xeno-Free?	References
Essential 8 (E8™)	DMEM/F-12, L-ascorbic acid, Selenium, Transferrin, NaHCO₃, Insulin, FGF2, TGFβ1,	Yes	(Life Technologies Corporation 2013)
mTeSR™	DMEM/F-12, L-ascorbic acid, Selenium, Transferrin, NaHCO₃, Glutathione, L-Glutamine, Defined lipids, Thiamine, Trace elements B, Trace Elements C, β-Mercaptoethanol, Albumin, Insulin, FGF2, TGFβ1, Pipecolic acid, LiCl, GABA, H₂O	mTeSR1™: No (with bovine serum albumin (BSA)) mTeSR2™: Yes	(Life Technologies Corporation 2013, STEMCELL Technologies 2017a, b)
StemMACS™	Proprietary composition	Yes	(Miltenyi Biotec 2017)
StemPro® hESC SFM	Proprietary composition	Yes	(Thermo Fisher Scientific 2017)

General components include basal media, such as Dulbecco's Modified Eagle Medium (DMEM)/F12, vitamins, antioxidants, salts, traces minerals, specific lipids, albumin, detergent, γ-aminobutyric acid (GABA), transforming growth factor (TGF) ligands, and basic fibroblast growth factor (bFGF, also known as FGF2). Extracellular matrix molecules, such as collagen, fibronectin, laminin, and vitronectin are also used as supporting substrata (Lu et al. 2006). Some formulations may also contain shear protectants, such as Pluronic variants (which are also antifoaming agents), while the addition of these agents may be necessary for other media intended for use in suspension culture.

Most studies on the effects of medium composition on human PSCs have been performed in monolayer and static suspension cultures. For example, Ludwig and colleagues used mTeSR1 to test the effects of pH, osmolarity, and gas atmosphere (CO_2/O_2) on several human ESC lines (H1, H7, H9, and H14) in monolayer cultures without feeder cells. The optimal conditions based on cell number, fraction of OCT4$^+$ cells and cloning efficiency, included a pH of 7.2, osmolarity of 350 mOsmol, and a 10% CO_2/5% O_2 atmosphere (Ludwig et al. 2006). In another report, eight different conditions and defined PSC culture media were tested side-by-side on static suspension cultures of human PSCs (human ESC lines, Royan H5 and hiPSC1, derived from human dermal fibroblasts). Cell proliferation rates were more than 12-fold higher in mouse embryonic fibroblast (MEF) feeder-cell conditioned medium. However, the inherent lot-to-lot variability of this type of medium makes it suboptimal for large-scale cultivation, and it is not xeno-free, which increases the chance for cross-contamination (Abbasalizadeh et al. 2012).

Furthermore, Chen and co-workers examined the effect of mTeSR™, or StemPro® hESC SFM (serum- and feeder-free medium), defined medium supplemented with 8 or 40 ng/mL bFGF on human ESCs (HES-2 and H9) in 6-well plates with orbital shaking. Starting at day 1, aggregates in StemPro® appeared more homogeneous with diameters of >200 μm, while those in mTeSR™ were larger and heterogeneous in size. Also, more aggregates appeared to form and cell numbers were significantly higher in StemPro® media by day 6 cell. HES-2 aggregates in StemPro® media started to show vacuolization (a sign of differentiation) by day 6, while this was not observed in H9 aggregates in any of the medium conditions. Cells from HES-2 aggregates were over 90% positive for pluripotency markers at day 3, but 70-90% at day 6, while cell from H9 aggregates maintained greater than 90% positive cells through day 6 (Chen et al. 2012). Similar studies testing various media for human PSC culture have also been performed in smaller scale suspension systems (e.g., see ref. [Elanzew et al. 2015]), but are not discussed here since the focus is on larger scale SSBs. Overall, the results for monolayer and static and dynamic suspension culture show that the composition of the medium affects cell proliferation, pluripotency and cloning efficiency and needs to be optimized for each cell line.

The development of medium for stem cell cultivation is particularly relevant when large-scale production of human PSC-based therapeutics is considered for clinical applications. Procedures for these applications should comply with Good Manufacturing Practice (GMP). This includes the use of xeno-free ingredients to avoid cross-contamination with pathogens from animal-derived components and limit batch-to-batch variability. To that end, recombinant versions of growth factors and cytokines, which are used for maintenance of pluripotency or induction of differentiation, serve as alternatives to corresponding animal-derived molecules. Yet, recombinant proteins are expensive, may exhibit variable activity and stability, and have lot-dependent half-lives. Consequently, derivation is sought of small molecules with improved or similar but more consistent bioactivity, higher stability, and lower cost to recombinant cytokines. Some examples of small molecules that have been created and successfully utilized for human PSC differentiation include CHIR99021 (a selective, potent inhibitor of GSK-3/activator of canonical Wnt signaling) (Cheng et al. 2017), SANT1 (inhibitor of Shh Sonic Hedgehog [Shh] signaling) (Chen et al. 2002), and LDN193189 (a bone morphogenetic protein (BMP) inhibitor) (Cuny et al. 2008). Development of other small molecules for stem cell culture would be extremely beneficial for large-scale culture especially since some growth factors, such as bFGF, are not thermally stable at 37°C.

The requirement for xeno-free components extends to substrates for human PSC culture. Recombinant human laminin-511/521, or vitronectin (Higuchi et al. 2015), can be used to coat culture surfaces instead of Matrigel, which is a mouse sarcoma extract with undefined composition or culture with MEFs. In this context, many of the existing human PSC lines are not GMP-compliant because they were cultured with animal-derived medium components and/or derived on MEFs, suggesting the need for generating lines using xeno-free substrates and components. Defined, xeno-free and consistent medium compositions and biomaterials both for maintaining PSC pluripotency and inducing differentiation are necessary for large-scale culture of human PSCs for clinical application, as well as for culture standardization. Recent and ongoing efforts in this area indicate the possibility of accomplishing this soon.

1.5   FEEDING STRATEGIES

Most reports on human PSC cultivation in bioreactors describe batch mode operation, that is, the cell suspension is initially loaded and the culture volume is kept constant, typically through partial exchanges of spent with fresh medium. In some instances, the starting medium volume is low to allow for cell aggregation, adhesion to microcarriers, and/or culture acclimation, and medium is added to the final working volume 24 hours post-seeding. While operating a bioreactor in batch mode is relatively straightforward, the cells may experience significant fluctuations in pH, osmolarity, concentration of metabolites (e.g., glucose, glutamine, lactate, glutamate), factors for human PSC pluripotency maintenance or specification, and dissolved gases upon addition/removal of fresh/spent medium. This suggests a variable environment with largely unexplored effects on cultured human PSCs to date.

Bioreactors can also be operated in fed-batch mode, as is done to produce protein therapeutics. Cells are generally cultured for longer periods leading to higher cell densities in comparison to batch bioreactor cultivation. Fresh medium (sometimes concentrated) enters the process continuously or intermittently up to a prescribed working volume. While this results in higher product titers, increases in cell concentration are less pronounced given that the increase of the biomass in absolute terms is concomitant with rises in culture volume.

Fed-batch strategies have been successful for cell culture for protein therapeutics and viral vaccine production, but they do not translate directly to human PSC cultivation. The use of pre-made medium with proprietary formulation for the maintenance of human PSCs is convenient, but affords little flexibility for intermittent, on-demand feeding of fed-batch bioreactor cultures. The dependence of human PSC pluripotency on factors (e.g., bFGF) with short half-lives adds a layer of complexity to potential fed-batch operation. Another concern is the accumulation of metabolic waste products that can be detrimental to human PSC growth, proliferation and may lead to aberrant differentiation. Overcoming these limitations will be a key to unlocking the potential of fed-batch (vs batch) cultivation to produce cells for regenerative medicine.

Perfusion processes have also been utilized in animal cell culture to produce recombinant therapeutics as well as human PSCs. This mode corresponds to a continuous operation involving the constant exchange of medium in the bioreactor, thereby removing cellular byproducts that potentially inhibit growth and/or other unwanted changes, such as induction of differentiation. Furthermore, cells are exposed to limited fluctuations in pH and nutrients, factors, and O_2 concentrations due to continuous addition and withdrawal of medium in conjunction with stirring. A cell retention system (e.g., settling cone, filter) for the exiting stream is necessary to minimize cell losses and recycle cells back to the culture. It should be noted that in the case of human PSCs, their cultivation in a perfusion system may lead to higher cell densities (see next paragraph), but will require greater amounts of medium, which is relatively expensive compared to general mammalian cell culture medium. This suggests a possible cost limitation for the wide-scale use of perfusion bioprocesses for PSC culture. Moreover, the presence of a cell retention device may make systems more prone to clogging/fouling, increasing maintenance needs and the risk for adverse effects on the cells.

Several studies have been conducted on the effects of feed strategies on human PSC culture. For example, Kropp and colleagues compared perfusion to repeated batch feeding in suspension cultures of human iPSC aggregates (hCBiPS2, from cord blood-derived endothelial cells, and hHSC_F1285T_iPS2, from hematopoietic stem cells). They utilized a system of stirred-tank mini-bioreactors with controlled temperature, pH and dissolved oxygen (DO) (Kropp et al. 2016). A cell retention apparatus was employed with perfusion at 4.2 mL/h, whereas batch feeding took place every 24 hours. Perfusion led to higher average cell densities of $2.85 \pm 0.34 \times 10^6$ versus $1.94 \pm 0.16 \times 10^6$ cells/mL for repeated batch cultivation or about a 6- and 4-fold expansion, respectively, after seven days. This difference could be explained, in part, by the higher levels of chemokine C-X-C motif ligand 5 (CXCL5) observed in perfusion supernatants, since the addition of CXCL5 showed some promotion of growth in control experiments. The average aggregate diameter on day 7 was similar at ~123 μm for repeated batch and ~134 μm for perfusion. Flow cytometry, qPCR, and global gene expression analysis confirmed the maintenance of human PSC pluripotency for both feeding strategies. Evidently, the repeated batch and perfusion modes were more efficient than a monolayer dish culture since they resulted in reduced medium consumption by 16% and 43% per 10^8 cells, respectively (Kropp et al. 2016).

Others also showed increases of up to 70% in the yield of human ESCs (HES-3) cultured as monolayers under perfusion *versus* non-perfusion (Fong et al. 2005). Furthermore, Serra and co-workers studied cultivation of human ESCs (SCED™461) on microcarriers, comparing perfusion with daily replacement of 50% of the medium. Cells in the perfusion reactor had faster cell growth, higher O_2 consumption, and lower lactate production, thus suggesting that a higher portion of the metabolism was aerobic (Serra et al. 2010).

Lastly, Kwok and collaborators demonstrated two different batch feeding systems applied to the suspension spinner culture of human iPSCs (AR1034ZIMA hiPSC clone 1, from adult fibroblasts, and FS hiPSC clone 2, from juvenile foreskin fibroblast). The first system started with 100 mL and the regular protocol comprised the replacement of medium up to 100 mL on days 2 and 3 post-inoculation. An optimized protocol was employed for the second system starting with a volume of 80 mL and the addition of 20 mL of medium on day 2 and 3 post-inoculation. Both protocols involved daily medium exchanges up to 100 mL from day 4 until day 7, keeping the total culture volume at 100 mL. The final density was $252 \pm 8 \times 10^6$ with the first protocol and $894 \pm 90 \times 10^6$ cells/mL with the second, that is, a 73% difference. Additionally, the optimized protocol had a 40.6% increase in yield compared to standard adherent culture (Kwok et al. 2017). Overall, these results demonstrate significant differences in cell growth based on the feeding protocols, suggesting the potential for further improvements in the scalable cultivation of human PSCs.

1.6   PRIMARY CULTURE VARIABLES/PARAMETERS

Traditional parameters for the SSB culture of mammalian cells include temperature, pH, pO_2, and pCO_2, agitation rate/shear stress, osmolarity, concentrations of nutrients and growth factors, seeding density, basal medium type, aggregate size, as well as size and geometry of the impeller and vessel. These parameters are also pertinent to the bioreactor culture of human PSCs and their progeny. While some parameters are relatively fixed (e.g., the cultivation temperature of human PSCs is 37°C and the pH 7–7.4), optimization of other parameters is necessary for different human PSC lines and conditions, especially considering the differences among PSC lines in propensity for commitment to specific fates and resistance to shear. This adds a layer of complexity to the translation of existing SSB culture strategies for human PSCs. Some of these culture parameters and general guidelines for the culture of human PSCs are discussed in the following.

1.6.1   OSMOLARITY

Osmolarity is a measurement of the osmotic pressure of a solution and ranges from 260 to 370 mOsm for commonly used culture media such as DMEM, DMEM/F-12, and Roswell Park Memorial Institute (RPMI) medium. Rajala and co-workers tested the effects of osmolarities of 260, 290, 320, and 350 mOsm on static monolayer culture of feeder-dependent human ESCs (HS346, HS401; and Regea 06/015, 06/040, 07/046, 08/013, and 08/017). An osmolarity of 320–330 mOsm was found to be optimal with KnockOut™ DMEM basal medium and 25 mM glucose for maintaining self-renewing human ESCs (Rajala et al. 2011). Others reported that exposure of human iPSC (from human foreskin fibroblasts) monolayers to high osmolarity (385 mOsm) by incubation with 30.5 mM glucose enhanced proliferation and upregulated pluripotency markers in comparison to 5.5 mM glucose (286 mOsm), as determined by western blot analysis and cytoskeletal remodeling (β-catenin, F- and G-actin, and tubing networks) (Madonna et al. 2014). Overall, these studies demonstrate the importance of osmolarity on the state of cultured PSCs, but do not indicate a specific osmolarity to use for human PSC culture or the effects of osmolarity for suspension culture.

1.6.2 SEEDING DENSITY

Suspension culture seeding density and extent of cell disassociation has been shown to affect cell proliferation and viability. For human PSC aggregate culture, seeding can be carried out with dispersed single cells, partially dissociated clumps, or preformed aggregates. When seeded as single cells, human PSC aggregate formation can be facilitated by the introduction of soluble extracellular matrix molecules, such as Matrigel or collagen. Alternatively, single cells can be coaxed to cluster in micropatterned materials or microwells. Considering the low survival of single human PSCs after dissociation from colonies, the use of a Rho-associated kinase (ROCK) inhibitor (e.g., Y-27632) is important to maintain cell viability (Watanabe et al. 2007). Seeding of single cells can generally result in a greater uniformity of aggregate size in comparison to culture inoculation with cell clumps. However, the use of a ROCK inhibitor has been associated with the higher incidence of aneuploidy, so cultures treated with ROCK should be closely monitored for such abnormalities (Riento and Ridley 2003). Current systems of micropatterned materials for coalescing dissociated human PSCs into aggregates are limited in their scalability for generating sufficient amounts of clusters for seeding large bioreactors. On the other hand, clump formation by partial disassociation of colonies is relatively straightforward and diminishes the need for ROCK, but the ensuing aggregates can be polydisperse.

In most reports of stirred-suspension culture of human PSC aggregates, the seeding concentration (or seeding density; the two terms are used interchangeably) ranges between 0.3 and 6×10^5 cells/mL (Cormier et al. 2006, Serra et al. 2010, Abbasalizadeh et al. 2012, Chen et al. 2012, Olmer et al. 2012, Elanzew et al. 2015, Kropp et al. 2016, Kwok et al. 2017). The selection of appropriate seeding densities pertains to the cell type and other culture characteristics, including the agitation rate and vessel/impeller geometries, which dictate the flow pattern inside the reactor. Thus, different concentrations of inoculated cells can lead to significantly different outcomes. For example, starting with 2×10^5 human iPSCs/mL resulted in less irregular clusters in spinner flask culture than seeding 5×10^5 iPSCs/mL (Kwok et al. 2017). Also, markedly different cell proliferation was reported for different seeding densities of human PSCs (human iPSC iLB-C-31f-rl, from dermal fibroblasts, and human ESC H9) in a BioLevitator™ (gently rotating 50 mL conical tubes), which is another benchtop culture system. Seeding 0.33×, 0.75× and 2×10^5 cells/mL resulted in 20-, 4.9- and 4.3-fold average increases, respectively (Elanzew et al. 2015).

Others reported the effect of seeding concentration on the yield of human PSCs in a 100 ml Cellspin suspension culture platform (spinner flask). Among different starting concentrations ($2\times$, $3\times$, $5\times$ and 10×10^5 cells/mL), the highest cell yield after 10 days of culture was achieved by seeding 3×10^5 cells/mL. Interestingly, the maximum cell density for the human ESC line was also achieved when cultures were started with 3×10^5 cells/mL, but for the human iPSCs the maximum cell density was observed when they were seeded at 10×10^5 cells/mL (Abbasalizadeh et al. 2012). These studies show that selection of the optimal concentration for seeding human PSCs for aggregate culture depends on the PSC line, the culture system, and the parameter(s) under consideration for optimization.

1.6.3 AGITATION

Agitation influences cells and the culture environment in multiple and frequently interrelated ways. It results in the mixing of the culture suspension, reduces concentration gradients that lead to heterogeneities within the bioreactor, and facilitates gas dispersion for systems featuring a gas sparger, which is typically located under the impeller. In addition to keeping the cells suspended, a rotating impeller creates shear, thereby influencing the formation of aggregates. However, excessive agitation-induced shear is detrimental, causing cell death or loss of pluripotency.

Cormier and colleagues grew mouse ESC aggregates in spinner flasks with maximum shear stresses of 4.5 and 6.1 dyne/cm^2 at 80 and 100 rpm, respectively. Proliferation was halted when the agitation was set to 120 rpm (7.8 dyne/cm^2 maximum shear stress) and a large amount of cell debris was noted. Cell expansion and peak density were greater at 100 rpm (31-fold expansion; 10^6 live cells/mL) versus 80 rpm (17-fold; 7×10^5 live cells/mL). Also, the average aggregate diameter was smaller at 100 rpm than at 80 rpm (~140 and 200 μm, respectively) (Cormier et al. 2006).

Along the same vein, Kwok and co-workers varied the stirring rate between 65 and 85 rpm in spinner flasks with human iPSC (adult fibroblast and juvenile foreskin fibroblast derived) in mTeSR1™ or StemMACS™ media (Kwok et al. 2017). At 65 rpm, the aggregates generally coalesced into larger clumps, while at 85 rpm more uniform round aggregates were observed. Another group (Nampe et al. 2017) used human ESCs (H9) or human iPSCs (adult fibroblast derived RIV9) to test four agitation rates (40, 60, 80, and 100 rpm) in 125 mL Corning® ProCulture® spinner flasks. After seven days of culture at 60 rpm, H9 and RIV9 cells grew ~38- and ~25-fold, respectively. The cells were >90% OCT4$^+$, TRA-1-60$^+$ and SSEA-4$^+$ for both lines at 60, 80, and 100 rpm, as determined by flow cytometry. Conversely, at 40 rpm (and in static cultures), the cells expressed lower levels of the pluripotency markers. Of note, the average aggregate diameter was ≥400 μm in static cultures and at 40 rpm, while it was 100–300 μm at 60–100 rpm. Moreover, cells within 100–300 μm aggregates were the most proliferative as determined by flow cytometry of cell fractions residing in the G$_2$/M cycle phase, while cells within aggregates ≥400 μm were the least proliferative. These findings demonstrate that the size of human PSC aggregates in culture is affected drastically by agitation with direct effects on cell pluripotency, proliferation, and viability (see also next section on dissolved O$_2$).

Stirring is linked not only to the vessel shape, the presence of baffles and probes or instrumentation and mixing patterns (typically, axial, or radial), but also to the impeller geometry and size. For example, Olmer and colleagues investigated different impeller sizes and geometries in a Cellferm® pro Parallel 125 ml Bioreactor System (DASGIP AG) for the cultivation of human iPSCs (cord blood-derived endothelial cell derived hCBiPSC2) in mTeSR1™. Triangular Teflon stirrers with diameters of 40 and 60 mm and blade impellers with diameters of 25 and 40 mm were tested. For the 40-mm blade impeller, three designs were used, 45°60°, 45°30°, and 60°60° (numbers denoting the pitch of the blade at the impeller center and edge, respectively). The different geometries at 30 rpm did not lead to homogeneous distribution of stained microcarriers (as substitutes for cell aggregates to monitor flow patterns). Greater homogeneity was observed at 60 rpm with the 60-mm stir bar and 40-mm impellers, but the other designs did not lead to a homogeneous stirring even with more vigorous agitation. Differences in average aggregate sizes were shown for the 40-mm impeller designs: 98 μm for 45°60°, 75 μm for 45°30°, and 77 μm for 60°60° at day 4 and, by day 7, 125, 93, and 106 μm, respectively. The expansion rates were equivalent for all the impellers and the cells remained pluripotent as determined by flow cytometry, immunostaining, qPCR, and differentiation of cultured cells to all three germ layers (Olmer et al. 2012).

It should be noted that the agitation rate (N) is related to the power (P_o) required for stirring and the power number (N_P) through an empirical correlation:

$$N_P = \frac{P_o}{\rho N^3 d_i^5} \tag{1.1}$$

where ρ and d_i are the medium density and impeller diameter, respectively. The N_P for ungassed bioreactors can be plotted against the Reynolds number:

$$Re = \frac{N d_i^2 \rho}{\mu} \text{ (1.2; } \mu \text{ is the liquid viscosity)}$$

for different impeller types. In a turbulent regime, the N_P is independent of Re, and thus the power requirement for a given impeller varies proportionally with N^3, D^5 and the liquid density (ρ), but not with the liquid viscosity (μ), that is,

$$P_o = k_1 N_P N^3 d_i^5 \rho \tag{1.3}$$

where k_1 is a constant. The power requirement (P_A) for aerated SSBs can also be calculated through empirical correlations, most often based on the power requirement for the ungassed bioreactor (P_o). For instance, the following relation has been derived from experimental data for six-flat blade turbines with a blade width equal to one-fifth of the impeller diameter:

$$\frac{P_A}{P_o} = \exp\left[-192\left(\frac{d_i}{d_v}\right)^{4.38} (Re)^{0.115} (Fr)^{1.96\left(\frac{d_i}{d_v}\right)}\left(\frac{Q}{N d_i^3}\right)\right] \tag{1.4}$$

$$Re = \frac{d_i^2 N \rho}{\mu} \tag{1.5}$$

$$Fr = \frac{d_i N^2}{g} \tag{1.6}$$

where Re is the Reynolds number and Fr represents the Froude number. Here, g is the gravitational acceleration, Q is the aeration rate and d_v is the diameter of the tank. Such relations are useful for the design and scale-up of bioreactors. The input power P (P_o or P_A) is related to the local rate of dissipation of turbulent kinetic energy per unit mass of fluid (ε):

$$P = \varepsilon \rho d_i^3 \tag{1.7}$$

The (Kolmogorov) eddy size λ can be estimated using the relation:

$$\lambda = \left(\frac{\mu^3}{\rho^3 \varepsilon} \right)^{0.25} \tag{1.8}$$

These eddies are rotational flow regions toward which the kinetic energy of turbulent fluid is directed due to agitation. Larger eddies disintegrate into smaller eddies, but below a certain size λ, the eddies are no longer sustained, causing their kinetic energy to dissipate. This dissipated energy can damage cells in close proximity. It has been shown experimentally that detrimental effects occur when λ drops below two-thirds to one-half of the aggregate (or microcarrier) diameter. Thus, for a target eddy size λ, one can calculate ε and using the correlation of P versus N_P that is typically available for a particular impeller, a suitable agitation rate N can be estimated. Of note, this workflow does not consider the effects of sparging on cells (e.g., damage due to bubble shear) (Nagata 1975).

Similar correlations can be developed for predicting the maximum average aggregate size ($\overline{d_a}$) as a function of dissipated power rate (ε and thus of agitation speed) and/or medium viscosity as elegantly shown for the culture of murine neural stems as neurospheres in stirred-suspension vessels (Sen et al. 2002). In another study, Hunt and colleagues investigated the effect of agitation rate and seeding density on human ESCs (H1 and H9) in 125 ml spinner flasks. Agitation rates 80, 100, and 120 rpm and seeding densities 2, 4, and 8 × 10⁴ were tested as a two-parameter, three-level (3^2) factorial experiment. The optimal culture parameters were dependent on whether fold increase, maximum cell density, or exponential growth rate was chosen as the primary output variable and there were significant interaction effects between the agitation rate and inoculation density indicating that each of these parameters cannot be considered alone (Hunt et al. 2014). Overall, shear stress due to agitation affects aggregate size, cell proliferation, pluripotency, and viability, all of which are important to human PSC culture.

1.6.4 DISSOLVED OXYGEN

Another important parameter is the oxygen (O_2) concentration in the liquid phase (also referred to as DO concentration), which is often expressed in mmole/L. For pure water saturated with O_2 under atmospheric pressure and 21% atmospheric O_2 at 37°C, the DO is 0.22 mmole/L. Since this DO concentration is in equilibrium with 159.6 mm Hg of O_2 or 21% in the gas phase, it can be expressed in mm Hg or %. Generally, human PSCs are cultured with 21% atmospheric O_2 supplemented with 5% CO_2. However, the O_2 tension *in vivo* during development ranges from 1.5% to 5.3% for rhesus monkeys, rabbits, and hamsters (Fischer and Bavister 1993). Therefore, it is important to determine whether or not hypoxic conditions are necessary for large-scale culture of human PSCs as aggregates.

Several studies have been conducted on the effects of O_2 concentration on human PSC monolayer cultures. Forsyth and colleagues (2006) compared the effects of physiologic (2%) or atmospheric (21%) levels of O_2 on the static, feeder-free culture of human ESCs (H1, H9, and RH1) in DMEM supplemented with KnockOut serum replacement (KSR). The human ESC clonal recovery was on average 6-fold greater in cultures with 2% O_2 than 21%. Moreover, the frequency of spontaneous chromosomal mutations significantly increased for cells maintained at 21% versus 2% O_2. Yet, there were no significant changes in the expression of SSEA4, TRA-1-60, and TRA-1-81 for the different conditions (Forsyth et al. 2006). Similar results were reported when human ESCs (CLS1, 2, 3, and 4) were grown for 18 and 42 months on a human foreskin fibroblasts feeder layer at 20% or 5% O_2. Spontaneous differentiation was observed at the center of normoxic human ESC colonies, whereas at 5% O_2 colonies were undifferentiated with homogenously flat morphology (Zachar et al. 2010). Another study by Forristal and co-workers looked at the role of hypoxia in pluripotency maintenance. HUES-7 human ESCs were cultured in 20% or 5% O_2 on Matrigel with MEF-conditioned medium. Twenty percent O_2 led to a decrease in cell proliferation with smaller colonies and lower cell numbers at days 2, 3, and 4 post-passaging compared to cultures with 5% O_2. Hypoxia-inducible factors 2A and 3A (HIF2A, HIF3A) were upregulated in cultures maintained at 5% versus 20% O_2. When HIF2A expression was knockdown with small (or short) interfering RNA (siRNA), cell proliferation and POU5F1 (Oct4), SOX2 and NANOG protein expression decreased, while SSEA1 (a marker of early differentiation) increased. This suggests that HIF2A is at least partially responsible for the beneficial effects of hypoxic conditions on human PSC pluripotency and proliferation in this cell line (Forristal et al. 2010). More generally, these studies demonstrated that hypoxia appears to promote clonal recovery and cell proliferation and minimize spontaneous chromosomal mutations. At the same time, the differences in pluripotency marker expression between atmospheric and hypoxic O_2 levels are trivial. Despite this, the data point to the conclusion that hypoxic conditions are preferable for long-term monolayer culture of human PSCs.

Beyond the availability of O_2 in the medium manifested by the DO, the ultrastructural characteristics of aggregates are determinant of the diffusional transfer of O_2 to the cells. The typical O_2 diffusion-limitation in tissues is about 100–200 µm. Wu and colleagues looked at the transfer of O_2 to the culture phase and through the human ESC (H9) aggregates in static and spinner flask cultures. A transient diffusion-reaction model combined with a population balance equation model was applied, coupled to experimental data considering aggregate features, such as the porosity and tortuosity, along with the evolving cluster size distribution due to agitation and cellular proliferation. This allowed the calculation of the effective diffusivity and kinetics of O_2 consumption within ESC aggregates under 45 rpm, 60 rpm or static culture. The results showed that human ESCs within aggregates <200 μm did not experience severe hypoxia in contrast to ~23% and 70% of cells in clusters with sizes greater than 400 and 1,000 μm, respectively. This report is aligned with the notion of faster O_2 transfer from the medium bulk to each aggregate in agitated vessels compared to static cultures (Wu et al. 2014). Importantly, the duration of exposure of the human PSCs to different O_2 levels during cultivation was estimated. This information is significant because of the effects of O_2 concentration on stem cell proliferation (as already discussed) and the propensity for specification to particular cell types.

Others investigated the effects of hypoxic (5%–6% O_2), uncontrolled (10%–13% O_2) and normoxic conditions (18%–21% O_2) on human PSC expansion and metabolic activity in suspension culture (100 mL CellSpin vessels). Cultivation under normoxia resulted in greater cell fold increase (20%–60%), reduced lag phase, and lengthened exponential growth than hypoxic and uncontrolled conditions (Abbasalizadeh et al. 2012). Serra and co-workers maintained human ESCs (SCED46) on microcarriers in MEF-conditioned medium supplemented with KSR in a semi-continuous fed controlled bioreactor (BIOSTAT® Qplus) to study the difference between 1% and 6% O_2. Cultures at 6% O_2 exhibited an almost three-times higher maximum cell concentration, two-fold increase, and three-times greater apparent growth rate, but also three-times higher apparent death rate in comparison to cultures at 1% O_2. The bioreactor cultures with 6% O_2 displayed a 2.5-fold increase in cell concentration compared to spinner flasks at ~20% O_2. However, this difference could be due to the distinctive culture systems and not O_2 concentrations since the spinner flask lacked pH control (Serra et al. 2010). Of note, cells in the perfusion reactor had higher O_2 consumption and lower lactate production, which suggests a change in human PSC metabolism, in comparison to semi-continuous culture (Serra et al. 2010). Similarly, examination of physiological and gene expression data from 3D suspension aggregate versus 2D cultures of human iPSCs (hCBiPS2 human cord blood derived or hHSC_F1285T_iPS2 hematopoietic stem cells derived) shows a switch from glycolysis to oxidative phosphorylation without inducing differentiation (Kropp et al. 2016). Whether this metabolic shift is due to the 3D versus 2D culture of the cells, changes in DO or both, remains unanswered.

These findings suggest that more studies are warranted to determine the optimal O_2 conditions for maintenance of human PSC pluripotency, proliferation, viability, and propensity for differentiation in SSB cultures. The culture of human PSCs under hypoxic conditions in static cultures (e.g., dishes) requires greater capital investment (e.g., special incubators) and labor than normoxia (Viswanathan et al. 2014). On the other hand, DO can be controlled in instrumented SSBs through sparging or by aerating the headspace with appropriate amounts of O_2, air, or mixtures of these with an inert gas, such as nitrogen. Tight control and monitoring of DO is critical for reducing batch-to-batch variability, cell differentiation, and cell death due to O_2 deprivation.

1.7 CHALLENGES AND CONCLUSIONS

Vast progress has been made in recent years moving from the 2D surface culture of human PSCs to scalable 3D cultivation in stirred-suspension platforms. Instrumented bioreactors provide the potential to produce clinically relevant quantities of human stem cells and their differentiated progeny and benefit from well-established methods and a significant body of work drawn from the area of cell culture for producing protein therapeutics. The focus is now shifting toward devising strategies for real-time, on-line assessment of the quality traits of cultured cells, including their epi/genetic stability, marker expression, and expected function(s), as well as control of nutrient concentrations. Additionally, significant strides will be required in the development of mathematical models to capture the heterogeneity of stem cell populations. The issue of heterogeneity is common in non-stem cell populations as well, but it becomes problematic in human PSC cultivation given that the cells (instead of, for example, secreted metabolites) are the desired product. Large-scale 3D culture of human PSCs is also hampered by the variability displayed by different human PSC lines with respect to their culture conditions and differentiation proclivities. With scale-up or scale-out strategies for the differentiation of stem cells, there is pressure to innovate on technologies for the rapid, high-throughput separation and purification of specific cell types. Overall, as the demand for stem cell products increases in the coming years, so will the need for developing robust methodologies for the scalable cultivation of human PSCs and for gaining a deeper understanding of the effects of bioprocessing on stem cell physiology.

REFERENCES

Abbasalizadeh, S., M. R. Larijani, A. Samadian, and H. Baharvand. 2012. Bioprocess development for mass production of size-controlled human pluripotent stem cell aggregates in stirred suspension bioreactor. *Tissue Eng Part C Methods* 18 (11):831–851. doi:10.1089/ten. TEC.2012.0161.

Amps, K., P. W. Andrews, G. Anyfantis, L. Armstrong, S. Avery, H. Baharvand, J. Baker et al. 2011. Screening ethnically diverse human embryonic stem cells identifies a chromosome 20 minimal amplicon conferring growth advantage. *Nat Biotechnol* 29 (12):1132–1144. doi:10.1038/nbt.2051.

Atkinson, S., and L. Armstrong. 2008. Epigenetics in embryonic stem cells: Regulation of pluripotency and differentiation. *Cell Tissue Res* 331 (1):23–29. doi:10.1007/s00441-007-0536-x.

Azarin, S. M., and S. P. Palecek. 2010. Development of scalable culture systems for human embryonic stem cells. *Biochem Eng J* 48 (3):378. doi:10.1016/j.bej.2009.10.020.

Barski, A., S. Cuddapah, K. Cui, T. Y. Roh, D. E. Schones, Z. Wang, G. Wei, I. Chepelev, and K. Zhao. 2007. High-resolution profiling of histone methylations in the human genome. *Cell* 129 (4):823–837. doi:10.1016/j.cell.2007.05.009.

Berry, J. D., P. Liovic, I. D. Šutalob, R. L. Stewart, V. Glattauerc, and L. Meagherc. 2016. Characterisation of stresses on microcarriers in a stirred bioreactor. *Appl Math Model* 40 (15–16):6787–6804.

Bhanu, N. V., S. Sidoli, and B. A. Garcia. 2016. Histone modification profiling reveals differential signatures associated with human embryonic stem cell self-renewal and differentiation. *Proteomics* 16 (3):448–458. doi:10.1002/pmic.201500231.

Buta, C., R. David, R. Dressel, M. Emgard, C. Fuchs, U. Gross, L. Healy et al. 2013. Reconsidering pluripotency tests: Do we still need teratoma assays? *Stem Cell Res* 11 (1):552–562. doi:10.1016/j.scr.2013.03.001.

Chan, E. M., S. Ratanasirintrawoot, I. H. Park, P. D. Manos, Y. H. Loh, H. Huo, J. D. Miller et al. 2009. Live cell imaging distinguishes bona fide human iPS cells from partially reprogrammed cells. *Nat Biotechnol* 27 (11):1033–1037. doi:10.1038/nbt.1580.

Chen, J. K., J. Taipale, K. E. Young, T. Maiti, and P. A. Beachy. 2002. Small molecule modulation of Smoothened activity. *Proc Natl Acad Sci U S A* 99 (22):14071–14076. doi:10.1073/pnas.182542899.

Chen, V. C., S. M. Couture, J. Ye, Z. Lin, G. Hua, H. I. Huang, J. Wu, D. Hsu, M. K. Carpenter, and L. A. Couture. 2012. Scalable GMP compliant suspension culture system for human ES cells. *Stem Cell Res* 8 (3):388–402. doi:10.1016/j.scr.2012.02.001.

Cheng, T., K. Zhai, Y. Chang, G. Yao, J. He, F. Wang, H. Kong et al. 2017. CHIR99021 combined with retinoic acid promotes the differentiation of primordial germ cells from human embryonic stem cells. *Oncotarget* 8 (5):7814–7826.

Chitty, L. S., K. O. Kagan, F. S. Molina, J. J. Waters, and K. H. Nicolaides. 2006. Fetal nuchal translucency scan and early prenatal diagnosis of chromosomal abnormalities by rapid aneuploidy screening: Observational study. *BMJ* 332 (7539):452–455. doi:10.1136/bmj.38730.655197.AE.

Come, J., X. Nissan, L. Aubry, J. Tournois, M. Girard, A. L. Perrier, M. Peschanski, and M. Cailleret. 2008. Improvement of culture conditions of human embryoid bodies using a controlled perfused and dialyzed bioreactor system. *Tissue Eng Part C* 14 (4):289–298.

Cormier, J. T., N. I. zur Nieden, D. E. Rancourt, and M. S. Kallos. 2006. Expansion of undifferentiated murine embryonic stem cells as aggregates in suspension culture bioreactors. *Tissue Eng* 12 (11):3233–3245.

Cuny, G. D., P. B. Yu, J. K. Laha, X. Xing, J. F. Liu, C. S. Lai, D. Y. Deng, C. Sachidanandan, K. D. Bloch, and R. T. Peterson. 2008. Structure-activity relationship study of bone morphogenetic protein (BMP) signaling inhibitors. *Bioorg Med Chem Lett* 18 (15):4388–4392. doi:10.1016/j.bmcl.2008.06.052.

Elanzew, A., A. Sommer, A. Pusch-Klein, O. Brustle, and S. Haupt. 2015. A reproducible and versatile system for the dynamic expansion of human pluripotent stem cells in suspension. *Biotechnol J* 10 (10):1589–1599. doi:10.1002/biot.201400757.

Fischer, B., and B. D. Bavister. 1993. Oxygen tension in the oviduct and uterus of rhesus monkeys, hamsters and rabbits. *J Reprod Fertil* 99 (2):673–679.

Fong, W. J., H. L. Tan, A. Choo, and S. K. Oh. 2005. Perfusion cultures of human embryonic stem cells. *Bioprocess Biosyst Eng* 27 (6):381–387. doi:10.1007/s00449-005-0421-5.

Forristal, C. E., K. L. Wright, N. A. Hanley, R. O. Oreffo, and F. D. Houghton. 2010. Hypoxia inducible factors regulate pluripotency and proliferation in human embryonic stem cells cultured at reduced oxygen tensions. *Reproduction* 139 (1):85–97. doi:10.1530/REP-09-0300.

Forsyth, N. R., A. Musio, P. Vezzoni, A. H. Simpson, B. S. Noble, and J. McWhir. 2006. Physiologic oxygen enhances human embryonic stem cell clonal recovery and reduces chromosomal abnormalities. *Cloning Stem Cells* 1 (8):16–23.

Garitaonandia, I., H. Amir, F. S. Boscolo, G. K. Wambua, H. L. Schultheisz, K. Sabatini, R. Morey et al. 2015. Increased risk of genetic and epigenetic instability in human embryonic stem cells associated with specific culture conditions. *PLoS One* 10 (2):e0118307. doi:10.1371/journal.pone.0118307.

GE Healthcare Life Sciences Wave Bioreactor™ System. 2008. Wave Bioreactor.

Gerecht-Nir, S., S. Cohen, and J. Itskovitz-Eldor. 2004. Bioreactor cultivation enhances the efficiency of human embryoid body (hEB) formation and differentiation. *Biotechnol Bioeng* 86 (5):493–502. doi:10.1002/bit.20045.

Goodwin, T. J., T. L. Prewett, D. A. Wolf, and C. F. Spaulding. 1993. Reduced shear stress: A major component in the ability of mammalian tissues to form three-dimensional assemblies in simulated microgravity. *J Cell Biochem* 51:301–311.

Higuchi, A., S. H. Kao, Q. D. Ling, Y. M. Chen, H. F. Li, A. A. Alarfaj, M. A. Munusamy et al. 2015. Long-term xeno-free culture of human pluripotent stem cells on hydrogels with optimal elasticity. *Sci Rep* 5:18136. doi:10.1038/srep18136.

Hulten, M. A., S. Dhanjal, and B. Pertl. 2003. Rapid and simple prenatal diagnosis of common chromosome disorders- advantages and disadvantages of the molecular methods FISH and QF–PCR. *Reproduction* 126:279–297.

Hunt, M. M., G. Meng, D. E. Rancourt, I. D. Gates, and M. S. Kallos. 2014. Factorial experimental design for the culture of human embryonic stem cells as aggregates in stirred suspension bioreactors reveals the potential for interaction effects between bioprocess parameters. *Tissue Eng Part C Methods* 20 (1):76–89. doi:10.1089/ten. tec. 2013.0040.

Ivanova, N. B., J. T. Dimos, C. Schaniel, J. A. Hackney, K. A. Moore, and I. R. Lemischka. 2002. A stem cell molecular signature. *Science* 298:601–604.

Kimura, H., M. Tada, N. Nakatsuji, and T. Tada. 2004. Histone code modifications on pluripotential nuclei of reprogrammed somatic cells. *Mol Cell Biol* 24 (13):5710–5720. doi:10.1128/mcb.24.13.5710-5720.2004.

Klockner, W., and J. Buchs. 2012. Advances in shaking technologies. *Trends Biotechnol* 30 (6):307–314. doi:10.1016/j.tibtech.2012.03.001.

Kotoula, V., S. I. Papamichos, and A. F. Lambropoulos. 2008. Revisiting OCT4 expression in peripheral blood mononuclear cells. *Stem Cells* 26 (1):290–291. doi:10.1634/stemcells.2007-0726.

Kropp, C., H. Kempf, C. Halloin, D. Robles-Diaz, A. Franke, T. Scheper, K. Kinast et al. 2016. Impact of feeding strategies on the scalable expansion of human pluripotent stem cells in single-use stirred tank bioreactors. *Stem Cells Transl Med* 5 (10):1289–1301. doi:10.5966/sctm.2015-0253.

Kumar, A., and B. Starly. 2015. Large scale industrialized cell expansion: Producing the critical raw material for biofabrication processes. *Biofabrication* 7 (4):044103. doi:10.1088/1758-5090/7/4/044103.

Kwok, C. K., Y. Ueda, A. Kadari, K. Gunther, A. Heron, A. C. Schnitzler, M. Rook, and F. Edenhofer. 2017. Scalable stirred suspension culture for the generation of billions of human induced pluripotent stem cells using single-use bioreactors. *J Tissue Eng Regen Med*. doi:10.1002/term.2435.

Lee, J. H., S. R. Hart, and D. G. Skalnik. 2004. Histone deacetylase activity is required for embryonic stem cell differentiation. *Genesis* 38 (1):32–38. doi:10.1002/gene.10250.

Lian, X., J. Zhang, S. M. Azarin, K. Zhu, L. B. Hazeltine, X. Bao, C. Hsiao, T. J. Kamp, and S. P. Palecek. 2013. Directed cardiomyocyte differentiation from human pluripotent stem cells by modulating Wnt/beta-catenin signaling under fully defined conditions. *Nat Protoc* 8 (1):162–175. doi:10.1038/nprot.2012.150.

Life Technologies Corporation. 2013. Essential 8™ Medium and Vitronectin FAQs. https://tools.thermofisher.com/content/sfs/manuals/FAQ_Essen8_Medium_vitronectin_man.pdf. Accessed May 1, 2018.

Lu, J., R. Hou, C. J. Booth, S. H. Yang, and M. Snyder. 2006. Defined culture conditions of human embryonic stem cells. *Proc Natl Acad Sci U S A* 103 (15):5688–5693. doi:10.1073/pnas.0601383103.

Ludwig, T. E., M. E. Levenstein, J. M. Jones, W. T. Berggren, E. R. Mitchen, J. L. Frane, L. J. Crandall et al. 2006. Derivation of human embryonic stem cells in defined conditions. *Nat Biotechnol* 24 (2):185–187. doi:10.1038/nbt1177.

Lund, R. J., T. Nikula, N. Rahkonen, E. Narva, D. Baker, N. Harrison, P. Andrews, T. Otonkoski, and R. Lahesmaa. 2012. High-throughput karyotyping of human pluripotent stem cells. *Stem Cell Res* 9 (3):192–195. doi:10.1016/j.scr.2012.06.008.

Madonna, R., Y. J. Geng, H. Shelat, P. Ferdinandy, and R. De Caterina. 2014. High glucose-induced hyperosmolarity impacts proliferation, cytoskeleton remodeling and migration of human induced pluripotent stem cells via aquaporin-1. *Biochim Biophys Acta* 1842 (11):2266–2275. doi:10.1016/j.bbadis.2014.07.030.

Mahdieh, N., and B. Rabbani. 2013. An overview of mutation detection methods in genetic disorders. *Iranian J Ped* 23 (4):375–388.

Merten, O. W. 2015. Advances in cell culture: Anchorage dependence. *Philos Trans R Soc Lond B Biol Sci* 370 (1661):20140040. doi:10.1098/rstb.2014.0040.

Miltenyi Biotec. 2017. StemMACS™ iPS-Brew XF, human. http://www.miltenyibiotec.com/en/products-and-services/macs-cell-culture-and-stimulation/media/stem-cell-media/stemmacs-ips-brew-xf-human.aspx.

Muller, F. J., J. Goldmann, P. Loser, and J. F. Loring. 2010. A call to standardize teratoma assays used to define human pluripotent cell lines. *Cell Stem Cell* 6 (5):412–414. doi:10.1016/j.stem.2010.04.009.

Nagata, S. 1975. *Mixing: Principles and Applications*. New York: Wiley.

Nampe, D., R. Joshi, K. Keller, N. I. Zur Nieden, and H. Tsutsui. 2017. Impact of fluidic agitation on human pluripotent stem cells in stirred suspension culture. *Biotechnol Bioeng*. doi:10.1002/bit.26334.

Olmer, R., A. Lange, S. Selzer, C. Kasper, A. Haverich, U. Martin, and R. Zweigerdt. 2012. Suspension culture of human pluripotent stem cells in controlled, stirred bioreactors. *Tissue Eng Part C Methods* 18 (10):772–784. doi:10.1089/ten.tec.2011.0717.

Phanstiel, D., J. Brumbaugh, W. T. Berggren, K. Conard, X. Feng, M. E. Levenstein, G. C. McAlister, J. A. Thomson, and J. J. Coon. 2008. Mass spectrometry identifies and quantifies 74 unique histone H4 isoforms in differentiating human embryonic stem cells. *Proc Natl Acad Sci U S A* 105 (11):4093–4098. doi:10.1073/pnas.0710515105.

Ponnuru, K., J. Wu, P. Ashok, E. S. Tzanakakis, and E. P. Furlani. 2014. Analysis of stem cell culture performance in a microcarrier bioreactor system. *Techconnect World Innovation Conference & Expo*, Washington, DC.

Prakash Bangalore, M., S. Adhikarla, O. Mukherjee, and M. M. Panicker. 2017. Genotoxic effects of culture media on human pluripotent stem cells. *Sci Rep* 7:42222. doi:10.1038/srep42222.

Rajala, K., H. Vaajasaari, R. Suuronen, O. Hovatta, and H. Skottman. 2011. Effects of the physiochemical culture environment on the stemness and pluripotency of human embryonic stem cells. *Stem Cell Studies* 1 (1):3. doi:10.4081/scs.2011.e3.

Ramalho-Santos, M., S. Yoon, Y. Matsuzaki, R. C. Mulligan, and D. A. Melton. 2002. "Stemness": Transcriptional profiling of embryonic and adult stem cell. *Science* 298:597–600.

Riento, K., and A. J. Ridley. 2003. Rocks: Multifunctional kinases in cell behaviour. *Nat Rev Mol Cell Biol* 4 (6):446–456. doi:10.1038/nrm1128.

Rodrigues, C. A., T. G. Fernandes, M. M. Diogo, C. L. da Silva, and J. M. Cabral. 2011. Stem cell cultivation in bioreactors. *Biotechnol Adv* 29 (6):815–829. doi:10.1016/j.biotechadv.2011.06.009.

Sen, A., M. S. Kallos, and L. A. Behie. 2002. Expansion of mammalian neural stem cells in bioreactors: Effect of power input and medium viscosity. *Brain Res Dev Brain Res* 134 (1–2):103–113.

Serra, M., C. Brito, M. F. Sousa, J. Jensen, R. Tostoes, J. Clemente, R. Strehl, J. Hyllner, M. J. Carrondo, and P. M. Alves. 2010. Improving expansion of pluripotent human embryonic stem cells in perfused bioreactors through oxygen control. *J Biotechnol* 148 (4):208–215. doi:10.1016/j.jbiotec.2010.06.015.

Squire, J.A., J. Bayani, C. Luk, L. Unwin, J. Tokunaga, C. MacMillan, J. Irish, D. Brown, P. Gullane, and S. Kamel-Reid. 2002. Molecular cytogenetic analysis of head and neck squamous cell carcinoma: By comparative genomic hybridization, spectral karyotyping, and expression array analysis. *Head Neck* 24 (9):874–887.

Stathopoulos, N. A., and J. D. Hellums. 1985. Shear stress effects on human embryonic kidney cells In Vitro. *Biotechnol Bioeng* 27:1021–1026.

STEMCELL Technologies. 2017a. mTeSR™1. https://www.stemcell.com/mtesr1.html.

STEMCELL Technologies. 2017b. TeSR™2. https://www.stemcell.com/tesr2.html.

Taapken, S. M., B. S. Nisler, M. A. Newton, B. L. Sampsell-Barron, K. A. Leonhard, E. M. McIntire, and K. D. Montgomery. 2011. Karyotypic abnormalities in human induced pluripotent stem cells and embryonic stem cells. *Nat Biotechnol* 29 (4):313–314.

Thermo Fisher Scientific. 2017. STEMPRO® hESC SFM—Human Embryonic Stem Cell Culture Medium. https://www.thermofisher.com/us/en/home/life-science/stem-cell-research/stem-cell-culture/stem-cell-research-misc/stempro-hesc-sfm.html.

Viswanathan, P., T. Gaskell, N. Moens, O. J. Culley, D. Hansen, M. K. Gervasio, Y. J. Yeap, and D. Danovi. 2014. Human pluripotent stem cells on artificial microenvironments: A high content perspective. *Front Pharmacol* 5:150. doi:10.3389/fphar.2014.00150.

Wang, Y. C., M. Nakagawa, I. Garitaonandia, I. Slavin, G. Altun, R. M. Lacharite, K. L. Nazor et al. 2011. Specific lectin biomarkers for isolation of human pluripotent stem cells identified through array-based glycomic analysis. *Cell Res* 21 (11):1551–1563. doi:10.1038/cr.2011.148.

Wang, Z., D. E. Schones, and K. Zhao. 2009. Characterization of human epigenomes. *Curr Opin Genet Dev* 19 (2):127–134. doi:10.1016/j.gde.2009.02.001.

Watanabe, K., M. Ueno, D. Kamiya, A. Nishiyama, M. Matsumura, T. Wataya, J. B. Takahashi et al. 2007. A ROCK inhibitor permits survival of dissociated human embryonic stem cells. *Nat Biotechnol* 25 (6):681–686. doi:10.1038/nbt1310.

Werner, S., S. C. Kaiser, M. Kraume, and D. Eibl. 2014. Computational fluid dynamics as a modern tool for engineering characterization of bioreactors. *Pharm Bioproc* 2 (1):85–99.

Wu, J., M. R. Rostami, D. P. Cadavid Olaya, and E. S. Tzanakakis. 2014. Oxygen transport and stem cell aggregation in stirred-suspension bioreactor cultures. *PLoS One* 9 (7):e102486. doi:10.1371/journal.pone.0102486.

Zachar, V., S. M. Prasad, S. C. Weli, A. Gabrielsen, K. Petersen, M. B. Petersen, and T. Fink. 2010. The effect of human embryonic stem cells (hESCs) long-term normoxic and hypoxic cultures on the maintenance of pluripotency. *In Vitro Cell Dev Biol Anim* 46 (3–4):276–283. doi:10.1007/s11626-010-9305-3.

2 Bioreactors for Human Pluripotent Stem Cell Expansion and Differentiation

*Carlos A.V. Rodrigues, Mariana Branco, Diogo
E.S. Nogueira, Teresa P. Silva, Ana Rita Gomes,
Maria Margarida Diogo, and Joaquim M. S. Cabral*

CONTENTS

2.1 CULTURE OF HUMAN PLURIPOTENT STEM CELLS—INTRODUCTION

Human pluripotent stem cells (hPSC) are capable of differentiating into cells from the three germ layers that give rise to somatic cells of the human body: ectoderm, endoderm, and mesoderm. hPSC also demonstrate indefinite self-renewal capacity, associated with high telomerase activity and undergo symmetric divisions in culture without differentiating (Takahashi et al. 2007). There are several types of hPSC, mainly embryonic stem cells (ESC), obtained from the inner cell mass of blastocyst or by somatic cell nuclear transfer (Thomson et al. 1998, Tachibana et al. 2013) and induced pluripotent stem cells (iPSC), which can be obtained from the reprogramming of somatic cells (Takahashi et al. 2007).

The first cultures of hESC were performed using mouse embryonic fibroblasts (MEF) feeder layers (Thomson et al. 1998), which secrete soluble factors and cytokines into the culture medium, supporting cell growth. However, the use of feeder cells has many drawbacks, including the creation of an undefined microenvironment, difficult to study and control, as well as the variability and technical complexity associated with culturing two types of cells (Discher et al. 2009).

Feeder-free culture strategies were afterwards introduced where feeder cells are replaced by Matrigel, the trade name for a basement membrane substrate extracted from Engelbreth-Holm-Swarm mouse tumors. Matrigel consists of a gelatinous mixture of proteins, including extracellular matrix (ECM) components that promote cell adhesion such as laminin, collagen IV or entactin, and several growth factors (Kleinman et al. 1982, Takahashi and Yamanaka 2006). However, the animal origin and variability between batches still limits the use of this substrate for clinical purposes. Other substrates have been shown to be efficient in expanding hiPSC in an undifferentiated state, including recombinant proteins, such as human recombinant laminin-511 (Rodin et al. 2010) or synthetic polymeric matrices (Nakagawa et al. 2014). Another field in progress is the development of engineered biomaterial surfaces, as Synthemax™ (a biologically active peptide-functionalized acrylate polymer), bringing a huge promise in terms of efficiency and scalability of components, potentially facilitating the translation for clinical application (Tong et al. 2015). Although they have great promise, these alternative substrates are still expensive and under-optimized.

Another important point that determines successful hPSC culture is cell-cell interactions, since these cells have a low survival rate when dissociated into single cells. A procedure that improves survival upon cell individualization consists in the incubation of the hPSC with a Rho-associated kinase (ROCK) inhibitor (Watanabe et al. 2007). Alternatively, hiPSC can be passaged as small aggregates and survive through re-aggregation, maximizing cell survival. For that purpose, an enzyme-free passaging method using ethylenediaminetetraacetic acid (EDTA) was developed (Beers et al. 2012), achieving high efficiency in terms of cell survival, without any additional drug treatment. Cadherins are proteins located on the cell surface, which have an important role on cell-cell adhesion and require calcium ions for their function (Hirano et al. 1987, Chen et al. 2012). EDTA acts as a chelating agent that removes calcium, leading to the dissociation of the human induced (hiPSC) colonies into small cell clumps (Beers et al. 2012). EDTA passaging is thus an efficient alternative to enzymatic methods (e.g., using Trypsin or Accutase), which is also gentler to the cells.

Different serum-free culture media have been formulated for long-term maintenance of hPSCs. mTeSR1, which has been widely used for the derivation and culture of hiPSC, contains a total of 18 components, including glucose, glutamine, basic fibroblast growth factor (bFGF), transforming growth factor β (TGFβ), bovine serum albumin (BSA), and others (Chen et al. 2011b). The role of bFGF is critical to prevent cells from differentiating, as it is an antagonist of the bone morphogenetic protein (BMP) signaling pathway. In addition, TGFβ is also important for pluripotency maintenance, as a cytokine that activates transcription factors. BSA is an animal origin protein that can introduce batch variability and therefore this medium is not xenogeneic (xeno)-free, and its composition is not fully defined. A new and completely defined culture medium formulation, named E8, was subsequently introduced (Chen et al. 2011b). Notably, this commercially available culture medium does not require BSA and is composed of only eight essential components—insulin, selenium, transferrin, L-ascorbic acid, bFGF and TGFβ in DMEM/F12 and NaHCO$_3$. E8 is a Good Manufacturing Practices (GMP)-compliant alternative to mTeSR1 and achieves equivalent results concerning hiPSC expansion and pluripotency maintenance (Chen et al. 2011b, Wang et al. 2013).

In order to scale-up the 2D monolayer culture system, other alternative systems have been developed, such as hiPSC culture on microcarriers (Badenes et al. 2016a) and 3D aggregates in suspension (Wang et al. 2013). These systems will be described with more detail in the following sections (see Sections 2.2 and 2.3).

The generation of patient-specific hiPSC from adult cells has the potential to revolutionize regenerative medicine, since these cells may be used to repair tissues affected by disease or injury. Furthermore, these cells circumvent the ethical issues related with the use of hESC and immune rejection (Yamanaka 2009). hiPSC-derived cells also allow the generation of *in vitro* disease models, which may be used for drug discovery, toxicology, and Precision Medicine (Grskovic et al. 2011, Hawgood et al. 2015). However, despite promising advances, these therapies are still far from being applied in the clinic. A more immediate application of hiPSC can be expected in the fields of drug discovery and safety pharmacology (Grskovic et al. 2011, Avior et al. 2016). In fact, only <10% of the drugs that enter the clinical phase of testing reach the stage of market approval (Inoue and Yamanaka 2011). One of the reasons for the failure of drug candidates is the fact that many compounds demonstrate benefit in animal models but fail to show effectiveness in humans (Inoue and Yamanaka 2011). The current pre-clinical models used to test the potential positive effects of some drugs and to study the molecular and cellular pathways of these diseases include animal models and immortalized human cell lines (Durnaoglu et al. 2011). Although these models help to understand the different mechanisms of degeneration, differences in anatomy, metabolism, and behavior between animals and humans make it difficult to fully recapitulate the human disease. Therefore, many candidate drugs that have significant effects on these models fail to show relevant positive effects in clinical trials. The potential ability to test new drugs on human cells, particularly cells that are difficult to access (e.g., neurons or cardiomyocytes), is expected to greatly enhance the success of new drug approval and reduce the costs associated with the discovery process and clinical studies.

Microfluidic approaches may be used to complement the potential of hPSC in pharmacological applications and help optimize the culture conditions to be used at larger scales. Microfluidic channels can be easily created using soft lithography techniques, where a 3D structure is printed and used as a mold to create the microfluidic channels with the desired material, and these channels allow to finely control the supply of nutrients, growth factors, small molecules, oxygen, or other components of the cell culture microenvironment, such as hydrodynamic shear. As such, they constitute a platform to perform high-throughput screening experiments, allowing for fast results with limited sample volumes. Another distinctive aspect of microfluidic devices is regarding the flow conditions they promote. Microfluidic devices present a low Reynolds number, and as such, liquid flow inside the channels is well organized (i.e., laminar flow) and allows for the creation of gradients of soluble factors inside the microchannels, thus allowing to screen compound dosage effects on the cells in one single device. Villa-Diaz and co-workers showed that hPSCs could be cultured as undifferentiated colonies in Matrigel-coated Polydimethylsiloxane (PDMS) microfluidic channels while maintaining a normal karyotype (Villa-Diaz et al. 2009). The flowing conditions inside the microchannel allowed for the formation of spatial gradients through the channel, thus targeting specific colonies, or even exposing one single colony to a gradient of molecules. Cimetta and co-workers used a perfusion microbioreactor to expose hPSC aggregates to gradients of growth factors to study the influence of their concentration in the mesodermal induction of the cells (Cimetta et al. 2013). Additional examples of the potential of microfluidics for hPSC applications, such as the effect of culture parameters (e.g., soluble factors, oxygen and shear stress) on cell fate and the cell-cell and cell-ECM interactions are reviewed elsewhere (Toh et al. 2010, Marques and Szita 2016).

2.2 BIOREACTORS FOR THE EXPANSION OF HUMAN PLURIPOTENT STEM CELLS

hPSC derivatives can be used for a wide variety of biomedical applications, as discussed earlier, but either of these applications require large amounts of these cells with the desired functionality. Traditional culture systems have relied on 2D surfaces coated with an ECM-like substrate, where cells would adhere to and proliferate as a monolayer. However, not only do these systems fail to properly translate the *in vivo* niche, which may negatively influence cell behavior (Birgersdotter et al. 2005, Baker and Chen 2012, Breslin and O'Driscoll 2013), they are also inefficient for the large-scale production of hPSC. Clinical translation of the human pluripotent stem cell-derived progenitors or fully differentiated cells implies substantial doses (10^9–10^{10}) of pure cell populations per patient to efficiently recover organ function (Kempf et al. 2015), and therefore many tissue culture plates would be required to attain the required surface area (Zweigerdt 2009). For that reason, the adaptation of the current known methodologies used in 2D culture dishes to bioreactors are crucial to reach the desired number of cells. Different strategies have been proposed for scaling-up hPSC culture. For example, the most explored strategy is the use of dynamic 3D culture in bioreactors, using microcarriers as support matrices or self-aggregation strategy. The following sections will review the challenges and the methods already described for hPSC culture in bioreactors, which are also summarized in Figure 2.1.

2.2.1 GENERAL CONSIDERATIONS

Cell culture, particularly when performed in bioreactors, requires the optimization of various parameters. pH, temperature, and dissolved oxygen may be monitored through the various probes in the reactor and controlled in line. For hPSC culture, pH and temperature are generally maintained at 7.4 and 37°C, respectively.

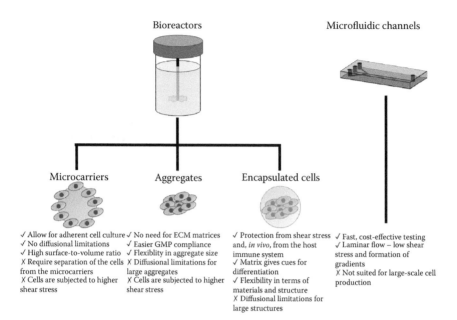

Bioreactors   Microfluidic channels

Microcarriers   Aggregates   Encapsulated cells

✓ Allow for adherent cell culture ✓ No need for ECM matrices   ✓ Protection from shear stress ✓ Fast, cost-effective testing
✓ No diffusional limitations   ✓ Easier GMP compliance   and, *in vivo*, from the host ✓ Laminar flow – low shear
✓ High surface-to-volume ratio ✓ Flexiblity in aggregate size immune system   stress and formation of
✗ Require separation of the cells ✗ Diffusional limitations for ✓ Matrix gives cues for   gradients
from the microcarriers   large aggregates   differentiation   ✗ Not suited for large-scale cell
✗ Cells are subjected to higher ✗ Cells are subjected to higher ✓ Flexibility in terms of   production
shear stress   shear stress   materials and structure
                                                                ✗ Diffusional limitations for
                                                                large structures

FIGURE 2.1   Bioreactor systems for hPSC expansion. hPSC can be cultured as suspension aggregates, attached to microcarriers or encapsulated in biomaterials. Microfluidic systems are powerful tools for process optimization.

Dissolved oxygen (% dO_2) in bioreactors is controlled as percentage of saturation when in contact with atmospheric air and, thus, 21% O_2 in the atmospheric air corresponds to 100% dO_2—a value that has been employed in many studies of hPSC expansion in bioreactors. More recently, however, some authors suggested that lower dO_2 may be favorable for the maintenance of pluripotency, since the early mammalian embryo is also subjected to hypoxic conditions (Ezashi et al. 2005, Serra et al. 2010).

The agitation rate is also a parameter that must be maintained in a tight interval. A minimum agitation rate is necessary to maintain the cells in suspension, but high rates should also be avoided to protect the cells from excessive shear stress, which can cause their uncontrolled differentiation or death (Nienow 2006, Leung et al. 2011). Cell damage can also be avoided by using gentle agitator geometries, such as a pitched-blade turbine (Olmer et al. 2012). Most notably, the recently introduced Vertical-Wheel™ Bioreactors (PBS Biotech) employ vertical agitation, allowing homogeneous mixing of the cell suspension with lower shear stress than conventional approaches, thus causing less damage to the cells (Croughan et al. 2016).

Many bioreactor studies employ a "repeated batch" feeding strategy, changing a percentage of the medium per time interval, similarly to 2D cultivation. Higher frequencies of medium exchange are correlated with higher cell yields, due to the more frequent replenishment of nutrients and depletion of harmful by-products, such as lactate (Lipsitz and Zandstra 2015). More recent studies have taken this to the limit, by applying a perfusion feeding strategy, where medium is constantly being replaced, while the cells are maintained inside the reactor via a retention device such as a spin filter (Kropp et al. 2016). Perfusion feeding was shown to be more beneficial for cell growth, avoiding exposing the cells to drastic changes in the medium composition and allowing for higher cell densities at the end of the culture.

All the detailed variables need to be optimized for the system under study, since different hPSC lines present different requirements. Although these variables are generally optimized independently, they interact with each other and, thus, this approach may not lead to the optimal set of conditions. Different groups developed a factorial experimental design to optimize multiple culture parameters simultaneously (e.g., bioreactor inoculation density and agitation rate) with a minimal number of individual runs (Hunt et al. 2014, Badenes et al. 2016b). In this approach, low, medium and high values of the variables to optimize are selected and all the combinations of those values are tested. Statistical analysis is then employed to generate surface plots, showing their influence on a selected readout (e.g., fold expansion, cell viability, aggregate size) and the optimal condition. Factorial experiment design is, thus, a powerful tool to understand and optimize hPSC cell culture in bioreactors.

2.2.2 MICROCARRIERS

Microcarriers are particles, normally of a spherical shape with a diameter between 100 and 230 µm (Markvicheva and Grandfils 2004), allowing for hPSC attachment and growth in suspension. This approach allows the transition from 2D monolayer cultures to 3D, as cells are similarly attached to a surface and directly exposed to the culture medium and oxygen, thus avoiding diffusional limitations. However, microcarriers allow for a much greater surface-to-volume ratio when compared to 2D, thus enabling large-scale expansion.

At the end of the expansion, the microcarriers must be removed from the cell suspension and for that purpose a size-based separation methodology, such as filtration, needs to be employed (Badenes et al. 2015). The requirement for this type of downstream processing increases the cost of the overall process and subjects the cells to straining procedures, thus impacting the number of viable cells that can be recovered from culture. This issue may be at least partially solved using dissolvable microcarriers that can be "digested" with little to no harm to the cells as shown by Fernandes and colleagues (Fernandes et al. 2007), for example, who described the culture of mouse ESC in spinner flasks using Cultispher® microcarriers, which can be dissolved enzymatically (e.g., with trypsin). On the other hand, microcarrier culture renders cells very susceptible to shear stresses. Since the agitation speed must be high to keep the microcarriers in suspension, the turbulent eddies will be smaller than the cell layer and, additionally, many bead-to-bead collisions will occur, both causing damage to the cells (Croughan et al. 1987).

There is a myriad of different commercial microcarriers currently available, differing in characteristics such as matrix composition, coating, and dimensions. Some examples include GE Healthcare®'s Cytodex® 1 and Cytodex® 3, which consist of a dextran matrix coated with either diethylaminoethyl (DEAE) or denatured type

I collagen, respectively; Whatman®'s DE-52 and DE-53 microcarriers, consisting of a cellulose matrix and DEAE coating; or Pall®'s SoloHill® microcarriers, which are available as xeno-free products. Most of these microcarriers were developed for large-scale cultures of mammalian cell lines for therapeutic protein production. For stem cell culture, most of these microcarriers must be adapted with an additional coating of ECM proteins. Uncoated plastic microcarriers are often used for this purpose, functionalized with a specific coating, such as Matrigel™ (Serra et al. 2010) or specific ECM proteins (Badenes et al. 2016b), according to the cell line or the desired application. Corning®'s Synthemax® II microcarriers, composed of a polystyrene matrix and a synthetic vitronectin-based peptide substrate (Chen et al. 2013, Chen et al. 2011a, Badenes et al. 2015) are also now commercially available and constitute a ready-to-use xeno-free alternative. All these examples refer to solid, or microporous, microcarriers. Alternatively, macroporous microcarriers, including Cytopore® or Cultispher®, have pore sizes large enough (10–30 μm) to accommodate cells in their interior, greatly increasing the surface area available per microcarrier. However, unlike solid microcarriers, macroporous microcarriers have not been widely explored for hPSC culture.

Microcarriers have been successfully applied in a variety of systems for hPSC culture. Initial reports focused on screening different microcarriers and demonstrating the feasibility of their use for hPSC culture on low-attachment tissue culture plates (Phillips et al. 2008, Nie et al. 2009, Oh et al. 2009, Chen et al. 2010, Chen et al. 2011a). These studies fostered the usage of microcarriers in dynamic, stirred systems, namely spinner flasks, with up to 13-fold expansion in viable cell numbers (Oh et al. 2009, Serra et al. 2010, Chen et al. 2011a, Leung et al. 2011, Bardy et al. 2013). Controlled stirred-tank bioreactors have also been used for cell expansion on microcarriers (Serra et al. 2010, Olmer et al. 2012, Silva et al. 2015), allowing for tight control of the process parameters (e.g., temperature, pH, dissolved oxygen), higher cell numbers, and demonstrating the high scalability of 3D cell culture. However, this setup has so far been relatively unexplored and limited to small volumes (≤300 mL) due to large requirements of expensive culture media. More recently, completely xeno-free systems have been described, envisaging the possibility of clinical translation. Badenes and co-workers obtained a 3.5-fold increase in cell density in spinner flasks by using polystyrene microcarriers coated with recombinant vitronectin for cell adhesion and E8 as the medium for cell growth (Badenes et al. 2016b).

2.2.3 3D CELL AGGREGATES

A different strategy for hPSC suspension culture consists in the formation of 3D spherical cell aggregates, without the need to adhere to external matrices, thus simplifying the downstream processing in relation to cell culture on microcarriers. Both of these factors contribute to an easier compliance with GMP, placing it as a more favorable approach for acceptance in the production of cells to be used in a clinical setting (Kropp et al. 2016). To generate the cell inoculum for aggregate suspension culture, hPSC colonies from 2D cultures can be dissociated into single cells using an enzymatic agent and incubated with culture medium supplemented with ROCK inhibitor for the first day of culture (Watanabe et al. 2007, Amit et al. 2011). The cells will then form aggregates, which will increase in size throughout the culture time, at an average diameter that depends on the stirring rate. The aggregates can then be dissociated and used for further hPSC expansion or a differentiation step can be integrated. The influence of aggregate size on differentiation trajectories has been widely demonstrated (Bauwens et al. 2008). In fact, the optimum initial aggregate size varies for differentiation into different cell types, and stirred bioreactors allow "fine-tuning" of the aggregate size obtained by adjusting the agitation speed, with greater agitation speeds resulting in a lower average aggregate size (Sen et al. 2002). Large aggregates (>200 μm), however, are not desirable due to a poor diffusion of oxygen and nutrients and could possibly lead to the formation of necrotic centers (Hoeben et al. 2004, Wu et al. 2014). On the other hand, small aggregates may also be problematic, since the high stirring rates necessary to obtain them exert strong shear stress over the cell and may affect cell fate and/or viability.

hPSC culture as aggregates has already been achieved in a variety of systems, while maintaining the expression of pluripotency markers and a normal karyotype. Aggregate culture in spinner flasks has resulted in expansion factors up to 25-fold in 10 days (Singh et al. 2010, Steiner et al. 2010, Zweigerdt et al. 2011, Amit et al. 2011). Kropp and co-workers have performed hPSC expansion as aggregates in 125 mL bioreactors, obtaining a four-fold increase in cell density after seven days of culture with a repeated batch feeding regimen and a seven-fold increase with a perfusion setup. Interestingly, this study reports a shift in cell metabolism from glycolysis to oxidative phosphorylation at later stages of the culture (Kropp et al. 2016). Using conditions optimized via factorial design, Hunt and co-workers (Hunt et al. 2014) obtained a maximum fold increase in total cell number of 12.2 following six days of expansion of hESCs in 125 mL bioreactors. All of the aforementioned studies used mTeSR1 as a culture medium, however, E8 medium has also been shown to be able to support undifferentiated hPSC growth as aggregates for over 10 passages, with an average 3.7-fold increase in cell number every three to four days in spinner flasks (Wang et al. 2013). Given the potential for GMP-compliant hPSC expansion as aggregates, some studies have explored this possibility. Chen and co-workers were able to expand three different hESC lines in the serum-free medium StemPro® in 125 mL spinner flasks. Cells were passaged every three to four days with an average 4.3-fold increase per passage and were demonstrated to maintain genomic integrity and high (>90%) viability and pluripotency during at least 21 passages. Notably, cells maintained their undifferentiated expansion when passaged from 125 to 500 mL spinner flasks, and were cryopreserved, generating GMP cell banks following expansion (Chen et al. 2012). More recently, Abecasis and co-workers expanded hPSC using Cellartis® DEF-CS™ xeno-free medium in 200 mL bioreactors, under hypoxic conditions (4% oxygen saturation) and with a perfusion rate optimized to maintain a non-inhibitory lactate concentration (<20 mM [mili molar]). This approach led to an average 18.7-fold cell expansion after three to four days of culture, with maintenance of karyotype and pluripotency during at least three passages (Abecasis et al. 2017).

2.2.4 MICROENCAPSULATION

Cell culture can also be performed by microencapsulating cells inside a biomaterial. The biomaterial confers protection against shear stress inside bioreactors and may provide biochemical and/or biomechanical cues to the cells based on its composition and stiffness, respectively. Hydrogels are the most common materials used for cell encapsulation, since these are biocompatible, elastic and their characteristics can be fine-tuned (Slaughter et al. 2009). Hydrogels responsive to stimuli like temperature or pH are also available (Li et al. 2016, Rodrigues et al. 2017). Materials that have been used for microencapsulation and further hPSC expansion as aggregates include polyethylene glycol (PEG), N-isopropylacrylamide (NIPAAm) (Li et al. 2012, Rodrigues et al. 2017), poly(lactic-*co*-glycolic acid) (PLGA), alginate (Serra et al. 2011), agarose or gelatin (Bratt-Leal et al. 2011), among others. Microencapsulation was also used for differentiation of hPSC into a variety of different cell lineages including cardiac (Li et al. 2012), neural (Rodrigues et al. 2017), hepatic (Fang et al. 2007) and vascular (Ferreira et al. 2007). The referenced approaches were tested only in static culture systems, but Serra and co-workers demonstrated the applicability of cell microencapsulation for culture inside 125 mL spinner flasks. Cells were encapsulated in alginate beads, either as single cells, aggregates or attached to microcarriers; while the first approach was not successful, the latter two allowed for improved cell expansion and maintenance of cell viability when compared to non-encapsulated cells on microcarriers. While cells in encapsulated aggregates only maintained pluripotency for two weeks, encapsulated cells on microcarriers could be expanded for over 19 days, and could be cryopreserved, ensuring better cell survival following thawing over non-encapsulated cells (Serra et al. 2011).

2.3 DIFFERENTIATION OF HUMAN PLURIPOTENT STEM CELLS IN BIOREACTORS

hPSC applications such as drug discovery or toxicology assays will likely require differentiation into specialized cell types. In the case of cellular therapies, hPSC cannot be used in their undifferentiated state as they can generate teratomas upon transplantation. hPSC must be differentiated into more committed progenitors, without tumorigenic potential, or into terminally differentiated cells before their use. Methods for hPSC differentiation can be performed as 2D monolayer cultures or as 3D cell aggregates and typically consist of complex multi-stage protocols where hPSC are manipulated and sequentially cultured in different culture media, containing different combinations of growth factors, until they acquire the desired phenotype. Differentiation systems where cells are cultured as a 2D monolayer are difficult to perform at a large scale, mostly if they are performed in culture platforms that must be extensively scaled-out to provide a sufficient surface area for large-scale production (e.g., culture flasks, multi-tray flasks). hPSC differentiation as 3D aggregates can be scaled-up by adapting the methodologies used in culture dishes to bioreactors with large volumes. These methods, however, rely on cell aggregation, which may not be straightforward due to the need of controlling aggregate size and homogeneity and, particularly under stirred conditions, to the influence of shear stress on cell survival and differentiation efficiency (Kempf et al. 2016).

Microcarriers can also be used for hPSC differentiation in suspension and facilitate the transition to larger scale bioprocesses. Most of the existing reports describe that hPSC cultured on microcarriers do not form single layers of cells but, instead, attach predominantly to each other, forming cell-microcarrier aggregates. In this context, the application of microcarrier technology to hPSC differentiation can be seen as a way to assist cell aggregation and to stabilize and minimize agglomeration of the aggregates (Lecina et al. 2010). The following section will describe examples of application of bioreactors for hPSC differentiation into cells of the three embryonic germ layers, either as cell aggregates or adherent to microcarriers.

2.3.1 Differentiation into Ectodermal Lineages—Neural Progenitor Cells

The human brain represents one of the most complex structures of the human body and efforts have been done for understanding its structural and functional complexity. The inaccessibility of fetal and adult human brain for experimental manipulation leads to the use of animal models for the study of embryonic brain development and adult brain related neurodegenerative diseases (Giandomenico and Lancaster 2017). Furthermore, most of the current knowledge about neurodegenerative disease-related neuronal phenotypes is based on *postmortem* studies (Durnaoglu et al. 2011).

Since the model systems currently used do not fully recapitulate human brain development and disease, it is essential to develop an alternative *in vitro* model that recapitulates key features of human brain regions. Importantly, manufacturing methods capable of generating large amounts of neural cells will be critical for the implementation of high-throughput drug screening and toxicity assays, disease modelling, and ultimately, cell therapies.

An improved understanding of the mechanisms of neural development has facilitated the establishment of efficient protocols for the generation of hPSC-derived neural progenitors. In 2009, Chambers and collaborators reported a protocol that enables the highly efficient differentiation of neural precursors from hPSC by using a monolayer culture system and two inhibitors of small Mothers Against Decapentaplegic (SMAD) signaling (Noggin and SB431542) (Chambers et al. 2009). Since 2D cultures do not fully reflect the complexity of tissues and do not allow the generation of large quantities of neural precursors, the original protocol has been adapted to dynamic 3D culture systems, using scalable suspension culture systems. In fact, an efficient microcarrier-based integrated process for hiPSC expansion and differentiation into neural progenitor cells (NPCs) in spinner flasks cultures has been reported (Bardy et al. 2013). The protocol published by Bardy and colleagues involved two different steps. First, cells were propagated using Matrigel-coated DE-53 Microcarriers, in 100 mL-stirred spinner flasks, obtaining a high density of hPSC (6×10^6 cells/mL, 20-fold expansion) during six to seven days of expansion. The expansion was followed by the neural differentiation, which yielded 333 NPCs per seeded hiPSC (78%–85% purity) as compared to 53 in the classical 2D tissue culture protocol. Furthermore, the generated NPC had the ability to further differentiate into neurons (TUJ1[+]), astrocytes (GFAP[+]) and oligodendrocytes (O4[+]) (Bardy et al. 2013).

Integrated culture platforms for neural commitment of hiPSC using 3D suspension conditions and chemically-defined culture media were also reported that do not require the use of adherent matrices or microcarriers. Miranda and co-workers reported an efficient integration of expansion and neural commitment of hiPSC as free-floating 3D aggregates based on the dual SMAD inhibition method and using defined medium conditions (Miranda et al. 2015). In this work, the authors optimized the initial inoculation for hiPSC expansion as 3D aggregates and determined the optimal size of aggregates to initiate neural commitment. The scaling-up of this protocol was also described, using 50 mL spinner flasks (Miranda et al. 2016). The process was optimized using a factorial design approach, involving parameters such as agitation rate and seeding density, resulting in an efficient protocol for the generation of large numbers of NPCs from hiPSC after only six days of neural induction (Miranda et al. 2016). Differentiation protocols for the generation of mature neurons in a highly reproducible manner, using 500 mL disposable spinner flasks, were reported by Rigamonti and colleagues (2016). The generated neurons show an effective degree of maturation, being able to form synapses, exhibit spontaneous electrical activity, and respond appropriately to depolarization (Rigamonti et al. 2016). Therefore, this protocol robustly generates large numbers of functional human neurons that can offer an important source of cells for disease modelling and high-throughput drug screening and toxicity assays (Rigamonti et al. 2016).

For disease modelling, it is important to recapitulate the complex 3D brain organization, which impacts both cell identity and function. Therefore, in order to better model the development of this tissue, new *in vitro* 3D models, named cerebral organoids, were established (Kelava and Lancaster 2016, Giandomenico and Lancaster 2017).

Lancaster and collaborators have described the development of cerebral organoids from hPSC, including ESC and patient-specific hiPSC. The protocol starts with the generation of floating embryoid bodies embedded in Matrigel, without the addition of any growth factor. In order to enhance nutrient and oxygen mass transfer, the organoids embedded in Matrigel were transferred into a spinning bioreactor. This bioreactor strategy improved the development of brain tissue, enabling the formation of longer continuous neuroepithelial-like zones, instead of the smaller rosette structures obtained under static conditions. As proof-of-concept, patient-derived hiPSC with microcephaly were generated, being a valuable tool used for modelling neurodevelopmental disorders (Lancaster et al. 2013). The cerebral cortical regions were similar to early stages of human brain development, namely showing characteristic progenitor zone organization with abundant Outer radial (oRG) stem cells. More recently, the authors used a bioengineering approach to improve tissue architecture. For that purpose, they use poly(lactide-co-glycolide) copolymer (PLGA) fiber microfilaments as a floating scaffold to generate elongated embryoid bodies, enhancing neuroectoderm formation and improving cortical development (Lancaster et al. 2017). Despite all the improvements leading to the generation of organoids for modelling multiple brain-like regions, these structures do not fully model the organization of the brain. The organoid protocol was unable to give rise to well-developed outer subventricular zone (oSVZ) layer, with specialized population of oRGCs, which is characteristic of human species brain development.

Quian and co-workers described the generation of brain-region-specific organoids from hiPSC, mainly forebrain, midbrain and hypothalamic organoids. This protocol was performed using a miniaturized spinning bioreactor (SpinΩ), a multi-well spinning device that fits a standard 12-well tissue culture plate. This cost-effective system allowed comparing several conditions in parallel while promoting cell viability and originating organoids that recapitulate features of human brain development, such as highly organized ventricular structures. As proof-of-concept, the forebrain platform was used for chemical compound testing and modeling of ZIKA virus infection, as these organoids recapitulate the developing human cortex (Qian et al. 2016).

2.3.2 DIFFERENTIATION INTO MESODERMAL LINEAGES—CARDIOMYOCYTES

Myocardial infarction and heart failure are the leading causes of death worldwide. The endogenous regenerative capacity of the myocardium is limited and thus not enough to compensate for the damage after heart failure. Despite the improvements in the medical field to overcome this problem, heart transplantation is still the only definitive treatment. However, immune rejection of the graft and the low survival rate after heart transplantation, increase the need for new alternatives. The generation of hPSC-derived cardiomyocytes (CMs) for cell replacement therapy constitutes a promising alternative approach. The translation of hPSC differentiation into CM to 3D culture has been achieved mainly by taking advantage of the temporal modulation of the Wnt pathway using small molecules (Lian et al. 2012). In order to obtain significant quantities of CMs, different groups have tried to adapt this method to scalable suspension culture systems. Kempf and colleagues (Kempf et al. 2014, Kempf et al. 2015) used a 100 mL stirred-tank bioreactor platform to combine the expansion and differentiation of hPSC into CMs from single cells. This suspension culture protocol allowed the production of an average of 69% cells expressing cardiac Troponin T (cTnT[+]) in defined medium without further enrichment after 10 days of culture, enabling an average production of 40 million CMs per bioreactor run. In a different study performed by Fonoudi and collaborators (Fonoudi et al. 2015), an integrated single-unit operation platform for the large-scale production of hPSC-CM aggregates was reported. Using a stirred bioreactor operated under a batch mode, with a 100 mL working volume, 85% αMHC[+] cells could be generated by day 10 of differentiation. This platform is expected to allow the production of the desired number of cells for clinical translation by either scaling-out or scaling-up. More recently, Chen and colleagues (Chen et al. 2015) were able to obtain 96% of CMs expressing cTnT in a 1,000 mL spinner flask with a CM production of more than 10^9 CMs in a single run, also by the integration of undifferentiated hPSC expansion and small molecule-induced cardiac differentiation.

Instead of self-aggregation, microcarriers were also reported as support matrix for hPSC expansion and CM differentiation. In a study by Ting and co-works (Ting et al. 2014), cardiac differentiation was performed using microcarriers, and different systems with different levels of shear stress were compared, particularly a "wave bioreactor"-like rocker platform and a 100 mL spinner flask. From the two analyzed systems, the highest differentiation efficiency was obtained with the rocking platform, being the system selected for the development of an integrated hPSC expansion/CM differentiation bioprocess. With this optimized process, a cell population with cTNT expression levels around 65.7% $\pm$ 10.7% was obtained, as well as a CM production of 31.75 CM per each hESC seeded. The agitation was demonstrated to be a parameter with a very high impact during the first three days of differentiation, suggesting that the initial differentiation process is very sensitive to shear stress. Comparing intermittent and continuous agitation during the first three days of differentiation, intermittent agitation during that period was crucial for the success of differentiation, showing CM differentiation efficiencies around 80% higher compared with continuous agitation.

For all of these previously reported studies, there were different parameters that required a challenging optimization. Among them, control of the aggregate size was considered, for the majority of these studies, a crucial aspect for cardiac differentiation success. In those cases, aggregate size was mainly controlled by cell seeding density, time in culture, and stirring conditions, specifically the agitation rate and stirring intervals. Other challenges faced by different studies were the inherent hPSC cell-line variability, which forced specific optimization for each cell line analyzed, such as what concerns the optimization of the concentrations of growth factors and/ or small molecules used for cardiac induction. Control of cell viability and aggregate integrity in the 3D dynamic culture system was mainly performed by controlling the agitation rate. The integration of hPSC expansion and differentiation with a purification or enrichment step was also addressed by Hemmi and colleagues (Hemmi et al. 2014) who used a 125 mL spinner flask to integrate these three steps, taking advantage of the distinct metabolic characteristics of hPSC-CMs for the enrichment step. This was achieved by changing the carbon source from glucose to lactate in the culture medium. With this process, the authors were able to achieve >99% α-actinin-positive CMs from an initial population of 25% α-actinin-positive CMs after a period of seven days in the presence of lactose-supplemented medium. This platform generated approximately 1×10^7 CMs from 5×10^6 hiPSC.

Another important aspect for clinical translation of cell culture is the cost associated with the entire process of production, as well as the use of a defined and xeno-free culture medium. In the case of cardiac differentiation, highly efficient CM generation from hPSC has been achieved using the Wnt signaling modulation method, which can be performed without the supplementation of expensive growth factors, since it only requires two small molecules (CHIR99021 and IWP2, respectively for activation and inhibition of the Wnt signaling pathway). A chemically-defined and albumin-free culture medium for CMs generation has also been described (Lian et al. 2015) but its use in bioreactor cultures has not been reported yet.

2.3.3 Differentiation into Endodermal Lineages—Hepatocyte-Like Cells

The development of cell therapies for liver failure would greatly benefit from the development of large-scale bioprocesses for the generation of hPSC-derived liver cells, particularly hepatocytes. To date, most hepatocyte differentiation systems rely on 2D culture platforms and on the use of animal derived components, which limit the large-scale production and the applicability of these hepatocyte-like cells (HLCs). Moreover, only a few publications have tried to demonstrate the feasibility of generating endoderm progenitor cells or HLCs, by a scalable process.

Lock and colleagues (Lock and Tzanakakis 2009) reported the differentiation of endoderm progenitor cells in a microcarrier stirred-suspension culture using a 50 mL spinner flask. With this platform, the authors were able to obtain $\approx$80% cells expressing both FOXA2 and SOX17 after 12 days of endoderm commitment and 2.14×10^5 definitive endoderm cells per mL of differentiation medium spent. A microcarrier-based system for combined expansion and differentiation of hPSC to HLCs, using a 250 mL spinner flask, was reported by Park and collaborators (Park et al. 2014). Shear stress was identified as a critical parameter for successful hepatic differentiation and, therefore, low agitation rates were used in the spinner flask cultures (20–25 rpm). However, the low agitation rates used led to agglomeration of the microcarriers creating a non-homogeneous differentiation environment and diffusion limitations, thus compromising the differentiation efficiency and maintenance of the cells in the early stages of differentiation expressing endodermal markers. After 18 days of differentiation, starting with a cell concentration of 1.0×10^6 , 5.9×10^5 cells/mL were obtained showing an albumin secretion rate of 1.77 and 31.8 μg/mL/10^6 cells of urea genesis, which still remained far from what is observed for human primary hepatocytes (urea secretion: $\approx$100 mg/dL/day/10^6 cells; albumin secretion: 1250 mg/dL/day/10^6 cells [Song et al. 2009]). An increased initial cell density (4.0×10^6 cells/mL) yielded 2.52×10^6 cells/mL at the end of differentiation, with hepatocyte-like cells showing comparable hepatic functionality, suggesting the potential for large-scale differentiation. Additionally, Vosough and co-workers (Vosough et al. 2013) reported the differentiation of HLCs in a 50 mL stirred bioreactor. The endoderm induction of spheroids started when these achieved an average diameter of 130 $\pm$ 40 μm, which was reported to be the ideal for endoderm induction. After 22 days of differentiation, 55% of cells expressed albumin. The derived hPSC-HLCs exhibited CYP450 metabolic activity and showed an albumin secretion rate of 0.15–0.45 mg/day/10^6 cell and urea production of 4–10 (mg/dL)/day/10^6 cells. However, once again, these values are still far from what is reported in literature for human primary hepatocytes and, therefore, additional maturation is needed. HLCs were enriched using a fluorescence-activated cell sorting (FACS) step that was based on a specific aspect of HLCs functionality, which is low-density lipoprotein (LDL) uptake. The spheroid-derived HLCs were pre-incubated with the acetylated LDL labeled with a fluorescent probe (DiI) for four hours and at the end, from the sorted cells, 75% of these were DiI-Ac-LDL-positive, which matched with albumin-positive cells. The authors referred that a clinically relevant cell number of hPSC-HLCs (approximately $5–8 \times 10^8$ cells) could be produced after a period of three weeks using four spinner flasks that contain a total of 400 mL of medium. However, this method must be further improved and performed under completely xeno-free conditions before clinical translation can be considered.

2.4 CONCLUSIONS AND FUTURE TRENDS

Despite the recent efforts to translate lab-scale expansion and differentiation of hPSC to achieve relevant numbers of cells for high-throughput drug screening and toxicity assays, disease modelling, and clinical translation, there are still some challenges that need to be addressed. For this purpose, the integration of hPSC expansion and differentiation seems to be the most efficient process to achieve relevant cell numbers. Shear stress, and how that affects the commitment of pluripotent stem cells and the differentiation process, is one of the main questions that arise when using dynamic culture conditions. The agitation should be sufficiently high to minimize diffusional and mass transfer problems that could be responsible for the formation of gradients of nutrients, oxygen or other important molecules inside the aggregates ending with inefficient and heterogeneous differentiation, but low enough to allow the maintenance of aggregate integrity. Encapsulation of the 3D structures, for example in hydrogels or Matrigel, has been one of the approaches used to protect and increase cell viability under dynamic conditions. However, encapsulation is an extremely labor-intensive process, being difficult to scale-up and some of the matrices used are undefined and from animal origin. The discovery of novel biomaterials and more automatized encapsulation techniques are expected to greatly widen the use of this approach. Besides that, new agitation designs, such as Vertical-Wheel™ bioreactors, emerged to provide a more homogeneous shear distribution inside the bioreactor, allowing a gentle and uniform fluid mixing, efficient particle suspension with low power input and agitation speeds.

Although many studies are still performed in simple stirred systems (e.g., spinner flasks), a more closed and automatized control over different parameters of cell culture will have to be implemented to generate high-quality cells and improve the robustness and reproducibility of the bioprocesses. On the other hand, when considering clinical implementation, hPSC must be expanded and differentiated under GMP conditions to ensure the safety of the cell product. The use of xeno-free and well-defined culture conditions will be a step forward for introducing a cell-based medicinal product on the pharmaceutical market, since the use of animal-origin components is associated with several obstacles including the risk of pathogens transition and immunological rejection. Disposable bioreactors can be also considered as a beneficial option when clinical translation is the final goal.

Another important and more recent topic related with tridimensional culture systems is the use of hPSC to generate 3D organ-like structures. These organoids have already been reported for gut, intestine, kidney, and brain recapitulating the cellular and molecular features associated with the *in vivo* phenotype, allowing the use of these structures for more fundamental studies including drug screening or disease modeling. The use of micro-fabricated bioreactor platforms to generate these organ-like structures can be very important for high-throughput screening of the cellular response in three-dimensional environments under combinatorial manipulated physical, mechanical, and biological conditions. Therefore, effects of different culture conditions, growth factors or biomaterials on proliferation and differentiation might be evaluated allowing the creation of more *in vivo*-like culture systems (Table 2.1).

TABLE 2.1

Summary of Bioreactor Processes for hPSC Differentiation

Cell Type	Culture System	Differentiation	Final Cell Concentration	Purity	Yield	References
hiPSC	Spinner flask/Microcarriers	Ectoderm (neural progenitors)	10.6×10^6 total cells/mL	78% PSA-NCAM positive cells	333 neural progenitors/hiPSC	Bardy et al. (2013)
hiPSC	Spinner flask/Aggregates	Ectoderm (neural progenitors)	1×10^6 cells/mL	80% Pax6 positive cells	NA	Miranda et al. (2016)
hiPSC	Spinner flask/Aggregates	Ectoderm (mature neurons)	30×10^6 cells/mL	70% TBR1/CTIP2/SATB2 positive cells	NA	Rigamonti et al. (2016)
hiPSC hESC	Spinner flask/Aggregates	Ectoderm (cerebral organoids)	NA	NA	NA	Lancaster et al. (2013, 2015)
hiPSC	Miniaturized spinning bioreactor (SpinΩ)/Aggregates	Ectoderm (cerebral organoids)	NA	NA	NA	Qian et al. (2016)
hiPSC	Bioreactor/Aggregates	Mesoderm (cardiomyocytes)	0.4×10^6 cells/mL	69% cTNT positive cells	NA	Kempf et al. (2014, 2015)
hiPSC hESC	Stirred Bioreactor/Aggregates	Mesoderm (cardiomyocytes)	NA	85% αMHC positive cells	NA	Fonoudi et al. (2015)
hiPSC hESC	Spinner flasks/Aggregates	Mesoderm (cardiomyocytes)	1×10^6 cells/mL	96% cTNT positive cells	NA	Chen et al. (2015)
hESC	"Wave bioreactor"-like rocker platform/Spinner flask Microcarriers	Mesoderm (cardiomyocytes)	NA	65.7% ± 10.7% cTNT positive cells	31.71 cardiomyocytes/ESC	Ting et al. (2014)
hiPSC	Spinner flask/Aggregates	Mesoderm (cardiomyocytes)	NA	>99% α actinin positive cells	10×10^6 cardiomyocytes from 5×10^6 hiPSC	Hemmi et al. (2014)
hESC	Spinner flask/Microcarriers	Endoderm (hepatocytes)	0.2×10^6 cells/mL	80% FOXA2/SOX17 positive cells	NA	Lock et al. (2009)
hESC	Spinner flask/Microcarriers	Endoderm (hepatocytes)	2.52×10^6 cells/mL	NA	NA	Park et al. (2014)
hiPSC hESC	Spinner flask/Aggregates	Endoderm (hepatocytes)	$1.25–2 \times 10^6$ cells/mL	55% albumin positive cells	NA	Vosough et al. (2013)

REFERENCES

Abecasis, B., T. Aguiar, E. Arnault, R. Costa, P. Gomes-Alves, A. Aspegren, M. Serra, and P. M. Alves. 2017. Expansion of 3D human induced pluripotent stem cell aggregates in bioreactors: Bioprocess intensification and scaling-up approaches. *J Biotechnol* 246:81–93.

Amit, M., I. Laevsky, Y. Miropolsky, K. Shariki, M. Peri, and J. Itskovitz-Eldor. 2011. Dynamic suspension culture for scalable expansion of undifferentiated human pluripotent stem cells. *Nat Protoc* 6(5):572–579.

Avior, Y., I. Sagi, and N. Benvenisty. 2016. Pluripotent stem cells in disease modelling and drug discovery. *Nat Rev Mol Cell Biol* 17 (3):170–182.

Badenes, S. M., T. G. Fernandes, C. A. Rodrigues, M. M. Diogo, and J. M. Cabral. 2015. Scalable expansion of human-induced pluripotent stem cells in xeno-free microcarriers. *Methods Mol Biol* 1283:23–29.

Badenes, S. M., T. G. Fernandes, C. A. Rodrigues, M. M. Diogo, and J. M. Cabral. 2016a. Microcarrier-based platforms for *in vitro* expansion and differentiation of human pluripotent stem cells in bioreactor culture systems. *J Biotechnol* 234:71–82.

Badenes, S. M., T. G. Fernandes, C. S. Cordeiro, S. Boucher, D. Kuninger, M. C. Vemuri, M. M. Diogo, and J. M. Cabral. 2016b. Defined essential 8 medium and vitronectin efficiently support scalable xeno-free expansion of human induced pluripotent stem cells in stirred microcarrier culture systems. *PLoS One* 11 (3):e0151264.

Baker, B. M., and C. S. Chen. 2012. Deconstructing the third dimension: How 3D culture microenvironments alter cellular cues. *J Cell Sci* 125 (Pt 13):3015–3024.

Bardy, J., A. K. Chen, Y. M. Lim, S. Wu, S. Wei, H. Weiping, K. Chan, S. Reuveny, and S. K. Oh. 2013. Microcarrier suspension cultures for high-density expansion and differentiation of human pluripotent stem cells to neural progenitor cells. *Tissue Eng Part C Methods* 19 (2):166–180.

Bauwens, C. L., R. Peerani, S. Niebruegge, K. A. Woodhouse, E. Kumacheva, M. Husain, and P. W. Zandstra. 2008. Control of human embryonic stem cell colony and aggregate size heterogeneity influences differentiation trajectories. *Stem Cells* 26 (9):2300–2310.

Beers, J., D. R. Gulbranson, N. George, L. I. Siniscalchi, J. Jones, J. A. Thomson, and G. Chen. 2012. Passaging and colony expansion of human pluripotent stem cells by enzyme-free dissociation in chemically defined culture conditions. *Nat Protoc* 7 (11):2029–2040.

Birgersdotter, A., R. Sandberg, and I. Ernberg. 2005. Gene expression perturbation in vitro—a growing case for three-dimensional (3D) culture systems. *Semin Cancer Biol* 15 (5):405–412.

Bratt-Leal, A. M., R. L. Carpenedo, M. D. Ungrin, P. W. Zandstra, and T. C. McDevitt. 2011. Incorporation of biomaterials in multicellular aggregates modulates pluripotent stem cell differentiation. *Biomaterials* 32 (1):48–56.

Breslin, S., and L. O'Driscoll. 2013. Three-dimensional cell culture: The missing link in drug discovery. *Drug Discov Today* 18 (5–6):240–249.

Chambers, S. M., C. A. Fasano, E. P. Papapetrou, M. Tomishima, M. Sadelain, and L. Studer. 2009. Highly efficient neural conversion of human ES and iPS cells by dual inhibition of SMAD signaling. *Nat Biotechnol* 27 (3):275–280.

Chen, A. K., S. Reuveny, and S. K. Oh. 2013. Application of human mesenchymal and pluripotent stem cell microcarrier cultures in cellular therapy: Achievements and future direction. *Biotechnol Adv* 31 (7):1032–1046.

Chen, A. K., X. Chen, A. B. Choo, S. Reuveny, and S. K. Oh. 2011a. Critical microcarrier properties affecting the expansion of undifferentiated human embryonic stem cells. *Stem Cell Res* 7 (2):97–111.

Chen, G., D. R. Gulbranson, Z. Hou, J. M. Bolin, V. Ruotti, M. D. Probasco, K. Smuga-Otto et al. 2011b. Chemically defined conditions for human iPSC derivation and culture. *Nat Methods* 8 (5):424–429.

Chen, V. C., J. Ye, P. Shukla, G. Hua, D. Chen, Z. Lin, J. C. Liu, J. Chai, J. Gold, J. Wu, D. Hsu, and L. A. Couture. 2015. Development of a scalable suspension culture for cardiac differentiation from human pluripotent stem cells. *Stem Cell Res* 15 (2):365–375.

Chen, V. C., S. M. Couture, J. Ye, Z. Lin, G. Hua, H. I. Huang, J. Wu, D. Hsu, M. K. Carpenter, and L. A. Couture. 2012. Scalable GMP compliant suspension culture system for human ES cells. *Stem Cell Res* 8 (3):388–402.

Chen, X., A. Chen, T. L. Woo, A. B. Choo, S. Reuveny, and S. K. Oh. 2010. Investigations into the metabolism of two-dimensional colony and suspended microcarrier cultures of human embryonic stem cells in serum-free media. *Stem Cells Dev* 19 (11):1781–1792.

Cimetta, E., D. Sirabella, K. Yeager, K. Davidson, J. Simon, R. T. Moon, and G. Vunjak-Novakovic. 2013. Microfluidic bioreactor for dynamic regulation of early mesodermal commitment in human pluripotent stem cells. *Lab Chip* 13 (3):355–364.

Croughan, M. S., J. F. Hamel, and D. I. Wang. 1987. Hydrodynamic effects on animal cells grown in microcarrier cultures. *Biotechnol Bioeng* 29 (1):130–141.

Croughan, M. S., D. Giroux, D. Fang, and B. Lee. 2016. Novel single-use bioreactors for scale-up of anchorage-dependent cell manufacturing for cell therapies. In *Stem Cell Manufacturing*, J. M.S. Cabral, C. L. da Silva, L. G. Chase, M. M. Diogo (Eds.), pp. 105–139. Boston, MA: Elsevier.

Discher, D. E., D. J. Mooney, and P. W. Zandstra. 2009. Growth factors, matrices, and forces combine and control stem cells. *Science* 324 (5935):1673–1677.

Durnaoglu, S., S. Genc, and K. Genc. 2011. Patient-specific pluripotent stem cells in neurological diseases. *Stem Cells Int* 2011:212487.

Ezashi, T., P. Das, and R. M. Roberts. 2005. Low O_2 tensions and the prevention of differentiation of hES cells. *Proc Natl Acad Sci U S A* 102 (13):4783–4788.

Fang, S., Y. D. Qiu, L. Mao, X. L. Shi, D. C. Yu, and Y. T. Ding. 2007. Differentiation of embryoid-body cells derived from embryonic stem cells into hepatocytes in alginate microbeads in vitro. *Acta Pharmacol Sin* 28 (12):1924–1930.

Fernandes, A. M., T. G. Fernandes, M. M. Diogo, C. L. da Silva, D. Henrique, and J. M. Cabral. 2007. Mouse embryonic stem cell expansion in a microcarrier-based stirred culture system. *J Biotechnol* 132 (2):227–236.

Ferreira, L. S., S. Gerecht, J. Fuller, H. F. Shieh, G. Vunjak-Novakovic, and R. Langer. 2007. Bioactive hydrogel scaffolds for controllable vascular differentiation of human embryonic stem cells. *Biomaterials* 28 (17):2706–2717.

Fonoudi, H., H. Ansari, S. Abbasalizadeh, M. R. Larijani, S. Kiani, S. Hashemizadeh, A. S. Zarchi et al. 2015. A universal and robust integrated platform for the scalable production of human cardiomyocytes from pluripotent stem cells. *Stem Cells Transl Med* 4 (12):1482–1494.

Giandomenico, S. L., and M. A. Lancaster. 2017. Probing human brain evolution and development in organoids. *Curr Opin Cell Biol* 44:36–43.

Grskovic, M., A. Javaherian, B. Strulovici, and G. Q. Daley. 2011. Induced pluripotent stem cells–Opportunities for disease modelling and drug discovery. *Nat Rev Drug Discov* 10 (12):915–929.

Hawgood, S., I. G. Hook-Barnard, T. C. O'Brien, and K. R. Yamamoto. 2015. Precision medicine: Beyond the inflection point. *Sci Transl Med* 7 (300):300ps17.

Hemmi, N., S. Tohyama, K. Nakajima, H. Kanazawa, T. Suzuki, F. Hattori, T. Seki et al. 2014. A massive suspension culture system with metabolic purification for human pluripotent stem cell-derived cardiomyocytes. *Stem Cells Transl Med* 3 (12):1473–1483.

Hirano, S., A. Nose, K. Hatta, A. Kawakami, and M. Takeichi. 1987. Calcium-dependent cell-cell adhesion molecules (cadherins): Subclass specificities and possible involvement of actin bundles. *J Cell Biol* 105 (6 Pt 1):2501–2510.

Hoeben, A., B. Landuyt, M. S. Highley, H. Wildiers, A. T. Van Oosterom, and E. A. De Bruijn. 2004. Vascular endothelial growth factor and angiogenesis. *Pharmacol Rev* 56 (4):549–580.

Hunt, M. M., G. Meng, D. E. Rancourt, I. D. Gates, and M. S. Kallos. 2014. Factorial experimental design for the culture of human embryonic stem cells as aggregates in stirred suspension bioreactors reveals the potential for interaction effects between bioprocess parameters. *Tissue Eng Part C Methods* 20 (1):76–89.

Inoue, H., and S. Yamanaka. 2011. The use of induced pluripotent stem cells in drug development. *Clin Pharmacol Ther* 89 (5):655–661.

Kelava, I., and M. A. Lancaster. 2016. Dishing out mini-brains: Current progress and future prospects in brain organoid research. *Dev Biol* 420 (2):199–209.

Kempf, H., B. Andree, and R. Zweigerdt. 2016. Large-scale production of human pluripotent stem cell derived cardiomyocytes. *Adv Drug Deliv Rev* 96:18–30.

Kempf, H., C. Kropp, R. Olmer, U. Martin, and R. Zweigerdt. 2015. Cardiac differentiation of human pluripotent stem cells in scalable suspension culture. *Nat Protoc* 10 (9):1345–1361.

Kempf, H., R. Olmer, C. Kropp, M. Ruckert, M. Jara-Avaca, D. Robles-Diaz, A. Franke et al. 2014. Controlling expansion and cardiomyogenic differentiation of human pluripotent stem cells in scalable suspension culture. *Stem Cell Reports* 3 (6):1132–1146.

Kleinman, H. K., M. L. McGarvey, L. A. Liotta, P. G. Robey, K. Tryggvason, and G. R. Martin. 1982. Isolation and characterization of type IV procollagen, laminin, and heparan sulfate proteoglycan from the EHS sarcoma. *Biochemistry* 21 (24):6188–6193.

Kropp, C., H. Kempf, C. Halloin, D. Robles-Diaz, A. Franke, T. Scheper, K. Kinast et al. 2016. Impact of feeding strategies on the scalable expansion of human pluripotent stem cells in single-use stirred tank bioreactors. *Stem Cells Transl Med* 5 (10):1289–1301.

Lancaster, M. A., M. Renner, C. A. Martin, D. Wenzel, L. S. Bicknell, M. E. Hurles, T. Homfray, J. M. Penninger, A. P. Jackson, and J. A. Knoblich. 2013. Cerebral organoids model human brain development and microcephaly. *Nature* 501 (7467):373–379.

Lancaster, M. A., N. S. Corsini, S. Wolfinger, E. H. Gustafson, A. W. Phillips, T. R. Burkard, T. Otani, F. J. Livesey, and J. A. Knoblich. 2017. Guided self-organization and cortical plate formation in human brain organoids. *Nat Biotechnol* 35:659–666.

Lecina, M., S. Ting, A. Choo, S. Reuveny, and S. Oh. 2010. Scalable platform for human embryonic stem cell differentiation to cardiomyocytes in suspended microcarrier cultures. *Tissue Eng Part C Methods* 16 (6):1609–1619.

Leung, H. W., A. Chen, A. B. Choo, S. Reuveny, and S. K. Oh. 2011. Agitation can induce differentiation of human pluripotent stem cells in microcarrier cultures. *Tissue Eng Part C Methods* 17 (2):165–172.

Li, Z., X. Guo, A. F. Palmer, H. Das, and J. Guan. 2012. High-efficiency matrix modulus-induced cardiac differentiation of human mesenchymal stem cells inside a thermosensitive hydrogel. *Acta Biomater* 8 (10):3586–3595.

Li, Z., Z. Fan, Y. Xu, W. Lo, X. Wang, H. Niu, X. Li, X. Xie, M. Khan, and J. Guan. 2016. pH-sensitive and thermosensitive hydrogels as stem-cell carriers for cardiac therapy. *ACS Appl Mater Interfaces* 8 (17):10752–10760.

Lian, X., C. Hsiao, G. Wilson, K. Zhu, L. B. Hazeltine, S. M. Azarin, K. K. Raval, J. Zhang, T. J. Kamp, and S. P. Palecek. 2012. Robust cardiomyocyte differentiation from human pluripotent stem cells via temporal modulation of canonical Wnt signaling. *Proc Natl Acad Sci U S A* 109 (27):E1848–E1857.

Lian, X., X. Bao, M. Zilberter, M. Westman, A. Fisahn, C. Hsiao, L. B. Hazeltine, K. K. Dunn, T. J. Kamp, and S. P. Palecek. 2015. Chemically defined, albumin-free human cardiomyocyte generation. *Nat Methods* 12 (7):595–596.

Lipsitz, Yonatan Y., and Peter W. Zandstra. 2015. Human pluripotent stem cell process parameter optimization in a small scale suspension bioreactor. *BMC Proceedings* 9 (9):O10.

Lock, L. T., and E. S. Tzanakakis. 2009. Expansion and differentiation of human embryonic stem cells to endoderm progeny in a microcarrier stirred-suspension culture. *Tissue Eng Part A* 15 (8):2051–2063.

Markvicheva, E., and C. Grandfils. 2004. Microcarriers for animal cell culture. In *Fundamentals of Cell Immobilisation Biotechnology*, V. Nedović and R. Willaert (Eds.), pp. 141–161. Dordrecht, the Netherlands: Springer.

Marques, M. P. C., and N. Szita. 2016. Microfluidic devices for the culture of stem cells. In *Stem Cell Manufacturing*, J. M. S. Cabral, C. L. da Silva, L. G. Chase, M. M. Diogo (Eds.), pp. 171–198. Boston, MA: Elsevier.

Miranda, C. C., T. G. Fernandes, J. F. Pascoal, S. Haupt, O. Brustle, J. M. Cabral, and M. M. Diogo. 2015. Spatial and temporal control of cell aggregation efficiently directs human pluripotent stem cells towards neural commitment. *Biotechnol J* 10 (10):1612–1624.

Miranda, C. C., T. G. Fernandes, M. M. Diogo, and J. M. Cabral. 2016. Scaling up a chemically-defined aggregate-based suspension culture system for neural commitment of human pluripotent stem cells. *Biotechnol J* 11 (12):1628–1638.

Nakagawa, M., Y. Taniguchi, S. Senda, N. Takizawa, T. Ichisaka, K. Asano, A. Morizane et al. 2014. A novel efficient feeder-free culture system for the derivation of human induced pluripotent stem cells. *Sci Rep* 4:3594.

Nie, Y., V. Bergendahl, D. J. Hei, J. M. Jones, and S. P. Palecek. 2009. Scalable culture and cryopreservation of human embryonic stem cells on microcarriers. *Biotechnol Prog* 25 (1):20–31.

Nienow, A. W. 2006. Reactor engineering in large scale animal cell culture. *Cytotechnology* 50 (1–3):9–33.

Oh, S. K., A. K. Chen, Y. Mok, X. Chen, U. M. Lim, A. Chin, A. B. Choo, and S. Reuveny. 2009. Long-term microcarrier suspension cultures of human embryonic stem cells. *Stem Cell Res* 2 (3):219–230.

Olmer, R., A. Lange, S. Selzer, C. Kasper, A. Haverich, U. Martin, and R. Zweigerdt. 2012. Suspension culture of human pluripotent stem cells in controlled, stirred bioreactors. *Tissue Eng Part C Methods* 18 (10):772–784.

Park, Y., Y. Chen, L. Ordovas, and C. M. Verfaillie. 2014. Hepatic differentiation of human embryonic stem cells on microcarriers. *J Biotechnol* 174:39–48.

Phillips, B. W., R. Horne, T. S. Lay, W. L. Rust, T. T. Teck, and J. M. Crook. 2008. Attachment and growth of human embryonic stem cells on microcarriers. *J Biotechnol* 138 (1–2):24–32.

Qian, X., H. N. Nguyen, M. M. Song, C. Hadiono, S. C. Ogden, C. Hammack, B. Yao et al. 2016. Brain-region-specific organoids using mini-bioreactors for modeling ZIKV exposure. *Cell* 165 (5):1238–1254.

Rigamonti, A., G. G. Repetti, C. Sun, F. D. Price, D. C. Reny, F. Rapino, K. Weisinger et al. 2016. Large-scale production of mature neurons from human pluripotent stem cells in a three-dimensional suspension culture system. *Stem Cell Reports* 6 (6):993–1008.

Rodin, S., A. Domogatskaya, S. Strom, E. M. Hansson, K. R. Chien, J. Inzunza, O. Hovatta, and K. Tryggvason. 2010. Long-term self-renewal of human pluripotent stem cells on human recombinant laminin-511. *Nat Biotechnol* 28 (6):611–615.

Rodrigues, G. M. C., T. Gaj, M. M. Adil, J. Wahba, A. T. Rao, F. K. Lorbeer, R. U. Kulkarni et al. 2017. Defined and scalable differentiation of human oligodendrocyte precursors from pluripotent stem cells in a 3D culture system. *Stem Cell Reports* 8 (6):1770–1783.

Sen, A., M. S. Kallos, and L. A. Behie. 2002. Expansion of mammalian neural stem cells in bioreactors: Effect of power input and medium viscosity. *Brain Res Dev Brain Res* 134 (1–2):103–113.

Serra, M., C. Brito, M. F. Sousa, J. Jensen, R. Tostoes, J. Clemente, R. Strehl, J. Hyllner, M. J. Carrondo, and P. M. Alves. 2010. Improving expansion of pluripotent human embryonic stem cells in perfused bioreactors through oxygen control. *J Biotechnol* 148 (4):208–215.

Serra, M., C. Correia, R. Malpique, C. Brito, J. Jensen, P. Bjorquist, M. J. Carrondo, and P. M. Alves. 2011. Microencapsulation technology: A powerful tool for integrating expansion and cryopreservation of human embryonic stem cells. *PLoS One* 6 (8):e23212.

Silva, M. M., A. F. Rodrigues, C. Correia, M. F. Sousa, C. Brito, A. S. Coroadinha, M. Serra, and P. M. Alves. 2015. Robust expansion of human pluripotent stem cells: Integration of bioprocess design with transcriptomic and metabolomic characterization. *Stem Cells Transl Med* 4 (7):731–742.

Singh, H., P. Mok, T. Balakrishnan, S. N. Rahmat, and R. Zweigerdt. 2010. Up-scaling single cell-inoculated suspension culture of human embryonic stem cells. *Stem Cell Res* 4 (3):165–179.

Slaughter, B. V., S. S. Khurshid, O. Z. Fisher, A. Khademhosseini, and N. A. Peppas. 2009. Hydrogels in regenerative medicine. *Adv Mater* 21 (32–33):3307–3329.

Song, Z., J. Cai, Y. Liu, D. Zhao, J. Yong, S. Duo, X. Song et al. 2009. Efficient generation of hepatocyte-like cells from human induced pluripotent stem cells. *Cell Res* 19 (11):1233–1242.

Steiner, D., H. Khaner, M. Cohen, S. Even-Ram, Y. Gil, P. Itsykson, T. Turetsky et al. 2010. Derivation, propagation and controlled differentiation of human embryonic stem cells in suspension. *Nat Biotechnol* 28 (4):361–364.

Tachibana, M., P. Amato, M. Sparman, N. M. Gutierrez, R. Tippner-Hedges, H. Ma, E. Kang et al. 2013. Human embryonic stem cells derived by somatic cell nuclear transfer. *Cell* 153 (6):1228–1238.

Takahashi, K., and S. Yamanaka. 2006. Induction of pluripotent stem cells from mouse embryonic and adult fibroblast cultures by defined factors. *Cell* 126 (4):663–676.

Takahashi, K., K. Tanabe, M. Ohnuki, M. Narita, T. Ichisaka, K. Tomoda, and S. Yamanaka. 2007. Induction of pluripotent stem cells from adult human fibroblasts by defined factors. *Cell* 131 (5):861–872.

Thomson, J. A., J. Itskovitz-Eldor, S. S. Shapiro, M. A. Waknitz, J. J. Swiergiel, V. S. Marshall, and J. M. Jones. 1998. Embryonic stem cell lines derived from human blastocysts. *Science* 282 (5391):1145–1147.

Ting, S., A. Chen, S. Reuveny, and S. Oh. 2014. An intermittent rocking platform for integrated expansion and differentiation of human pluripotent stem cells to cardiomyocytes in suspended microcarrier cultures. *Stem Cell Res* 13 (2):202–213.

Toh, Y. C., K. Blagovic, and J. Voldman. 2010. Advancing stem cell research with microtechnologies: Opportunities and challenges. *Integr Biol (Camb)* 2 (7–8):305–325.

Tong, Z., A. Solanki, A. Hamilos, O. Levy, K. Wen, X. Yin, and J. M. Karp. 2015. Application of biomaterials to advance induced pluripotent stem cell research and therapy. *EMBO J* 34 (8):987–1008.

Villa-Diaz, L. G., Y. S. Torisawa, T. Uchida, J. Ding, N. C. Nogueira-de-Souza, K. S. O'Shea, S. Takayama, and G. D. Smith. 2009. Microfluidic culture of single human embryonic stem cell colonies. *Lab Chip* 9 (12):1749–1755.

Vosough, M., E. Omidinia, M. Kadivar, M. A. Shokrgozar, B. Pournasr, N. Aghdami, and H. Baharvand. 2013. Generation of functional hepatocyte-like cells from human pluripotent stem cells in a scalable suspension culture. *Stem Cells Dev* 22 (20):2693–2705.

Wang, Y., B. K. Chou, S. Dowey, C. He, S. Gerecht, and L. Cheng. 2013. Scalable expansion of human induced pluripotent stem cells in the defined xeno-free E8 medium under adherent and suspension culture conditions. *Stem Cell Res* 11 (3):1103–1116.

Watanabe, K., M. Ueno, D. Kamiya, A. Nishiyama, M. Matsumura, T. Wataya, J. B. Takahashi et al. 2007. A ROCK inhibitor permits survival of dissociated human embryonic stem cells. *Nat Biotechnol* 25 (6):681–686.

Wu, J., M. R. Rostami, D. P. Cadavid Olaya, and E. S. Tzanakakis. 2014. Oxygen transport and stem cell aggregation in stirred-suspension bioreactor cultures. *PLoS One* 9 (7):e102486.

Yamanaka, S. 2009. A fresh look at iPS cells. *Cell* 137 (1):13–17.

Zweigerdt, R. 2009. Large scale production of stem cells and their derivatives. *Adv Biochem Eng Biotechnol* 114:201–235.

Zweigerdt, R., R. Olmer, H. Singh, A. Haverich, and U. Martin. 2011. Scalable expansion of human pluripotent stem cells in suspension culture. *Nat Protoc* 6 (5):689–700.

3 Differentiation of Human Pluripotent Stem Cells for Red Blood Cell Production

*Mark C. Allenby, Susana Brito dos Santos,
Nicki Panoskaltsis, and Athanasios Mantalaris*

CONTENTS

3.1 CELL SOURCE: INDUCED PLURIPOTENT STEM CELLS AND EMBRYONIC STEM CELLS

Embryonic stem cells (ESCs) and induced pluripotent stem cells (iPSCs) are defined as immortal cells with the capacity to self-renew and the potential to differentiate into cells of all three germ layers. Human ESCs were first isolated from three to four-day-old blastocysts (Thomson et al. 1998), and subsequent studies described their capability to be maintained indefinitely in cultures due to high telomerase activity. Mouse iPSCs were first produced by reprograming adult and embryonic fibroblasts in mice by transferring four genes, which encode for transcription factors known to maintain pluripotency in ESCs (Oct3/4, Sox2, c-Myc, and Klf4; Takahashi and Yamanaka 2006). Human fibroblasts were reprogrammed almost simultaneously by Yamanaka and Thomson's groups shortly thereafter. While Yamanaka's group used the exact same gene combination reported for mice, Thomson's group chose the combination of Oct3/4, Sox2, Nanog and Lin28, thereby eliminating the need for forced expression of the oncogene c-Myc (Takahashi et al. 2007; Yu et al. 2007).

The reprograming process was first accomplished by lentivirus-mediated transfection, but subsequently, other methods of reprograming were reported, such as ectopic expression of the same set of transcription factors (Stadtfeld et al. 2008), the use the excisable transposons (Lacoste et al. 2009), or external supplementation of these factors attached to a cell-penetrating peptide sequence (Zhou et al. 2009). iPSCs possess similar characteristics to human ESCs: they are capable of *in vitro* self-renewal and differentiation into all cell types of the three germ layers and, when transplanted into immunodeficient mice, they generate teratomas, providing evidence for their pluripotency. Human iPSCs have two major advantages over ESCs: they have no ethical concerns regarding their origin and they can be used for autologous treatments (e.g., to match a rare blood type; Dorn et al. 2015).

3.2   2D SYSTEMS FOR HUMAN PLURIPOTENT STEM CELLS DIFFERENTIATION TOWARDS RED BLOOD CELLS

In vitro erythroid differentiation from ESCs or iPSCs is accomplished by following two main stages: lineage restriction followed by erythroid differentiation. The first step, which consists of differentiation of PSC to hematopoietic-committed progenitors, may be performed by using two alternative approaches, as shown in Figure 3.1, (a) formation of embryoid bodies (EBs) in suspension cultures or (b) co-culture on murine or human stromal-feeder layers, such as S17 or FH-B-hTERT cell lines. The first approach relies on a cocktail of cytokines to form cell aggregates that can be differentiated into multipotent hematopoietic stem/progenitor cells (HSCs) or hemato-endothelial progenitors (hemangioblasts). In the second approach, undirected differentiation is induced by undefined factors produced by stromal cells, and an extra step is required to select the cells which committed to the hematopoietic lineage (HSCs). This selection is usually performed by CD34$^+$ sorting using magnetic-activated cell sorting (MACS). Although the use of a low-oxygen environment (1%–5%), compared to atmospheric levels, to induce and maintain pluripotency has been reported (Ezashiet al. 2005; Yoshida et al. 2009), differentiation protocols are mainly performed in normoxic conditions (20% O_2).

Proof of concept of the differentiation of hPSCs into erythroid cells was first established for ESCs. In 2006, Olivier and colleagues produced primitive erythroid cells starting from undifferentiated human ESCs, which were first differentiated into CD34$^+$ cells by co-culture with a human fetal liver cell line (FH-B-hTERT). The derived CD34$^+$ cells (sorted by MACS) were expanded and differentiated in serum-free liquid culture during 24 days using a three-step method, which included addition of hydrocortisone (10^{-6}M), interleukin (IL)-3 (13 ng/mL), bone morphogenetic protein (BMP)-4 (13 ng/mL), stem cell factor (SCF, 40 ng/mL), erythropoietin (EPO, 3.3 U/mL), and insulin-like growth factor (IGF)-1 (40 ng/mL), followed by co-culture with an MS-5 feeder layer. This protocol yielded 5 million basophilic erythroblasts from 50,000 CD34$^+$ cells that did not enucleate and expressed a mixture of embryonic and fetal globins.

Ma and co-workers (2008) generated hematopoietic progenitors by co-culture with murine fetal liver-derived stromal cells in serum containing medium (15%). Erythroid differentiation was performed in a methylcellulose medium supplemented

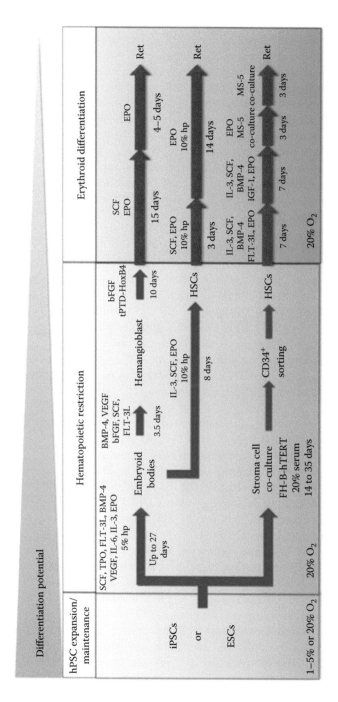

FIGURE 3.1 Principal approaches and cell culture parameters to induce erythroid differentiation of hPSCs. Ret—reticulocyte, hp—human plasma.

with SCF (100 ng/mL), IL-3 (10 ng/mL), IL-6 (100 ng/mL), thrombopoietin (TPO, 10 ng/mL), granulocyte colony-stimulating factor (G-CSF, 10 ng/mL), and EPO (4 U/mL). The produced erythroid cells expressed β-globin, simultaneously with γ and ε-globin, and were proved functional based on equilibrium oxygen curves and glucose-6-phosphate dehydrogenase activity.

Lu and colleagues (2008) obtained 10%–30% enucleated erythrocytes through formation of hemangioblasts using a 21-day four-step protocol. This approach started with embryoid bodies and hemangioblast formation in serum-free medium supplemented with BMP-4 (50 ng/mL), vascular endothelial growth factor (VEGF, 50 ng/mL), and basic fibroblast growth factor (bFGF, 20 ng/mL), SCF (20 ng/mL), and FMS-like tyrosine kinase 3 ligand (FLT-3L, 20 ng/mL), followed by hemangioblast expansion. Erythroid differentiation and enrichment was accomplished in Stemline II-based medium containing SCF (100 ng/mL), EPO (3 U/mL), and methylcellulose. The produced erythrocytes synthesized fetal and embryonic, but not adult, forms of hemoglobin. Higher efficiency of enucleation (60%) was achieved in the presence of OP9 feeder cells.

Lapillonne and collaborators (2010) reported the first two-step protocol for erythroid differentiation and maturation of human iPSC lines and compared them with the ESC line H1. Their protocol started with 20 days differentiation of human iPSCs by the formation of embryoid bodies in the presence of SCF (100 ng/mL), FLT-3L (100 ng/mL), TPO (100 ng/mL), BMP-4 (10 ng/mL), VEGF (5 ng/mL), IL-3 (5 ng/mL), IL-6 (5 ng/mL), EPO (3 U/mL), and 5% human plasma (hp) to obtain early erythroid commitment. Erythroid differentiation was then performed during 26 days in human plasma-containing medium (10%) with addition of IL-3 (5 ng/mL), SCF (100 ng/mL), and EPO (3 U/mL). Efficiency of enucleation was 52%–66% for ESCs and 4%–10% for iPSCs, and the produced erythrocytes synthesized fetal hemoglobin.

Dias and co-workers (2011) generated red blood cells (RBCs) from human iPSC and ESCs using two different methods (60–125 days): (a) co-culture with OP9 followed by CD34+ sorting and erythroid expansion and differentiation in serum-free medium supplemented with dexamethasone (10^{-6}M), SCF (50–100 ng/mL), EPO (2 U/mL), TPO (50 ng/mL), IL-3 (5 ng/mL), and IL-6 (10 ng/mL); and (b) formation of cell aggregates in presence of the same cytokines plus FLT-3L (200 ng/mL) to induce hematopoietic differentiation followed by MS-5 feeder-layer co-culture to achieve erythroid maturation. In both methods, cells were co-cultured with a MS-5 feeder layer up to 20 days to induce maturation. Produced erythroid cells were 2%–10% enucleated and expressed predominately fetal and embryonic globin forms. Kobari and colleagues (2012) adapted the previous protocol and demonstrated the capacity of iPSC-derived erythroblasts to switch from fetal to adult hemoglobin after infusion of nucleated erythroid precursors into mice.

Only in 2016, a serum-free, feeder-free 31-day protocol for erythroid differentiation of ESCs and iPSCs was developed by Olivier and others. This protocol relies on several steps that combine cocktails of cytokines with small molecules (such as StemRegenin [SR1], isobutyl methyl xanthine [IBMX], GSK3β inhibitor VIII, and Pluripotin [SC1]) and achieved 99.1% CD235a+ cells with predominant expression of fetal globin chains and 10% enucleation.

Using an alternative approach, Szabo and colleagues (2010) demonstrated the ability to generate progenitors of hematopoietic fate directly from human dermal fibroblasts without establishing pluripotency. Ectopic expression of Oct4 activated hematopoietic transcription factor in combination with a specific cytokine treatment (IGF-II, bFGF, FLT-3L, and SCF), which allowed for the generation of cells expressing the pan-hematopoietic marker CD45. The capacity of these fibroblast-derived cells to originate granulocytic, monocytic, megakaryocytic, and erythroid lineages and to engraft *in vivo* was also demonstrated.

In conclusion, the field of erythroid cell formation from ESCs and iPS cells has expanded and different methods have been used that show impressive results in terms of reproducibility of the method, quantity of cells produced, and functional activity of these cells (Kaufman 2009; Dias et al. 2011) (Figure 3.1).

3.3 3D SYSTEMS FOR HUMAN PLURIPOTENT STEM CELLS EXPANSION AND PRODUCTION OF RED BLOOD CELLS

The expansion and differentiation of hPSCs into RBCs remains a promising cell source for the unlimited production of blood for transfusion. However, no method exists which produces clinically relevant numbers of functional RBCs, limiting the clinical use of *in vitro* expanded hPSCs as a RBC source. These limitations stem from a lack of robust control during expansion and differentiation processes, as the standard use of feeder (most of xenogeneic origin) co-cultures, xenogeneic serum, and Matrigel cultures introduce sources of irreproducibility, while the formation of embryoid bodies, teratomas, and hematopoietic colonies restrict user control as paracrine interactions dominate over exogenously added soluble factors. Recent bioprocess advances have engineered predictable culture platforms through the study of substrate mechanics, surface modifications, and scaffold topology at multiple points throughout the production of hPSC-derived RBCs.

3.3.1 Human Pluripotent Stem Cells Expansion/Maintenance

The expansion of hPSCs is an essential bioprocess to scale-up production of differentiated cell therapies but spontaneous apoptosis and differentiation limits quality control at an early stage. Initially, these undesired cell-fate decisions were avoided by supplementing exogenous extracellular signals in excess using mouse embryonic fibroblast (MEF) layers or Matrigel, which retard hPSC doubling time but preserve hPSC phenotypes in comparison with 2D (Thomson et al. 1998; Xu et al. 2001). More recently, this use of xenogeneic and undefined MEFs and Matrigel coatings has been avoided by adjusting (a) extracellular matrix chemistry and/or (b) mechanical stiffness to provide support for undifferentiated hPSC expansion in the absence of feeder cells and undefined factors.

Extracellular matrix stiffness promotes the self-renewal of hPSCs through maintaining thin monolayer colonies of cells on the surface of various scaffold topologies. Porous scaffolds have been engineered to permit the inclusion of medium and nutrients but not cell colonies. Colonies formed on scaffolds are constantly exposed to culture medium and user-supplemented factors, eliminating paracrine differentiation cues and zones of metabolic starvation. These scaffolds have been formed by the weaving, electrospinning, or UV-casting polymers which can be coated with extracellular proteins (laminin, vitronectin).

Extracellular surface chemistry, imparted by adhering coatings of resin methacrylates or matrix proteins fibronectin, laminin, vitronectin, also promote hPSCs to spread into a homogenous monolayer, instead of forming colonies, which would inevitably start differentiation as embryoid bodies or teratomas and could be kept for more than four months at 95% purity (Braam et al. 2008; Villa-Diaz et al. 2010; Rodin et al. 2010). These coatings could be implemented on microcarriers to increase culture surface area for larger-density cell cultures. In a comprehensive report implementing 10 and 8 microcarrier coatings, it was found that 100-μm diameter spherical or cylindrical microcarriers coated with Matrigel or laminin allowed for high pluripotent cell growth and could be effectively scaled from 4 mL 6-well plates to 50 mL spinner flasks and further optimized for microcarrier cell seeding density and shear agitation (Oh et al. 2009; Chen et al. 2011; Heng et al. 2012; Badenes et al. 2016; Li et al. 2017).

A recent study found that the incorporation of electrospun gelatin nanofibers onto woven polyglycolic acid (PGA) microfibers significantly enhanced hPSC cell attachment in comparison with Matrigel and PGA microfibers alone. These scaffolds were typically less efficient at expanding cell numbers than Matrigel, but operated within a few-fold difference while illustrating increased viability (>90% vs 80%) and a better maintenance of immature hPSC phenotypes (>95% OCT4[+] SSEA4[+] TRA-1-60[+] SSEA1[-]), even when scaled-up to 55 mL, seven-day or three-passage, fed-batch cultures (Liu et al. 2017). Biomaterial-supported hPSC expansion also implemented a micro-tripod array of casted PDMS coated in gelatin, or chemically modified nanocrystalline graphene, which studied under long-term culture conditions (>10-passage or 30-day culture), performed similarly to Matrigel with respect to hPSC doubling rate and phenotypic maintenance at a six-well plate scale (Lee et al. 2016, 2017). Material parameters controlling matrix stiffness and hPSC colony size can also be optimized for hPSC expansion or for differentiation into specific germ layers; with increased mesodermal differentiation occurring with high scaffold stiffness (50% brachyury[+] at 313 kPa vs 18% during suspension embryoid body formation; Maldonado et al. 2016).

3.3.2 HEMATOPOIETIC RESTRICTION

Germ layer induction is typically achieved through the formation of EBs, a spontaneous process once hPSCs are cultured as single cells in suspension without feeder layers or differentiation-inhibiting factors (Xu et al. 2001). Ultra-low attachment plates coated with a hydrophilic layer are utilized to avoid hPSC settling into adherent monolayer colonies, forming spheroids in suspension instead. Suspension culture provides the largest quantity and size of EBs, but it inefficiently controls germ layer induction due to chemical and soluble factor diffusion from bulk culture liquid and paracrine cellular communication. While feeder or biomaterial-based methods for bypassing EB formation have been reported (Dias et al. 2011; Liu et al. 2014), controlling EB size is a common technique to limit EB cell expansion while increasing mesoderm purity. Also forced aggregation (hanging drop, microwell, or encapsulates of specific sizes) or mechanical shear (spinner flask, rotating vessel, microfluidic platform; Rungarunlert et al. 2009; Sheridan et al. 2012) can help control EB size.

Suspension and hanging drop cultures of hPSC aggregates were compared to suspension cultures of a single cell after dissociation (termed forced aggregation) for hematopoietic specification after 15 days. Suspension cultures of hPSC aggregates contained the largest number of hematopoietic colony-forming unit (CFU) cells, while forced aggregation improved EB homogeneity; higher seeded hESC input (8×10^5) significantly increased day 14 content of CD34$^+$CD45$^+$ and CD45$^+$ cells within EBs, most of which was distributed in larger EBs ($>8 \times 10^5$). Interestingly, the use of 20% (v/v) supplementation of fetal calf serum during EB formation biased hematopoietic specification toward a granulocytic fate (four-fold more CFU-granulocyte), while serum-free EB cultures promoted an erythroid fate (4.7-fold more CFU-erythrocyte; Hong et al. 2010).

To promote uniform EB formation and expansion through forced aggregation of dissociated hPSCs, patterned culture microwells (≥ 200 cells/well) have been engineered that limit maximum EB size and maintain or differentiate hPSCs (Mohr et al. 2010; Choi et al. 2010; Sato et al. 2016). In a recent study, mesodermal specification and colony-forming-cell output was maximized when 100 hPSCs were seeded onto very small pyramidal microwells (approx. 750 μm^3, 10^{11} hPSC/mL density) forcing tight aggregation within a much lower bulk medium dilution (10^5 hPSC/mL). Differentiation was further enhanced at reduced cost when employing local soluble factor delivery from microparticles (Purpura et al. 2012). Commercial cellulose microcarriers have improved the feeder-free liquid suspension expansion of EB cells from hESC cells from 1 to 1.5-fold on MEF layers, 0.11-fold on Matrigel, to 1.19-fold on microcarriers. These microcarriers also produced a significantly higher number of blast cells (4.41 per input hESC) in comparison with 2D Matrigel (0.21 per input hESC; Lu et al. 2013).

In dynamic systems, EB formation size has been controlled by using mechanical shear within 250 mL stirred-tank reactors to achieve a 15-fold expansion for 21 days of culture, as compared to 4-fold in static culture while maintaining a similar purity of hematopoietic CD34$^+$CD45$^+$ cells. The number of EBs in static culture decreased within the initial four days (perhaps due to agglomeration), while the spinner flask maintained a consistent EB number. On the other hand, at day 21, the number of cells per EB was almost two-fold higher in the spinner flask (73×10^3 cells) compared to the static system (Cameron et al. 2006). Other studies investigated the role of shear on EB growth by using high aspect rotating vessels and slow turning lateral vessels to study the effect of hypoxia and aggregation methods on murine models (Dang et al. 2002; Dang et al. 2004; Gerecht-Nir et al. 2004). Once an appropriate state of hematopoietic lineage specification has been reached, cells are typically dissociated from their biomaterial scaffolding, purified by immunophenotypic separation (CD34$^+$) and further differentiated towards an erythroid fate.

3.3.3 Erythroid Differentiation

Many biomaterial platforms have been proposed for the culture of HSCs (e.g., CD34$^+$) derived from multipotent sources (e.g., UCB, BM, PB) by mimicking aspects of adult hematopoietic niches. However few have investigated *ex vivo* biomaterials for erythropoiesis, and none of which utilize hPSC-derived HSCs. As hESCs undergo blood formation in a different environment than fetal/neonatal (UCB-HSCs) and adult (BM-HSCs), optimal biomaterial platform properties, such as scaffold mechanics (porosity, stiffness, shear) and surface modification (functional groups, peptides, protein coatings), might vary for embryonic, fetal, and adult RBCs. This process is similar to how the optimal soluble factors used for RBC production from embryonic hPSC-derived HSCs (Olivier et al. 2016) vary from isolated UCB HSCs (Timmins et al. 2011). Even so, biomaterial platforms and parameters successfully used for neonatal and adult HSC production of RBCs can be investigated to inspire future models of hPSC-derived HSC erythropoiesis.

Substrate elasticity and shear stress have been investigated to promote hematopoietic proliferation and erythropoiesis. The implementation of tropoelastin, the most elastic biomaterial available, provided a two- to threefold expansion of human UCB-HSCs in comparison to common tissue culture controls, providing an endothelial-like contact for emerging hematopoietic cells (Holst et al. 2010). HSCs have demonstrated greater proliferation, viability, and adherence when cultured on soft surfaces, and greater migratory properties on stiffer surfaces, which might corroborate with the range of matrix stiffness values found in the murine marrow (0.05 kPa to 1 kPa; Choi and Harley 2012; Lee-Thedieck et al. 2012; Chitteti et al. 2015; Nelson and Roy 2016). Wall shear stress has been shown to promote embryonic hematopoiesis from disassociated murine EB cells by increasing *Runx1* expression in CD41$^+$c-KIT$^+$ hematopoietic progenitors and increasing their hematopoietic colony-forming potential, perhaps providing a physiological stress as found after initiation of the heartbeat in vertebrates (Adamo et al. 2009).

Biomaterials implemented for erythroid cultures frequently leverage cell adhesive molecules (e.g., VCAMs, VLAs) present on hematopoietic and erythroid progenitors (Chow et al. 2013; Silberstein et al. 2016) by modifying biomaterial surfaces chemically (amine functional groups; Chua et al. 2006; Sakthivel et al. 2009) or with cell-adherent peptides (CS-1, RGD; Jiang et al. 2006; Chen et al. 2012), and proteins (fibronectin, collagen type-1; Dao et al. 1998; Feng et al. 2006; Mortera-Blanco et al. 2011; Lee et al. 2014). As erythroid cell mature *in vivo*, cell adhesive proteins are lost after enucleation and departure from erythroblastic islands (Manwani and Bieker 2008). Therefore, cell-adherent biomaterials can maintain hemato-erythroid progenitor cells while continuously separating mature red cell egress (Severn et al. 2016). Nanorough substrate topologies increase the coatable surface area and the number of HSC-adhesive sites to increase HSC expansion and maintenance of potency (Chen et al. 2012; Muth et al. 2013). HSC and erythroid-adhesive contacts have not only been mimicked *ex vivo* by adhesive coatings, but also by promoting intercellular associations. Expanded late-stage erythroblasts were plated at high densities (10^7/mL) in a variety of 2D and 3D platforms and found to be making cellular contacts. These cells exhibited enhanced terminal differentiation (13% versus 22%) and enucleation (23% versus 46%) toward red blood cells when compared with low cell density controls (10^6/mL; Lee et al. 2014).

Many scaffold and carrier constructs utilizing the different properties detailed earlier have been implemented for the expansion of HSCs, but very few have investigated their use in erythropoiesis. In 2014, Arteriocyte™ (now Isto Biologics™) patented an aminated PAAc-grafted PES nanofiber mesh scaffold for the production of RBCs from UCB CD34$^+$CD133$^+$ HSCs. Within 18 days of culture, the scaffold reached a 2630-fold increase in total cell number with 95% enucleation rate, producing 35×10^6 RBCs per HSC input (Sakthivel et al. 2009). In 2013, a perfusion 3D hollow fiber bioreactor design was patented and allowed for the high-density production of RBCs from UCB. The reactor implemented medium-perfused ceramic and polymeric hollow fibers within a collagen type-1 coated polyurethane scaffold, which was inoculated with 10^8 mononucleated cells (MNCs)/mL and produced a total number of 36×10^9 across 31 days, including 830×10^6 enucleate red blood cells within one 10 mL reactor (Macedo 2011). In these two culture platforms, the use of shear stress, nanoroughness, and different surface modifications dealt with three major challenges restricting commercialization of RBC production *ex vivo*: expansion, culture density, and quality, all of which influence cost per RBC product.

3.4 CHALLENGES TO OVERCOME

Pluripotent stem cells represent an advantageous cell source in conditions where no other alternative is available, such as to support autologous transfusions for allo-immune patients or to produce a rare blood group such as O^-Rh^-. Their capacity to be propagated and expanded *in vitro* indefinitely turn those to a potentially inexhaustible and donorless source of cells for human therapy. The ability to generate and use iPSCs avoids the ethical concerns associated with human ESCs.

Considerable progress has been made toward the *ex vivo* erythroid differentiation of hPSCs by Olivier and colleagues (2016). Their seminal work combined conventional PSC expansion, EB formation, hematopoietic restriction, and erythroid differentiation culture stages within serum-free, feeder-free cell suspension cultures, providing a comprehensive experimental approach to produce RBCs from hPSCs. Alas, this work highlighted many challenges that must be overcome before RBC-derived hPSC will be feasible and appropriate for clinical transfusion, which biomaterial-based approaches may provide answers.

An inefficient mesodermal induction and hematopoietic restriction during embryoid body formation remains a stumbling step between unlimited hPSC expansion and proliferous RBC production from HSCs, typically eliciting <20% CD34+ HSCs within EBs (Cameron 2006; Olivier et al. 2016). Embryoid body growth remains without bioprocess control with the concentrations of nutrients/metabolites and other soluble factors varying greatly from the outside surface of the EB to its core. Therefore, culture platforms, which optimize EB differentiation aim to limit EB size, and recent protocols are finding new ways to reach hematopoietic lineage specification while avoiding EB formation altogether. Although this possibility has existed with the use of murine feeder layers, recent developments in biomaterials culture allow for a more defined, clinically-relevant methods (Dias et al. 2011; Liu et al. 2014).

The production of RBCs from HSCs has reached a clinically sufficient fold-expansion, but is limited by low cell density requirements, making medium the most prohibitive culture cost towards clinical translation (Rousseau et al. 2014). Recently, scaffold-based bioreactors have allowed for hematopoietic culture at 10- to 100-fold higher cell densities than liquid suspension without decreased cell viabilities, including stirred-tank reactors with microcarriers (10^7 cells/mL; Ratcliffe et al. 2012) and hollow fiber bioreactors (2×10^8 cells/mL; Housler et al. 2012), while erythroid culture platforms at high density ($>10^7$ cells/mL) have demonstrated an enhanced RBC product quality by producing a higher expression of adult phenotypes and enucleation frequency (Lee et al. 2014).

RBC product quality must also be improved, as cells express mainly embryonic and fetal forms of hemoglobin (ε and γ globin chains), which do not resemble definitive adult erythropoiesis. A recently described approach, involving the immortalization of an erythroblast cell line, has allowed for the indefinite expansion of progenitor cells that can be released to differentiate into red blood cells with adult globin, whose deformability index was comparative to native reticulocytes (Trakarnsanga et al. 2017). However, these cells were kept at low densities $\leq 3 \times 10^5$/mL, the protocol used included high concentrations of expensive protein supplements, and only 30% enucleation was achieved, suggesting similar difficulties might arise during scale-up.

hPSCs represent an unlimited cell source for RBC production whose bioprocess has been developed across the last 20 years, and includes: hPSC expansion, hematopoietic specification, and erythropoiesis. Biomaterials have been utilized to improve product purity, yield, and compliance with GMP requirements (serum-free, feeder-free) for hPSC expansion and hPSC hematopoietic specification, but not yet for hPSC-derived erythropoiesis to maturation, a feat only recently accomplished in liquid suspension. Therefore, the future for hPSC-derived RBCs remains ripe for exploration and optimization, with very promising preliminary results (Table 3.1).

TABLE 3.1

Comparison of Feeder-Free 2D (White) and 3D (Gray) Platforms for Different Stages of hPSC Expansion and Maintenance (Top), Hematopoietic Restriction (Middle), and RBC Production (Bottom) with Current Liquid Suspension State of the Art for hPSC-Derived Cells (Pink)

hPSC Expansion/Maintenance

Biomaterial	Single-cell liquid suspension w/ ROCK inhibitor	Laminin-511 or methacrylate monolayers	Vitronectin-coated polystyrene microcarriers in stirred flask w/ ROCK inhibitor	Gelatin micro-nanofibers in perfused bag w/ ROCK inhibitor
Input Density	2×10^5 hESCs/mL	10^5/mL hPSCs	8×10^5/mL hPSCs	3×10^4/mL hPSCs
Output Density [days]	6×10^5/mL [4d] hESCs	10^6/mL [7d] hPSCs	1.6×10^6/mL [10d] hPSCs	1.5×10^6/mL [7d] hPSCs
Fold Expansion	3	10	4	50
Volume	4 mL	2 mL	50 mL	55 mL
Reference	Harb et al. (2008)	Ludwig et al. (2006), Rodin et al. (2010), Villa-Diaz et al. (2010)	Badenes et al. (2016)	Liu et al. (2017)

(Continued)

TABLE 3.1 (*Continued*)
Comparison of Feeder-Free 2D (White) and 3D (Gray) Platforms for Different Stages of hPSC Expansion and Maintenance (Top), Hematopoietic Restriction (Middle), and RBC Production (Bottom) with Current Liquid Suspension State of the Art for hPSC-Derived Cells (Pink)

Hematopoietic Restriction

Biomaterial	Olivier et al. liquid suspension	Suspension stirred-flask culture	Microfabricated plastic vessels w/ ROCK inhibitor	Cellulose microcarriers
Input Density	5×10^4/mL hPSCs	10^5/mL hPSCs	2×10^5/mL hPSCs	10^6/mL hIPSCs
Output Density [days]	10^6/mL [4d] 92% CD34+	2.5×10^6/mL [21d] EB cells	1×10^6/mL [5d] EB cells	2×10^7/mL [4d], 20% CFC cells
Fold Expansion	100	25	5	20
Volume	4 mL	100 mL	10^{-4} mL	4 mL
Reference	Olivier et al. (2016)	Cameron et al. (2006)	Sato et al. (2016)	Lu et al. (2013)

Erythroid Differentiation

Biomaterial	Olivier et al. liquid suspension	Liquid suspension followed by collagen scaffold in shaker	Aminated PAAc-grafted PES nanofiber mesh	Collagen-coated PU scaffold with hollow fiber perfusion
Input Density	10^5/mL hPSC derived CD34+	10^6/mL UCB CD34+ followed by 10^7/mL erythroblasts	10^3/mL UCB CD133+	10^8/mL UCB MNCs
Output Density [days]	10^6/mL [26d] 10% enuc	9×10^8/well [19d] 46% enuc	5×10^7/mL [18d] 6% enuc	4×10^9/mL, [31d] 5% enuc
Fold Expansion	1,000	1,000	2,630	34
Volume	4 mL	125 mL	4 mL	10 mL
Reference	Olivier et al. (2016)	Lee et al. (2014)	Sakthivel et al. (2009)	Housler et al. (2012), Panoskaltsis et al. (2012)

REFERENCES

Adamo, L., O. Naveiras, P. L. Wenzel, S. McKinney-Freeman, P. J Mack, J. Gracia-Sancho, A. Suchy-Dicey et al. 2009. Biomechanical forces promote eHaematopoiesis. *Nature* 459 (7250): 1131–1135. doi:10.1038/nature08073.

Badenes, S. M., T. G. Fernandes, C. S. M. Cordeiro, S. Boucher, D. Kuninger, M. C. Vemuri, M. Margarida Diogo, and J. M. S. Cabral. 2016. Defined essential 8 medium and vitronectin efficiently support scalable xeno-free expansion of human induced pluripotent stem cells in stirred microcarrier culture systems. *PLoS One* 11 (5): 1–19. doi:10.1371/journal.pone.0151264.

Braam, S. R., L. Zeinstra, S. Litjens, D. Ward-van Oostwaard, S. van den Brink, L. van Laake, F. Lebrin et al. 2008. Recombinant vitronectin is a functionally defined substrate that supports human embryonic stem cell self-renewal via alphavbeta5 Integrin. *Stem Cells* 26 (9): 2257–2265. doi:10.1634/stemcells.2008-0291.

Cameron, C. M., W.-S. Hu, D. S. Kaufma. 2006. Improved development of human embryonic stem cell-derived embryoid bodies by stirred vessel cultivation. *Biotechnology and Bioengineering*. doi:10.1002/bit.

Chen, A. K. L., X. Chen, A. B. Hwa Choo, S. Reuveny, and S. Kah Weng Oh. 2011. Critical microcarrier properties affecting the expansion of undifferentiated human embryonic stem cells. *Stem Cell Research* 7 (2): 97–111. doi:10.1016/j.scr.2011.04.007.

Chen, L. Y., Y. Chang, J.-S. Shiao, Q.-Dong Ling, Y. Chang, Y. H. Chen, D.-C. Chen et al. 2012. Effect of the surface density of nanosegments immobilized on culture dishes on ex vivo expansion of hematopoietic stem and progenitor cells from umbilical cord blood. *Acta Biomaterialia* 8 (5): 1749–1758. doi:10.1016/j.actbio.2012.01.002.

Chitteti, B. R., M. A. Kacena, S. L. Voytik-Harbin, and E. F. Srour. 2015. Modulation of hematopoietic progenitor cell fate in vitro by varying collagen oligomer matrix stiffness in the presence or absence of osteoblasts. *Journal of Immunological Methods* 425: 108–113. doi:10.1016/j.jim.2015.07.001.

Choi, J. S., and B. A. C. Harley. 2012. The combined influence of substrate elasticity and ligand density on the viability and biophysical properties of hematopoietic stem and progenitor cells. *Biomaterials* 33 (18): 4460–4468. doi:10.1016/j.biomaterials.2012.03.010.

Choi, Y. Y., B. G. Chung, D. H. Lee, A. Khademhosseini, J. H. Kim, and S. H. Lee. 2010. Controlled-size embryoid body formation in concave microwell arrays. *Biomaterials* 31 (15): 4296–4303. doi:10.1016/j.biomaterials.2010.01.115.

Chow, A., M. Huggins, J. Ahmed, D. Hashimoto, D. Lucas, Y. Kunisaki, S. Pinho et al. 2013. CD169+ Macrophages provide a niche promoting erythropoiesis under homeostasis and stress. *Nature Medicine* 19 (4): 429–436. doi:10.1038/nm.3057.

Chua, K.-N., C. Chai, P.-C. Lee, Y.-N. Tang, S. Ramakrishna, K. W. Leong, and H.-Q. Mao. 2006. Surface-aminated electrospun nanofibers enhance adhesion and expansion of human umbilical cord blood hematopoietic stem/progenitor cells. *Biomaterials* 27 (36): 6043–6051. doi:10.1016/j.biomaterials.2006.06.017.

Dang, S. M., M. Kyba, R. Perlingeiro, G. Q. Daley, and P. W. Zandstra. 2002. Efficiency of embryoid body formation and hematopoietic development from embryonic stem cells in different culture systems. *Biotechnology and Bioengineering* 78 (4): 442–453. doi:10.1002/bit.10220.

Dang, S. M., S. Gerecht-Nir, J. Chen, J. Itskovitz-Eldor, and P. W. Zandstra. 2004. Controlled, scalable embryonic stem cell differentiation culture. *Stem Cells* 22 (3): 275–282. doi:10.1634/stemcells.22-3-275.

Dao, M. A., K. Hashino, I. Kato, and J. A. Nolta. 1998. Adhesion to fibronectin maintains regenerative capacity during ex vivo culture and transduction of human hematopoietic stem and progenitor cells. *Blood* 92 (12): 4612–4621.

Dias, J., M. Gumenyuk, H. Kang, M. Vodyanik, J. Yu, J. A. Thomson, and I. I. Slukvin. 2011. Generation of red blood cells from human induced pluripotent stem cells. *Stem Cells and Development* 20 (9): 1639–1647. doi:10.1089/scd.2011.0078.

Dorn, I., K. Klich, M. J. Arauzo-Bravo, M. Radstaak, S. Santourlidis, F. Ghanjati, T. F. Radke et al. 2015. Erythroid differentiation of human induced pluripotent stem cells is independent of donor cell type of origin. *Haematologica* 100 (1): 32–41. doi:10.3324/haematol.2014.108068.

Douay, L., and G. Andreu. 2007. Ex vivo production of human red blood cells from hematopoietic stem cells: What is the future in transfusion? *Transfusion Medicine Reviews* 21 (2): 91–100. doi:10.1016/j.tmrv.2006.11.004.

Ezashi, T., P. Das, and R. M. Roberts. 2005. Low O_2 tensions and the prevention of differentiation of hES cells. *Proceedings of the National Academy of Sciences of the United States of America* 102 (13): 4783–4788. doi:10.1073/pnas.0501283102.

Feng, Q., C. Chai, X. S. Jiang, K. W. Leong, and H. Q. Mao. 2006. Expansion of engrafting human hematopoietic stem/progenitor cells in three-dimensional scaffolds with surface-immobilized fibronectin. *Journal of Biomedical Materials Research A* 78 (4): 781–791.

Gerecht-Nir, S., S. Cohen, and J. Itskovitz-Eldor. 2004. Bioreactor cultivation enhances the efficiency of human embryoid body (hEB) formation and differentiation. *Biotechnology and Bioengineering* 86 (5): 493–502. doi:10.1002/bit.20045.

Harb, N., T. K. Archer, and N. Sato. 2008. The rho-rock-myosin signaling axis determines cell-cell integrity of self-renewing pluripotent stem cells. *PLoS One* 3 (8). doi:10.1371/journal.pone.0003001.

Heng, B. C., J. Li, A. K.-Li. Chen, S. Reuveny, S. M. Cool, W. R. Birch, and S. K.-W. Oh. 2012. Translating human embryonic stem cells from 2-dimensional to 3-dimensional cultures in a defined medium on laminin- and vitronectin-coated surfaces. *Stem Cells and Development* 21 (10): 1701–1715. doi:10.1089/scd.2011.0509.

Holst, J., S. Watson, M. S. Lord, S. S. Eamegdool, D. V. Bax, L. B. Nivison-Smith, A. Kondyurin et al. 2010. Substrate elasticity provides mechanical signals for the expansion of hemopoietic stem and progenitor cells. *Nature Biotechnology* 28 (10): 1123–1128. doi:10.1038/nbt.1687.

Hong, S. H., T. Werbowetski-Ogilvie, V. Ramos-Mejia, J. B. Lee, and M. Bhatia. 2010. Multiparameter comparisons of embryoid body differentiation toward human stem cell applications. *Stem Cell Research* 5 (2): 120–130. doi:10.1016/j.scr.2010.04.007.

Housler, G. J., T. Miki, E. Schmelzer, C. Pekor, X. Zhang, L. Kang, V. Voskinarian-Berse, S. Abbot, K. Zeilinger, and J. C. Gerlach. 2012. Compartmental hollow fiber capillary membrane-based bioreactor technology for in vitro studies on red blood cell lineage direction of hematopoietic stem cells. *Tissue Engineering. Part C, Methods* 18 (2): 133–142. doi:10.1089/ten.tec.2011.0305.

Jiang, X. S., C. Chai, Y. Zhang, R. X. Zhuo, H. Quan Mao, and K. W. Leong. 2006. Surface-immobilization of adhesion peptides on substrate for ex vivo expansion of cryopreserved umbilical cord blood CD34+ cells. *Biomaterials* 27 (13): 2723–2732. doi:10.1016/j.biomaterials.2005.12.001.

Kaufman, D. S. 2009. Toward clinical therapies using hematopoietic cells derived from human pluripotent stem cells. *Stem Cells* 114 (17): 3513–3523. doi:10.1182/blood-2009-03-191304.

Kobar, L., F. Yates, N. Oudrhiri, A. Francina, L. Kiger, C. Mazurier, S. Rouzbeh et al. 2012. Human induced pluripotent stem cells can reach complete terminal maturation: In vivo and in vitro evidence in the erythropoietic differentiation model. *Haematologica* 97: 1795–1803. doi:10.3324/haematol.2011.055566.

Lacoste, A., F. Berenshteyn, and A. H. Brivanlou. 2009. An efficient and reversible transposable system for gene delivery and lineage-specific differentiation in human embryonic stem cells. *Cell Stem Cell* 5 (3): 332–342. doi:10.1016/j.stem.2009.07.011.

Lapillonne, H., L. Kobari, C. Mazurier, P. Tropel, M. C. Giarratana, I. Zanella-Cleon, L. Kiger et al. 2010. Red blood cell generation from human induced pluripotent stem cells: Perspectives for transfusion medicine. *Haematologica* 95 (10): 1651–1659. doi:10.3324/haematol.2010.023556.

Lawes, R. 2011. NHS Blood and Transplant Commercial Review.

Lee, E., S. Y. Han, H. Sook Choi, B. Chun, B. Hwang, and E. Jung Baek. 2014. Red blood cell generation by three-dimensional aggregate cultivation of late erythroblasts. *Tissue Engineering. Part A* 21: 1–30. doi:10.1089/ten. TEA.2014.0325.

Lee, H., D. Nam, J.-K. Choi, M. J. Arauzo-Bravo, S.-Y. Kwon, H. Zaehres, T. Lee et al. 2016. Establishment of feeder-free culture system for human induced pluripotent stem cell on DAS nanocrystalline graphene. *Scientific Reports* 6: 20708. doi:10.1038/srep20708.

Lee-Thedieck, C., N. Rauch, R. Fiammengo, G. Klein, and J. P. Spatz. 2012. Impact of substrate elasticity on human hematopoietic stem and progenitor cell adhesion and motility. *Journal of Cell Science* 125 (16): 3765–3775. doi:10.1242/jcs.095596.

Li, J., F. Zhang, L. Yu, N. Fujimoto, M. Yoshioka, X. Li, J. Shi et al. 2017. Culture substrates made of elastomeric micro-tripod arrays for long-term expansion of human pluripotent stem cells. *Journal of Materials Chemistry B* 5 (2): 236–244. doi:10.1039/C6TB02246D.

Liu, L., K.-I. Kamei, M. Yoshioka, M. Nakajima, J. Li, N. Fujimoto, S. Terada et al. 2017. Nano-on-micro fibrous extracellular matrices for scalable expansion of human ES/iPS cells. *Biomaterials* 124: 47–54. doi:10.1016/j.biomaterials.2017.01.039.

Liu, Y., V. Fox, Y. Lei, B. Hu, K. Il Joo, and P. Wang. 2014. Synthetic niches for differentiation of human embryonic stem cells bypassing embryoid body formation. *Journal of Biomedical Materials Research—Part B Applied Biomaterials* 102 (5): 1101–1112. doi:10.1002/jbm.b.33092.

Lu, S. J., T. Kelley, Q. Feng, A. Chen, S. Reuveny, R. Lanza, and S. K. W. Oh. 2013. 3D microcarrier system for efficient differentiation of human pluripotent stem cells into hematopoietic cells without feeders and serum. *Regenerative Medicine* 8 (4): 413–424. doi:10.2217/rme.13.36.

Lu, S.-J., Q. Feng, J.S. Park, L. Vida, B.-S. Lee, M. Strausbauch, P. J. Wettstein, G. R. Honig, and R. Lanza. 2008. Biologic properties and enucleation of red blood cells from human embryonic stem cells. *Blood* 112 (12): 4475–4484. doi:10.1182/blood-2008-05-157198.

Ludwig, T. E., V. Bergendahl, M. E. Levenstein, J. Y. Yu, M. D. Probasco, and J. A. Thomson. 2006. Feeder-independent culture of human embryonic stem cells. *Nature Methods* 3 (8): 637–646. doi:10.1038/nmeth1006-867.

Ma, F., Y. Ebihara, K. Umeda, H. Sakai, S. Hanada, H. Zhang, Y. Zaike et al. 2008. Generation of functional erythrocytes from human embryonic stem cell-derived definitive hematopoiesis. *National Academic Science.* 105 (35): 13087–13092. doi:10.1073/pnas.0802220105.

Macedo, H. 2011. A novel 3D dual hollow fibre bioreactor for the production of human red blood cells. PhD Theseis, Imperial College London, London, UK.

Maldonado, M., G. Ico, K. Low, R. J. Luu, and J. Nam. 2016. Enhanced lineage-specific differentiation efficiency of human induced pluripotent stem cells by engineering colony dimensionality using electrospun scaffolds. *Advanced Healthcare Materials* 5 (12): 1408–1412. doi:10.1002/adhm.201600141.

Manwani, D., and J. J. Bieker. 2008. The erythroblastic island. *Current Topics in Development Biology* 82: 23–52. doi:10.1016/S0070-2153(07)00002-6.

Mohr, J. C., J. Zhang, S. M. Azarin, A. G. Soerens, J. J. de Pablo, J. A. Thomson, G. E. Lyons, S. P. Palecek, and T. J. Kamp. 2010. The microwell control of embryoid body size in order to regulate cardiac differentiation of human embryonic stem cells. *Biomaterials* 31 (7): 1885–1893. doi:10.1016/j.biomaterials.2009.11.033.

Mortera-Blanco, T., A. Mantalaris, A. Bismarck, N. Aqel, and N. Panoskaltsis. 2011. Long-term cytokine-free expansion of cord blood mononuclear cells in three-dimensional scaffolds. *Biomaterials* 32 (35): 9263–9270.

Muth, C. A., C. Steinl, G. Klein, and C. Lee-Thedieck. 2013. Regulation of hematopoietic stem cell behavior by the nanostructured presentation of extracellular matrix components. *PLoS One* 8 (2). doi:10.1371/journal.pone.0054778.

Nelson, M. R., and K. Roy. 2016. Bone-marrow mimicking biomaterial niches for studying hematopoietic stem and progenitor cells. *Journal of Materials Chemistry B* 3490 (4): 3490–3503. doi:10.1039/c5tb02644j.

Oh, S. K. W., A. K. Chen, Y. Mok, X. Chen, U. M. Lim, A. Chin, A. B. H. Choo, and S. Reuveny. 2009. Long-term microcarrier suspension cultures of human embryonic stem cells. *Stem Cell Research* 2 (3): 219–230. doi:10.1016/j.scr.2009.02.005.

Olivier, E. N., C. Qiu, M. Velho, R. E. Hirsch, and E. E. Bouhassira. 2006. Large-scale production of embryonic red blood cells from human embryonic stem cells. *Experimental Hematology* 34 (12): 1635–1642. doi:10.1016/j.exphem.2006.07.003.

Olivier, E. N., L. Marenah, A. McCahill, A. Condie, S. Cowan, and J. C. Mountford. 2016. High-efficiency serum-free feeder-free erythroid differentiation of human pluripotent stem cells using small molecules. *Stem Cells Translational Medicine*, 1–12. doi:10.5966/sctm.2015-0371.

Panoskaltsis, N., H. Macedo, M. Teresa, A. Mantalaris, and A. G. Livingston. 2012. No Title. Patent WO 2012/06.

Purpura, K. A., A. M. Bratt-Leal, K. A. Hammersmith, T. C. McDevitt, and P. W. Zandstra. 2012. Systematic engineering of 3D pluripotent stem cell niches to guide blood development. *Biomaterials* 33 (5): 1271–1280. doi:10.1016/j.biomaterials.2011.10.051.

Ratcliffe, E., K. E. Glen, V. L. Workman, A. J. Stacey, and R. J. Thomas. 2012. A novel automated bioreactor for scalable process optimisation of haematopoietic stem cell culture. *Journal of Biotechnology* 161 (3): 387–390. doi:10.1016/j.jbiotec.2012.06.025.

Rodin, S., A. Domogatskaya, S. Ström, E. M. Hansson, K. R. Chien, J. Inzunza, O. Hovatta, and K. Tryggvason. 2010. Long-term self-renewal of human pluripotent stem cells on human recombinant laminin-511. *Nature Biotechnology* 28 (6): 611–115. doi:10.1038/nbt.1620.

Rousseau, G. F., M. C. Giarratana, and L. Douay. 2014. Large-scale production of red blood cells from stem cells: What are the technical challenges ahead? *Biotechnology Journal* 9 (1): 28–38. doi:10.1002/biot.201200368.

Rungarunlert, S., M. Techakumphu, M. K. Pirity, and A. Dinnyes. 2009. Embryoid body formation from embryonic and induced pluripotent stem cells: Benefits of bioreactors. *World Journal of Stem Cells* 1 (1): 11–21. doi:10.4252/wjsc.v1.i1.11.

Sakthivel, R., D. J. Brown, H.-Q. Mao, L. Douay, V. J. Pompili, K. McIntosh, H. Das, and Y. Zhao. 2009. Erythrocytes differentiated in vitro from nanofiber expanded CD133+ cells. issued 2009.

Sato, H., A. Idiris, T. Miwa, and H. Kumagai. 2016. Microfabric vessels for embryoid body formation and rapid differentiation of pluripotent stem cells. *Scientific Reports* 6 (April): 31063. doi:10.1038/srep31063.

Severn, C. E., H. Macedo, M. J. Eagle, P. Rooney, A. Mantalaris, and A. M. Toye. 2016. Polyurethane scaffolds seeded with CD34(+) cells maintain early stem cells whilst also facilitating prolonged egress of haematopoietic progenitors. *Scientific Reports* 6: 32149. doi:10.1038/srep32149.

Sheridan, S. D., V. Surampudi, and R. R. Rao. 2012. Analysis of embryoid bodies derived from human induced pluripotent stem cells as a means to assess pluripotency. *Stem Cells International* 2012. doi:10.1155/2012/738910.

Silberstein, L., K. A. G.oncalves, P. V. Kharchenko, R. Turcotte, Y. Kfoury, F. Mercier, N. Baryawno et al. 2016. Proximity-based differential single-cell analysis of the niche to identify stem/progenitor cell regulators. *Cell Stem Cell* 19 (4): 530–543. doi:10.1016/j.stem.2016.07.004.

Stadtfeld, M., M. Nagaya, J. Utikal, G. Weir, and K. Hochedlinger. 2008. Induced pluripotent stem cells generated without viral integration. *Nature* 322: 945–949.

Szabo, E., S. Rampalli, R. M. Risueño, A. Schnerch, R. Mitchell, A. Fiebig-comyn, M. Levadoux-Martin et al. 2010. Direct conversion of human fibroblasts to multilineage blood progenitors. *Nature* 468 (7323): 521–526. doi:10.1038/nature09591.

Takahashi, K., and S. Yamanaka. 2006. Induction of pluripotent stem cells from mouse embryonic and adult fibroblast cultures by defined factors. *Cell* 126 (4): 663–676. doi:10.1016/j.cell.2006.07.024.

Takahashi, K., K. Tanabe, M. Ohnuki, M. Narita, T. Ichisaka, K. Tomoda, and S. Yamanaka. 2007. Induction of pluripotent stem cells from adult human fibroblasts by defined factors. *Cell* 131 (5): 861–872. doi:10.1016/j.cell.2007.11.019.

Thomson, J. A., J. Itskovitz-eldor, S. S. Shapiro, M. A. Waknitz, J. J. Swiergiel, V. S. Marshall, and J. M. Jones. 1998. Embryonic stem cell lines derived from human blastocysts. *Science* 1145 (12): 1145–1148. doi:10.1126/science.282.5391.1145.

Timmins, N. E., S. Athanasas, M. Gü, P. Buntine, and L. K. Nielsen. 2011. Ultra-high-yield manufacture of red blood cells from hematopoietic stem cells. *Tissue Engineering Part C Methods* 17 (11): 1131–1137. doi:10.1089/ten.tec.2011.0207.

Trakarnsanga, K., R. E. Griffiths, M. C. Wilson, A. Blair, T. J. Satchwell, M. Meinders, N. Cogan et al. 2017. An immortalized adult human erythroid line facilitates sustainable and scalable generation of functional red cells. *Nature Communications* 8 (14750): 1–7. doi:10.1038/ncomms14750.

Villa-Diaz, L. G., H. Nandivada, J. Ding, N. C. Nogueira-de-Souza, P. H. Krebsbach, K. S. O'Shea, J. Lahann, and G. D. Smith. 2010. Synthetic polymer coatings for long-term growth of human embryonic stem cells. *Nature Biotechnology* 28 (6): 581–583. doi:10.1038/nbt.1631.

Xu, C., M. S. Inokuma, J. Denham, K. Golds, P. Kundu, J .D. Gold, and M. K. Carpenter. 2001. Feeder-free growth of undifferentiated human embryonic stem cells. *Nature Biotechnology* 19 (10): 971–974. doi:10.1038/nbt1001-971.

Yoshida, Y., K. Takahashi, K. Okita, T. Ichisaka, and S. Yamanaka. 2009. Hypoxia enhances the generation of induced pluripotent stem cells. *Cell Stem Cell* 5 (3): 237–241. doi:10.1016/j.stem.2009.08.001.

Yu, J., M. A. Vodyanik, K. Smuga-Otto, J. Antosiewicz-Bourget, J. L. Frane, S. Tian, J. Nie et al. 2007. Induced pluripotent stem cell lines derived from human somatic cells. *Science* 318: 1917–1919. doi:10.1126/science.1151526.

Zhou, H., S. Wu, J. Y. Joo, S. Zhu, D. W. Han, T. Lin, S. Trauger et al. 2009. Generation of induced pluripotent stem cells using recombinant proteins. *Cell Stem Cell* 8 (4): 381–384.

4 3D Strategies for Expansion of Human Cardiac Stem/ Progenitor Cells

Maria João Sebastião, Bernardo Abecasis,
Manuel J.T. Carrondo, Paula M. Alves,
Patrícia Gomes-Alves, and Margarida Serra

CONTENTS

4.1 INTRODUCTION

Ischemic Heart Disease (IHD) is the most common type of heart condition as well as the most prevalent cause of death worldwide, with an estimation of 7.4 million deaths in 2012 ("WHO|Cardiovascular Diseases Fact Sheet" 2016). IHD consists of plaque build-up in heart arteries inner walls, narrowing and reducing blood flow to the heart, ultimately leading to myocardial infarction (MI), with tissue damage and irreversible loss of cardiomyocytes (CMs). Current treatments have success in reducing immediate mortality but fail to recover the injured tissue. The subsequent degeneration of myocardium with loss of contractile function leads to Chronic Heart Failure (CHF), a potential fatal condition in which the only available clinic option is a heart transplant, which is impaired by the scarcity of available hearts donors, high costs, and need for immunosuppression (Yusen et al. 2015).

For many years, the adult mammalian heart has been considered a post-mitotic organ without regenerative potential. In 2003, Beltrami and colleagues identified a population positive for the stem cell marker tyrosine kinase (c-kit, also known as CD117) in the adult mouse heart (Beltrami et al. 2003), followed by the same discovery in other species, including humans (Linke et al. 2005; Bearzi et al. 2007).

Endogenous cardiac stem/progenitor cells (CPCs) seem to play an important role in cardiac homeostasis and in response to physiological stress and cardiac injury (Figure 4.1). CPCs can self-renew and differentiate *in vitro* into the three main cardiac

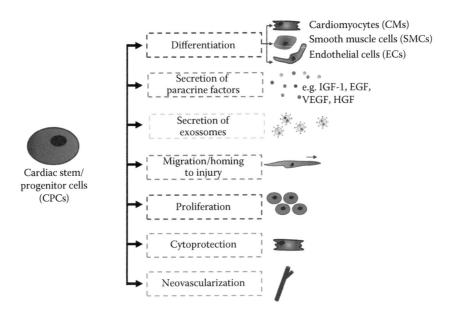

FIGURE 4.1 Overview of the quality attributes described for cardiac stem/progenitor cells (CPCs) in cardiac regeneration and repair. CPCs are described to be able to differentiate into the three main cardiac lineages, secrete an array of paracrine factors (e.g., insulin-like growth factor 1 [IGF-1], epidermal growth factor [EGF], vascular endothelial growth factor [VEGF], hepatocyte growth factor [HGF]) and exosomes, proliferate, migrate to the site of injury, protect the endogenous myocardial tissue, and promote neovascularization.

lineages (CMs, vascular smooth muscle cells (SMCs), and endothelial cells (ECs)) (Beltrami et al. 2003). Although some doubts and controversy still exists regarding the ability of endogenous and transplanted CPCs to differentiate *in vivo* upon injury (Ellison et al. 2013; van Berlo et al. 2014; Nadal-Ginard, Ellison, and Torella 2014), the regenerative paracrine potential of these cells has already been well documented in several animal studies (Urbanek et al. 2005a; Torella et al. 2007; Crisostomo et al. 2015a, Menárd et al. 2005). CPCs have been shown to secrete important paracrine factors involved in the modulation of cell proliferation, angiogenesis, vasculogenesis, and pro-survival of CMs (Torella et al. 2007; Miyamoto et al. 2010; Li et al. 2012; Sharma et al. 2016) and to express receptors recognizing factors highly released in the myocardium upon ischemic injury and other physiologic stresses (Urbanek et al. 2005b; Ellison et al. 2012; Li et al. 2014b; Gomes-Alves et al. 2015). Several of these cytokines and growth factors have shown to induce CPCs proliferation *in vitro* and *in vivo*, such as insulin-like growth factor 1 (IGF-1) (Ellison et al. 2011; Ellison et al. 2012; Koudstaal et al. 2014; Waring et al. 2014), epidermal growth factor (EGF) (Aghila Rani and Kartha 2010), and connective tissue growth factor (CTGF) (Stastna and Van Eyk 2012). Indeed, the number of CPCs has been shown to increase markedly upon acute MI in mouse (Ryzhov et al. 2012; Valiente-Alandi et al. 2016) and human (Urbanek et al. 2005a) hearts. Upon injury, CPCs also home and migrate to the site of injury, a process activated by ischemia through hypoxia-inducible factor 1-alpha (HIF-1α) transcription factor, chemokine stromal cell-derived factor 1 (SDF-1) (Ceradini et al. 2004; Rota et al. 2008), and by EGF and hepatocyte growth factor (HGF) (Urbanek et al. 2005b; Boucek et al. 2015). CPCs are also proposed to exert potent cytoprotective effects in cardiomyocytes through factors such as IGF-1 (Linke et al. 2005; Miyamoto et al. 2010; Kawaguchi et al. 2010), HGF (Linke et al. 2005), and vascular endothelial growth factor (VEGF) (Miyamoto et al. 2010). Other factors secreted by CPCs have also been shown to promote neovascularization in infarcted hearts, including pro-angiogenic factors VEGF (Miyamoto et al. 2010; Wang et al. 2014), transforming growth factor beta (TGF-β) (Ellison et al. 2012; Park et al. 2016), as well as several interleukins (Valiente-Alandi et al. 2016).

Exosomes have recently emerged as a novel player in CPCs regenerative properties, with observed cardiac functional improvement after administration of CPC-derived exosomes in mouse (Ibrahim, Cheng, and Marbán 2014) and rat (Barile et al. 2014) infarcted myocardium.

The documented regenerative mechanisms of CPCs upon myocardial injury have led, in the last 15 years, to a boost in the research for the development of clinical therapies involving the transplantation of CPCs onto ischemic patients. In order to generate CPCs in sufficient number and high quality for clinical trials and for further therapy development, several strategies have been pursued to cultivate and expand CPCs *in vitro*, including the use of controlled bioreactors and three-dimensional (3D) culture configurations.

In this chapter, we review and discuss the clinical relevance of CPCs, the ongoing clinical trials, different sources of CPCs, as well as methodologies for isolation, culture, and characterization of these cells, with emphasis on the application of 3D culture strategies and bioreactor technologies, as well as on the analytical tools used for the characterization of CPC quality attributes and process monitoring.

4.2   CLINICAL TRANSLATION/RELEVANCE OF CARDIAC STEM/PROGENITOR CELLS

The promising regenerative potential demonstrated for endogenous CPCs and the beneficial heart functionality effects of CPCs-based transplantation therapies in preclinical models of IHD (average of 10.7% improvement in left ventricular ejection [Zwetsloot et al. 2016]) led to a rapid translation of CPCs application in clinical trials (Table 4.1). Several other cell types have been applied in clinical trials of ischemic heart pathologies, including "first generation" stem cell sources such as bone marrow mononuclear cells (BM-MNCs) (e.g., REPAIR-AMI trial, NCT00279175), bone marrow-derived mesenchymal stem/stromal cells (BM-MSCs) (e.g., BOOST trial, NCT00224536), and adipose tissue-derived mesenchymal stem/stromal cells (AT-MSCs) (e.g., APOLLO trial, NCT00442806). More focus is now being devoted to therapies with more purified and homogeneous "second generation" stem cell sources, such as CPCs, which are considered by several authors as the best candidate cell therapy for these diseases, mainly due to their physiologic location and function in the heart, their potential to differentiate into myocardial lineages, and the promising regenerative effects of CPCs transplantation in MI preclinical models (Tang et al. 2010; Crisostomo et al. 2015a). In fact, in a direct comparison of BM-MNCs, BM-MSCs, AT-MSCs, and CPCs, the latter demonstrated superior functional benefit after transplantation in a MI mouse model, as well as a superior growth factor secretion profile (Li et al. 2012). Moreover, comparing transplantation of embryonic stem cell-derived CPCs (ESC-CPCs), embryonic stem cell-derived CMs (ESC-CMs), and BM-MSCs in a rat model of MI, ESC-CMs and ESC-CPCs outperformed BM-MSCs in terms of heart muscle function and contractility (Fernandes et al. 2015). However, such a trend in the superior effects of CPCs compared with other cell types seems to be lost when moving to large animal models, as demonstrated by a recent meta-analysis of CPC transplantation in preclinical MI models (Zwetsloot et al. 2016).

4.2.1   CARDIAC STEM/PROGENITOR CELLS IN CELL THERAPY: AUTOLOGOUS VERSUS ALLOGENEIC

The first clinical trials using CPCs were based on autologous therapy (Chugh et al. 2012; Makkar et al. 2012; Ishigami et al. 2015) (Table 4.1), since it holds the advantages of avoiding major ethical concerns as well as not presenting immunogenicity risks to the patients. However, autologous cell therapy is associated with serious limitations that compromise widespread "off-the-shelf" clinical applications, such as the difficult logistic, economic and time constraints in patient specific tissue harvesting and expansion. Moreover, cell's phenotype, regenerative potential, and quality will be highly variable and dependent on patient's age, co-morbidities, and genetic background (Dimmeler and Leri 2008; Sharma et al. 2016; Wu et al. 2016). To overcome such limitations, the field has been moving towards allogeneic

TABLE 4.1
Published Clinical Trials with Transplantation of hCPCs

	Phase	Clinicaltrials.gov Identifier	CPC Type	Patients	Route of Administration	Dosage	Status	Functional Outcomes	nr of Patients	References
SCIPIO - Cardiac Stem Cell Infusion in Patients With Ischemic Cardiomyopathy	I	NCT00474461	Autologous c-kit+ CPCs	MI (pre-CABG) (LVEF<40%)	Intracoronary injection	Single dose (0.5–1 × 10⁶)	Completed	↑LVEF; ↓infarct size; ↑contractile function	33	Bolli et al. (2011), Chugh et al. (2012)
CADUCEUS - Cardiosphere-Derived Autologous Stem Cells to Reverse Ventricular Dysfunction	I	NCT00893360	Autologous CDCs	Recent MI (LVEF<45%)	Intracoronary injection	Single dose (12–25 × 10⁶)	Completed	↓infarct size, ↓viable mass; ↑contractile function	25	Makkar et al. (2012)
ALLSTAR - Allogeneic Heart Sem Cells to Achieve Myocardial Regeneration	I/II	NCT01458405	Allogeneic CDCs	Recent and chronic MI (LVEF≤45%)	Intracoronary injection	Single dose (12.5 and 25 × 10⁶)	Phase I completed; Phase II ongoing	Phase I: ↓infarct size; ↑ infarcted segments wall thickening	Phase I: 14	Chakravarty et al. (2017)
HOPE - Halt Cardiomyopathy Progression in Duchenne	I/II	NCT02485938	Allogeneic CDCs	Duchenne muscular dystrophy cardiomyopathy	Intracoronary injection in each of the three left ventricle territories (anterior, lateral, inferior)	Single dose (75 × 10⁶)	Ongoing	–	Phase I: 25	(i)

(Continued)

TABLE 4.1 (*Continued*)
Published Clinical Trials with Transplantation of hCPCs

	Phase	Clinicaltrials.gov Identifier	CPC Type	Patients	Route of Administration	Dosage	Status	Functional Outcomes	nr of Patients	References
DYNAMIC - Dilated Cardiomyopathy Intervention with Allogeneic Myocardially-Regenerative Cells	I	NCT02293603	Allogeneic CDCs	Severe heart failure (LVEF≤35%)	Intracoronary injection	Single dose	Ongoing	–	–	(i)
ALCADIA - Autologous Human Cardiac-Derived Stem Cell to Treat Ischemic Cardiomyopathy	I	NCT00981006	Autologous CPCs + bFGF	Ischemic cardiomyopathy (LVEF ≤35%)	Intramyocardial injection; gelatin hydrogel patch with bFGF implanted on epicardium injection sites	Single dose (0.5 × 10⁶/ kg); 200 mg bFGF	Completed	↑LVEF; ↓infarct size; ↑Maximal aerobic exercise capacity	6	(i)
CAREMI - Safety and Efficacy Evaluation of Intracoronary Infusion of Allogeneic Human Cardiac Stem Cells in Patients With AMI	I/II	NCT02439398	Allogenous c-kit⁺ CPCs	Segment elevation myocardial infarction	Intracoronary injection	Single dose (35 × 10⁶)	Ongoing	–	55	(i)

(Continued)

TABLE 4.1 (*Continued*)
Published Clinical Trials with Transplantation of hCPCs

	Phase	Clinicaltrials.gov Identifier	CPC Type	Patients	Route of Administration	Dosage	Status	Functional Outcomes	nr of Patients	References
ESCORT - Transplantation of Human Embryonic Stem Cell-derived Progenitors in Severe Heart Failure	1	NCT02057900	ESC derived CD15+ Isl-1+ CPCs	Severe heart failure (LVEF≤35%)	Epicardial delivery of cells embedded in fibrin patch	Single dose (4×10^6)	Recruiting participants	–	–	Menasché et al. (2005a)
TICAP - Transcoronary Infusion of Cardiac Progenitor Cells in Patients With Single Ventricle Physiology	1	NCT01273857	Autologous CDCs	Pediatric patients with hypoplastic left heart syndrome	Intracoronary injection	Single dose (0.3×10^6/ kg)	Completed	↑RVEF; ↑somatic growth	18	Ishigami et al. (2015)
PERSEUS - Cardiac Progenitor Cell Infusion to Treat Univentricular Heart Disease	II	NCT01829750	Autologous CDCs	Univentricular heart disease	Intracoronary injection	Single dose (0.3×10^6 cells/kg)	Completed	↓infarct size; ↑somatic growth; ↑ trophic factors production	41	Ishigami et al. (2017)

(Continued)

TABLE 4.1 (Continued)
Published Clinical Trials with Transplantation of hCPCs

	Phase	Clinicaltrials.gov Identifier	CPC Type	Patients	Route of Administration	Dosage	Status	Functional Outcomes	nr of Patients	References
CONCERT-HF - Combination of Mesenchymal and C-kit+ Cardiac Stem Cells as Regenerative Therapy for Heart Failure	II	NCT02501811	Autologous BM-MSCs, c-kit+ CPCs both alone and in combination	Ischemic cardiomyopathy	Transendocardial injection	Single dose (150×10^6 BM-MSCs; 5×10^6 CSCs)	Recruiting participants	–	–	(i)
Transplantation of Autologous Cardiac Stem Cells in Ischemic Heart Failure	II	NCT01758406	Autologous CPCs	Ischemic heart failure	Intracoronary injection	Single dose ($5–100 \times 10^6$)	Recruiting participants	–	–	(i)
APOLLON - Cardiac Stem/Progenitor Cell Infusion in Univentricular Physiology	III	NCT02781922	Autologous CPCs	Pediatric patients with heart failure	Intracoronary injection	Single dose (0.3×10^6 cells/kg)	Recruiting patients	–	–	(i)

(Continued)

TABLE 4.1 (Continued)
Published Clinical Trials with Transplantation of hCPCs

	Phase	Clinicaltrials.gov Identifier	CPC Type	Patients	Route of Administration	Dosage	Status	Functional Outcomes	nr of Patients	References
TAC-HFT-II - The Transendocardial Autologous Cells (hMSC) or (hMSC) and (hCSC) in Ischemic Heart Failure Trial	I/II	NCT02503280	Combination of autologous MSCs and c-kit$^+$ CPCs or MSCs alone	Chronic ischemic left ventricular dysfunction and/ or heart failure secondary to MI	Transendocardial injection	Single dose (200 × 10^6 BM-MSCs or 190 × 10^6 BM-MSC + 1 × 10^6 CSCs)	Recruiting patients	–	–	(i)

Abbreviations: Cardiac stem/ progenitor cells (CPCs); Cardiosphere-derived cells (CDCs); Embryonic stem cells (ESC); Bone marrow-derived mesenchymal stem/stromal cells (BM-MSCs); mesenchymal stem/stromal cells (MSCs); Myocardial infarction (MI); Coronary artery bypass grafting (CABG); Left ventricular ejection fraction (LVEF); Right ventricular ejection fraction (RVEF).

(i) *Information were obtained from www.clinicaltrials.gov, accessed June 2017.*

clinical approaches, such as the clinical study ALLSTAR (*Allogeneic Heart Stem Cells to Achieve Myocardial Regeneration*) and more recently, the CAREMI trial (*Safety and Efficacy Evaluation of Intracoronary Infusion of Allogeneic Human Cardiac Stem Cells in Patients With AMI*), which had success in demonstrating safety and lack of rejection of transplanted cells, as well as improvements in infarct size (Table 4.1) (preliminary results of CAREMI in press release at *www.tigenix.com* [*accessed in June 2017*]).

Moreover, a cross-talk of transplanted allogeneic c-kit[+] human CPCs (hCPCs) with innate natural killer cells has been shown to result in attenuation of myocardium inflammation and prevention of adverse cardiac remodeling (Boukouaci et al. 2014). The same group has also identified programmed death ligand 1 (PDL-1) interaction with T regulatory cells as one of the mechanisms involved in this immunomodulatory capacity and as a marker to identify and select low immunogenic risk allogeneic c-kit[+] CPCs (Lauden et al. 2013).

4.2.2 Pre-Clinical and Clinical Outcomes

For all different CPC subpopulations tested, clinical trials have demonstrated some physiological improvements, increase in viable tissue and in heart functional outcome (Table 4.1), but very limited cell retention and engraftment in the heart, regardless of the route of administration and cell dosage. Within 24 hours of delivery, less than 10% of injected cells remain at the targeted location, and most of the successfully retained cells die, probably due to the inflammatory and ischemic environment in the infarct and infarct border zones of the myocardium (Hong and Bolli 2014; Mathur et al. 2017). In patients with heart failure, the lack of signals emitted from the target tissue also does not facilitate the homing of transplanted cells. In order to further improve the physiologic benefit of cell transplantation, several strategies have been pursued to increase cell retention, including preconditioning of cells to be transplanted, for example, with a hypoxia cultivation priming phase (Hosoyama et al. 2015), preconditioning the target tissue (Assmus et al. 2013), repeated cell dosage (Tokita et al. 2016), and biomaterial-based approaches (Hosseinkhani et al. 2010; Rajabi-Zeleti et al. 2014; Kryukov et al. 2014; Menasché et al. 2015a; Gaetani et al. 2015). The low engraftment in preclinical and clinical studies strongly supports the hypothesis that the beneficial physiological effect of transplanted CPCs is due to paracrine modulation rather than tissue engraftment and differentiation. The paracrine factors secreted by the transplanted CPCs include VEGF, HGF, IGF, and SDF-1 among others. These factors play an important role in promoting angiogenesis, inhibiting resident myocardium cells apoptosis, as well as activating endogenous CPCs populations (Chimenti et al. 2010; Sharma et al. 2016). Novel strategies for heart regeneration involve the activation of endogenous CPCs populations directly with IGF-1 and HGF administration (Urbanek et al. 2005b; Ellison et al. 2011; Koudstaal et al. 2014), or combinatorial approaches, such as the one employed in the ALCADIA trial, in which a sustained release of basic fibroblast growth factor (bFGF) from a gelatin hydrogel sheet was used to augment the effect exerted by the transplanted CPCs (Takehara et al. 2012) (Table 4.1).

4.3 MANUFACTURING OF CARDIAC STEM/ PROGENITOR CELLS FOR CELL-BASED THERAPY

As a direct result of the increasing application of CPCs in clinical therapies and the low number of CPCs that can be isolated from tissue biopsies, extensive efforts have been made focusing on the development of methods for the isolation or derivation of CPCs, as well as on *in vitro* cell expansion (schematic representation in Figure 4.2). Several studies have been performed with the aim of developing culture protocols that ensure the efficient proliferation of CPCs while maintaining, or further improving, their regenerative properties, and minimizing the potential risks for cell transplantation.

4.3.1 ISOLATION OF ENDOGENOUS CARDIAC STEM/PROGENITOR CELLS

The mammalian adult heart harbors a small percentage of endogenous CPCs (about one stem cell per 8000–20000 CMs [Anversa et al. 2006]). CPCs are in organized hypoxic niches within the myocardium, more abundant in lower hemodynamic stress areas such as the atrium and apex. Within the niches, lineage-negative CPCs are typically clustered together with early committed cells and adult CMs (Leri et al. 2014).

Several CPCs subpopulations have been described and isolated from the adult heart according to their phenotypic profile and differential expression of surface molecular markers. These subpopulations include c-kit$^+$, Sca1$^+$ and Isl-1$^+$ CPCs (for a more detailed review see Santini et al. 2016).

4.3.1.1 C-Kit$^+$ Cardiac Stem/Progenitor Cells

As described before, c-kit$^+$ cells were the first population of CPCs to be identified in the adult mouse heart in 2003 (Beltrami et al. 2003). These cells are characterized by the expression of stemness marker c-kit and expression of cardiac lineage transcription factors, such as GATA4, Nkx2.5, and Mef2C, and the absence of hematopoietic lineage markers such as CD34 and CD45 (Santini et al. 2016). C-kit$^+$ CPCs are one of the most extensively studied CPCs subpopulations, already employed in several clinical trials, such as SCIPIO, CAREMI, CONCERT and TAC-HFT-II (Table 4.1).

4.3.1.2 Sca-1$^+$ Cardiac Stem/Progenitor Cells

Sca-1$^+$ CPCs are characterized by the expression of the endothelial marker Sca1 (stem cells antigen-1). These cells have been identified in mice and human adult heart and express the cardiac transcription factors GATA4, Mef2C, and Tef1, but lack other cardiac lineage markers, such as Nkx2.5, hematopoietic markers CD34 and CD45, and mature endothelial markers such as CD31 (Santini et al. 2016).

4.3.1.3 Isl-1$^+$ Cardiac Stem/Progenitor Cells

Another population of CPCs, characterized by the expression of the LIN-homeodomain transcription factor insulin gene enhancer protein (Isl-1) has also been described in murine and human hearts. These cells are negative for c-kit, Sca-1, and CD31, while expressing the cardiac transcription factors Nkx2.5 and GATA4. The expression of Isl-1 is closely associated with age: Isl-1$^+$ cells can be

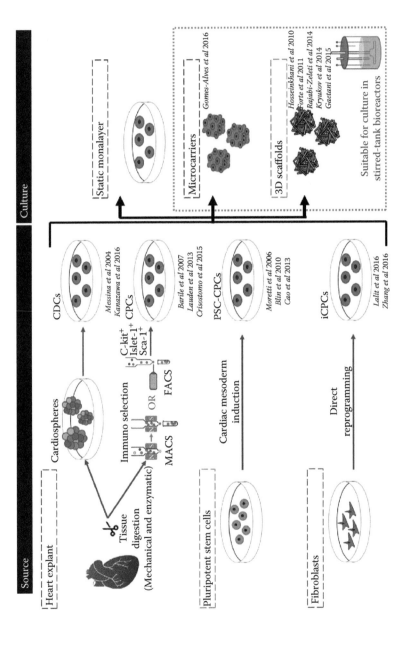

FIGURE 4.2 Schematic representation of sources and culture systems currently available for cardiac stem/progenitor cells. Abbreviations: Magnetic-activated cell sorting (MACS); Fluorescence-activated cell sorting (FACS); Cardiosphere-derived cells (CDCs); Cardiac stem/progenitor cells (CPCs); Pluripotent stem cell-derived cardiac stem cells (PSC-CPCs); induced cardiac stem cells (iCPCs).

found predominantly in fetal and neonatal myocardium, yet very low levels of this population can be found in adult hearts, which limits their clinical application (Weinberger et al. 2012).

4.3.2 ISOLATION METHODS

CPC subpopulations are usually isolated from the adult human heart tissue either from cadavers or from biopsies of patients undergoing heart surgery. Experimental approaches for the isolation of hCPCs can be separated into two categories: cardiosphere-based and immunoselection-based on the expression of specific molecular markers (Figure 4.2).

The first method to isolate CPCs from human heart biopsies was reported in 2004 (Messina et al. 2004) (Patent number WO2005012510). Cardiac tissue explants were enzymatically digested with trypsin and collagenase and cultured as adherent explants. These explants gave rise to a layer of fibroblast-like cells over which small, phase-bright cells migrated. These phase-bright cells were collected and cultured in medium supplemented with bFGF, EGF, cardiotrophin-1, thrombin, and B27 in coating substrate poly-D-lysine. In these conditions, cells spontaneously aggregate, giving rise to 3D structures, named cardiospheres, which are composed of mixed cell populations with undifferentiated cells expressing stem cell markers in the core and differentiating cells expressing markers characteristic of cardiac, vascular endothelial, and stromal commitment in the periphery (Davis et al. 2010). Cardiospheres are then collected, replated in fibronectin adherent culture dishes, and further expanded as monolayers to yield cardiosphere-derived cells (CDCs). These cells are uniformly positive for CD105 and negative for the pan-hematopoietic marker CD45, with a 25%–60% population positive for mesenchymal marker CD90 and a minority of cells (≈3%) expressing c-kit (Cheng et al. 2014; Kapelios et al. 2016).

Several preclinical studies in myocardial injury models (Kanazawa et al. 2016) have demonstrated the beneficial effects of transplantation of CDCs in improving cardiac function and decreasing the size of scar tissue (for detailed reviews consult Marbán 2014; Kapelios et al. 2016). These studies have already been translated into several human clinical trials, such as CADUCEUS and ALLSTAR (Table 4.1). However, the cell population heterogeneity inherent to the cardiosphere isolation method has been pointed as a disadvantage in terms of obtaining robust and predictable clinical outcomes.

Another method for isolating hCPCs from heart tissue is by the enzymatic digestion of tissue explants followed by immuno-selection for stem cell markers, such as the surface receptors c-kit, Sca-1, Isl-1, and immunodepletion of hematopoietic and mesenchymal markers, such as CD45 and Tryp, by magnetic-activated cell sorting (MACS) or fluorescence-activated cell sorting (FACS) (Barile et al. 2007; Lauden et al. 2013; Goichberg et al. 2014). In particular, CPCs immunoselected for c-kit[+] represent the most extensively characterized CPC population. Although controversy exists regarding the differentiation capability of c-kit[+] CPCs (Ellison et al. 2013; van Berlo et al. 2014; Nadal-Ginard et al. 2014) and the stability of c-kit marker expression during *in vitro* culture (Forte et al. 2011), numerous preclinical studies (for a more detailed review consult Nigro et al. 2015) and clinical trials (Table 4.1) demonstrate that c-kit[+] CPCs have relevant regenerative properties, since transplantation of this cell population results in improved cardiac tissue function and reduction of scar tissue size.

It is not yet clear what is the more advantageous CPC cell source and isolation method, probably due to the lack of direct comparison studies and lack of standardization of preclinical trial studies in terms of animals used, cell dosage, method of administration, and efficacy end-point measurements (Fernández-Avilés et al. 2017). Nonetheless, the transplantation of CDCs was shown to outperform Sca-1[+] CPCs in a meta-analysis in small animal models of AMI (Zwetsloot et al. 2016) in terms of improvement in left ventricular ejection fraction.

4.3.3 DERIVATION OF HUMAN CARDIAC STEM/PROGENITOR CELLS FROM OTHER CELL SOURCES

Besides donor hearts, other sources for CPC isolation have emerged as a putative resource for clinical hCPC translation. Pluripotent stem cells (PSCs), including induced pluripotent stem cells (iPSCs) and embryonic stem cells (ESCs), have unlimited self-renewal capacity and the potential to differentiate *in vitro*, holding great promise for the clinical translation of cell therapies. In particular, hiPSCs hold the advantage of allowing for autologous therapies without invasive isolation procedures. In fact, as development of less invasive surgical intervention techniques is further explored, the availability of biopsies for isolation of autologous and allogeneic hCPCs will be scarcer in a near future. As so, efforts have been made throughout the years to obtain cardiac stem/progenitor cells from pluripotent stem cells (PSC-CPCs).

The first study reporting PSC-CPCs (Isl-1[+]) was performed using murine ESCs and demonstrated the proliferative capacity of these derived CPCs on a feeder layer of mesenchymal cells (Moretti et al. 2006). Similarly, in the following studies, human PSC (hPSC) were differentiated into CPCs either through an embryoid body (EB)-based spontaneous differentiation step originating Isl-1[+] hCPCs (Bu et al. 2009) or a bone morphogenetic protein 2 (BMP-2) directed differentiation originating stage specific embryonic antigen-1 (SSEA1[+]) hCPCs (Tomescot et al. 2007; Blin et al. 2010). Both types of hPSC-CPC have shown similar regenerative potential to the CPCs isolated from human fetal hearts and have proven to be able to repopulate the infarcted heart after *in vivo* differentiation into CMs in both rat and non-human primate models. More recently, these advances in the differentiation of hPSC-CPCs led to the translation from bench scale to a cell-based medicinal product (Menasché et al. 2015b) culminating in the ongoing clinical trial with hESC-CPCs for patients with severe heart failure (ESCORT Trial, Table 4.1). More recently, efforts have been made to achieve more efficient protocols for differentiation and expansion of CPCs from hPSCs in a chemically defined medium under feeder- and serum-free culture conditions resulting in high purities of SSEA1[+] mesoderm posterior protein 1 positive (MESP1[+]) cells (Cao et al. 2013).

Several questions arise today on how these hPSC-CPCs compare to the CPCs isolated from the human heart. Within this context, several studies have been trying to bring light on the different cardiac progenitors generated during heart development, and to which signaling pathways may be manipulated to obtain those (Birket et al. 2015).

In a more recent study, global transcriptomic analysis of patient epicardium-derived CPCs and hPSC-CPCs was carried out showing that more than three thousand genes were differentially expressed between the two cell types, with hierarchical clustering analysis denoting a pronounced separation between the two types of CPCs (Synnergren et al. 2016) and confirming the phenotypic differences between these two populations.

Other innovative strategies for the generation of CPCs have emerged through direct reprogramming of fibroblasts (iCPCs). These iCPCs have been described as having (1) expression of cardiac signature genes, (2) extensive proliferative capacity, and (3) ability to differentiate into the three main cardiac lineages *in vitro* and *in vivo* after transplantation into infarcted mouse hearts (Lalit et al. 2016; Zhang et al. 2016). Although these iCPCs have shown promising *in vivo* regeneration capacity, they still lack direct comparison to isolated endogenous CPCs or hPSC-CPCs. In addition, these studies were performed using murine cells and further studies with human cells are still required, as well as protocols for their safe and efficient application in a clinical setting (Chen and Wu 2016).

4.3.4 *Ex-Vivo* Expansion of Human Cardiac Stem/Progenitor Cells

In current clinical trials, high cell dosages, ranging from 0.5 to 75 million cells are being used for cardiac regeneration (Table 4.1). Since endogenous CPCs represent only a minor population of the human heart (Section 3.2) and very low cell numbers are recovered after isolation, efficient protocols allowing the expansion of these cells *ex-vivo* are required to ensure the production of clinically relevant numbers of hCPC in high quality.

Current manufacturing protocols for hCPCs for clinical settings use standard static planar (or two-dimensional, 2D) culture systems. Although there is little information about the proliferation potential of endogenous hCPCs *in vitro* on these systems, it has already been described that these cells are able to expand approximately 4-fold in seven days (Gomes-Alves et al. 2016) or reach a population doubling level (PDL) of six (estimated 64-fold expansion) in 40 days of culture (Smith et al. 2007). The differences observed in population doubling time between the two studies (approx. 3.5 days in Gomes Alves et al. 2016 study and 6.7 in Smith et al. 2007 study) may reflect the distinct cell origins (c-kit CPCs vs CDCs) as well as the different culture conditions used for cell growth (Gomes-Alves study used growth factors 10 ng/mL bFGF, 20 ng/mL EGF, 30 ng/mL IGF-II, B-27 (1X), and 10% (v/v) fetal bovine FBS, while the Smith et al CDC growth medium only included 20% FBS (v/v) supplementation). On the other hand, hiPSC have unlimited proliferative potential and hiPSC-CPCs have also been described to be able to expand through 15 consecutive passages with consistent growth rates and high fold expansions in 2D culture systems (10^7-fold in 45–60 days) (Cao et al. 2013).

In the last decade, efforts have been made to invest in the development of scalable and efficient strategies to culture these cells in more controlled culture conditions, which have shown to be advantageous over static 2D culture systems (Serra et al. 2012; Gomes-Alves et al. 2016). In the following sections, the advances on 3D cell culturing of hCPCs will be described with a focus on the use of bioreactor technology, as well as on culture critical process parameters and characterization (and monitoring) of cells' critical quality attributes.

4.3.4.1 3D Strategies for the Expansion of Human Cardiac Stem/Progenitor Cells

Given the importance of cell–cell and cell-matrix interactions on stem cell fate decisions (self-renewal, differentiation, apoptosis, senescence), a variety of 3D culture strategies has been applied for stem cell bioprocessing, including self-aggregated spheroids and cell immobilization on microcarriers or hydrogels (Serra et al. 2012). The technological progresses made in the last decade show that there is no optimal 3D culture strategy capable of embracing all the applications of these cells. Each case faces unique challenges and evaluating them prior processing is crucial to decide on the appropriate method to be used.

As mentioned earlier (Section 3.1), self-aggregated spheroids (or 3D cellular structures) have long been used for the isolation and cultivation of hCPCs (Messina et al. 2004; Bartosh et al. 2008; Davis et al. 2010). These 3D structures benefit from the advantages typically reported for 3D culture systems: (1) mimic cardiac tissue morphology; (2) provide a suitable environment for cell–cell and cell-matrix contact; (3) enable self-organization of the tissue construct; and (4) allow for partial differentiation to occur (Barile et al. 2007). In particular, self-aggregated spheroids of hCPCs, such as cardiospheres, showed to promote the growth of different hCPCs populations (c-kit$^+$, Isl-1$^+$, etc.) within the same culture, all having potential for cardiac regeneration (Davis et al. 2010) as opposed to other immuno-based isolation protocols, which only enable the growth of a specific hCPC population. This high cell heterogeneity contributes to a functional advantage of cardiospheres over CDCs, namely regarding their differentiation potential (Davis et al. 2010), paracrine properties (Machida et al. 2011), and resistance to oxidative stress (Bartosh et al. 2008). However, the application of cardiospheres for intracoronary infusion, which from a clinical perspective represents the most practical method for cell delivery since it is less invasive and ensures cell administration to the entire myocardium (Crisostomo et al. 2015b), is limited due to the complications associated with microembolism. Therefore, cells have been delivered both intra-myocardially or trans-endocardially (Crisostomo et al. 2015b). In fact, previous data from Bauer and co-workers showed that the delivery of CPC in the form of 3D aggregates into the myocardium in a murine model of cardiac ischemia-reperfusion injury improves cell retention and survival when compared to single cell suspension therapy (Bauer et al. 2012), confirming the higher efficacy of these 3D aggregates of CPC.

Another strategy for the culture of hCPCs in a 3D-like structure is the use of microcarriers. A wide variety of microcarriers (microporous, macroporous, nonporous) is available for the culture of stem cells. Moreover, different materials (e.g., polystyrene, dextran, cellulose) have been used to produce different types of microcarriers and further functionalization/coating of the microcarrier with extracellular matrix-derived proteins (e.g., collagen, fibronectin, Matrigel™) may be required prior to culture. The selection of the microcarrier type and further functionalization must be performed according to the cell type, its characteristics (size, morphology, etc.), and the process requirement (expansion, differentiation, harvesting, etc.) (further reviewed in Badenes et al., 2016 and Serra et al., 2012). Microcarrier technology has several advantages for cell expansion, namely its adjustable surface to volume ratio (with diameters ranging from 100 to 400 µm), allowing an efficient cell expansion and facilitating the scale-up process. Another advantage when comparing to self-aggregated 3D spheroids is the minimization of oxygen and nutrient diffusion gradients in the non-porous or microporous microcarriers, ensuring a more homogenous culture microenvironment. Microcarrier technology has been already applied for the expansion of hCPCs. A screening of five different types of microcarriers (Cytodex 1, Cytodex coated with CELLstart CTS, Cytodex 3, Cultispher S, and Cytopore 2) was performed and hCPC were able to adhere and grow in all microcarriers tested (Gomes-Alves et al. 2016). Noteworthy, Cytodex 1 coated with CELLstart CTS was selected for further studies in stirred tank bioreactors as it allowed efficient cell attachment, microcarrier colonization, and highest cell growth (compared to the other microcarriers without further coating). This technology still holds some disadvantages such as the occurrence of microcarrier clumping during cell culture, the need for cell detachment from the microcarriers, and the additional costs with downstream processing for microcarrier removal. These disadvantages have been tackled in the last few years with the development of new strategies for: (1) the detachment of human MSC from microcarriers in bioreactors (Nienow et al. 2014; Cunha et al. 2017) and (2) the separation of microcarriers from human MSC suspension using filtration-based methodologies (Cunha et al. 2015).

Cell immobilization in biomaterials/hydrogels also guarantees a 3D environment for stem cell expansion (Serra et al. 2012). Different designs and biomaterials may be used as scaffolds, allowing to recreate a customized microenvironment. However, the impact of this type of immobilized cultures in the expansion of hCPCs has not been thoroughly evaluated. Studies using biomaterial-based approaches have been mostly focused on improving the delivery and therapeutic effect of hCPCs. A 3D-printed patch composed of hCPCs in a hyaluronic acid/gelatin-based matrix has shown to significantly reduce the adverse cardiac remodeling and preserve cardiac performance after MI by supporting the long-term *in vivo* survival and engraftment of the cells (Gaetani et al. 2015). Alternatively, extracellular matrix-based approaches, as a pericardium-derived decellularized 3D macroporous scaffold, have been shown to enable hCPCs to migrate, survive, proliferate, and differentiate at higher rates compared with decellularized pericardium membranes and collagen scaffolds (Rajabi-Zeleti et al. 2014). In addition, scaffold-free engineered tissues have been fabricated using temperature-responsive surfaces in which hCPCs remain embedded in self-produced extracellular matrix (Forte et al. 2011). After safe pericardial delivery of these tissues into murine hearts, cells migrate and integrate into the myocardium and in the vascular walls.

4.3.4.2 Bioreactors for the Expansion of Human Cardiac Stem/Progenitor Cells

The development of scalable bioprocesses for hCPC manufacturing in controlled culture systems, such as bioreactors, is crucial to produce clinically relevant numbers of high quality cells (in terms of identity, purity, and functionality/potency). These bioprocesses should also be cost-effective and compliant with good manufacturing practices (GMP-compliant) to ensure rapid and efficient transfer of cell-based products to late stage clinical phases and commercialization. Several bioreactors types have been proposed for the bioprocessing of either adult or pluripotent stem cells (for more details please see Serra et al. 2012 and Abraham et al. 2017). For the expansion of hCPCs, two different types of bioreactors have been reported: stirred-tank bioreactors (Gomes-Alves et al. 2016) and fixed-bed (also referred to as packed-bed) perfusion bioreactors (Hosseinkhani et al. 2010; Kryukov et al. 2014) (please see more details described in the following sections). Regarding hPSC-CPCs, although the expansion of undifferentiated hPSC in bioreactor culture systems have been extensively studied (Serra et al. 2012; Abecasis et al. 2017; Abraham et al. 2017), no integrated differentiation bioprocess for the generation of hPSC-CPCs in bioreactors have been proposed thus far.

4.3.4.2.1   Stirred-Tank Bioreactors

Stirred culture vessels are very attractive tools since they allow cell cultivation in a dynamic and homogeneous environment. These culture systems have been used across different chemical and biological industries for several years, and their scale-up is facilitated due to their extensive study and characterization. In addition, these bioreactors hold several advantages for stem cell bioprocessing, such as: efficient gas and nutrient transfer; precise control and monitoring of the culture environment; and non-destructive sampling (Serra et al. 2012). In addition, they are highly flexible as they can operate under different culture operation modes (batch, fed-batch, perfusion) and can be adapted to different 3D culturing strategies (self-aggregated spheroids or immobilized cells on microcarriers or hydrogels), presenting great potential for the manufacturing of cell-based products. There are still some disadvantages associated with cell culture in stirred culture vessels, namely the hydrodynamic stress promoted by stirring and the high volumes typically required to set up the experiments (50–200 mL), requiring higher starting cell numbers and increasing the cost in the bioprocess development studies. But large efforts have been made towards the development of smaller scale systems (working volume of 10–15 mL) and several options are available today such as the ambr® systems (from Sartorius).

With the aim to design a robust, scalable, cost-effective, and GMP-compliant strategy for hCPCs cultivation, a microcarrier-based stirred culture system was established in our laboratory using stirred-tank bioreactors (Gomes-Alves et al. 2016). As previously stated (Section 4.3.4.1), after optimization of the type of microcarrier to use in these cultures (Cytodex 1 coated with CELLstart CTS) (Figure 4.3), results have shown that the microcarrier-based stirred culture conditions led to more than three-fold higher cell expansion as compared to standard static 2D culture. To evaluate the robustness of the developed bioprocess, c-kit$^+$ hCPCs isolated from three different donors where tested, revealing comparable cell growth profiles, maximum growth rates of 0.43–0.50 day^{-1}, 9- to 11-fold increase in cell concentration, and 3.2–3.5 number of population doublings for the seven-day culture period. Some differences were observed in the efficiency of cell attachment to the microcarriers (attributed to the donor variability), which did not compromise cell expansion factors. Importantly, cell identity and potency were evaluated and compared to the hCPCs expanded in static planar culture systems, showing maintenance of phenotypic expression, growth factor/cytokine secretory capability, and differentiation potential. This study was part of the CAREMI project, which resulted in the namesake clinical trial

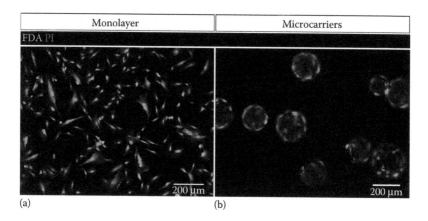

Monolayer	Microcarriers

(a) (b)

FIGURE 4.3 c-kit⁺ human cardiac stem/progenitor cells (hCPCs) cultured in monolayer (a) and adherent to microcarriers (b). Cell staining with fluorescein diacetate (FDA, live cells, green) and propidium iodide (PI, dead cells, red), scale bars 200 μm.

(Table 4.1). If the present therapeutic cell dosage for this clinical trial (35 million cells) and cell expansion factors were to be maintained in future clinical settings, it is estimated that this bioprocess could ensure the production of 10–12 cell doses per liter of culture.

4.3.4.2.2 Perfusion Bioreactors

Perfused bioreactor culture systems hold the advantage of continuously providing nutrients and oxygen and removing metabolic waste, while also enabling the use of 3D culturing strategies. Immobilization of cells in biomaterials, such as porous scaffolds, facilitates the use of these perfusion culture systems with anchorage-dependent cells, resulting in a fixed-bed perfusion culturing system. Compared to stirred-tank, fixed-bed perfusion bioreactors have the advantage of decreasing the hydrodynamic stress to which cells are exposed while maintaining the controlled levels of nutrients and oxygen. However, these bioreactors still hold some disadvantages as the heterogeneous cell distribution across the scaffolds, the difficulties in non-destructive sampling (hampering the precise monitoring of the culture), and low cell recovery yields usually obtained upon cell harvesting from the scaffolds.

For the culture of hCPCs, porous scaffolds, as collagen sponges (Hosseinkhani et al. 2010) or alginate scaffolds (Kryukov et al. 2014), have been combined with perfused operation to design fixed-bed perfusion bioreactor culture systems. These studies have used bioreactors with controlled circulating perfusion systems, which ensure culture medium oxygenation and heating. In these systems, the cells were seeded in the respective scaffolds and held in a fixing net in the interior of the bioreactor vessel to allow for the culture medium perfusion through the scaffold during the culture period. Results have shown that the attachment and proliferation of CPCs in the 3D culture was enhanced by 3.5-fold in the bioreactor perfusion system compared with 3D static and monolayer culture systems (Hosseinkhani et al. 2010). This dynamic system reduced the mass transfer limitations within the internal pores of the scaffold, which consequently promoted overall cell proliferation. Although these fixed-bed culture systems have limited potential for scalability when compared to the stirred-tank bioreactors, they showed suitability for robust expansion of human CPCs while maintaining their progenitor state and differentiation potential into the cardiovascular cell lineages (Kryukov et al. 2014).

4.3.4.3 Critical Process Parameters of Human Cardiac Stem/Progenitor Cells

As previously stated, environmentally controlled bioreactors enable the maintenance of physiological levels of gases (e.g., oxygen), nutrients, metabolites, and growth factors throughout culture time. These process parameters are controlled by the bioreactor through the different controlled variables (agitation, gas composition, perfusion rate, and so on) and will influence the outcome of the cell culture both in terms of quantity and quality. In particular, oxygen concentration and culture medium composition have been identified as critical process parameters (CPP) that influence the final outcome of hCPC bioprocessing.

Cell culture under physiological oxygen levels (3–4% O_2) has proven to be beneficial for the expansion of several stem cell types (Krinner et al. 2009; Serra et al. 2010; Li et al. 2011; Tiwari et al. 2016; Abecasis et al. 2017). For hCPC cultivation, it is already reported that these low oxygen partial pressure conditions affect not only final cell yield, but more importantly, their genomic stability and potency for myocardial repair (Li et al. 2011; Lauden et al. 2013).

The composition of the medium used for hCPCs culture also plays an important role in their expansion *in vitro*. The culture medium used for hCPCs cultivation is typically rich in glucose and supplemented with growth factors, such as EGF and bFGF, which are reported to promote the proliferation of different progenitor cells (Tropepe et al. 1999; Gharibi and Hughes 2012). Different studies have shown that activation of canonical Wnt signaling plays an important role in the embryonic heart development and, consequently, promotes the expansion of CPCs (Cohen et al. 2007; Kwon et al. 2007). Also, TGF-β signaling inhibition has been shown to improve yield and function of c-kit[+] CPCs from cardiac explants (Zakharova et al. 2013). Endothelin-1 (ET-1) has also been identified as a key component in cardiac development and regeneration, acting through both autocrine and paracrine means to maintain and expand clonal Isl-1[+] CPCs derived from human ESC (Soh et al. 2016). In fact, CPCs exposed to ET-1 were shown to differentiate primarily into endothelial and smooth muscle cell lineages, forming a functional circulatory system *in vivo*. Consequently, precise control of the concentration of metabolites and growth factors in the culture medium throughout time of culture is essential to maintain hCPC quality during bioprocessing and can be further improved using perfusion culture systems.

4.3.5 Critical Quality Attributes of Human Cardiac Stem/Progenitor Cells

As a cell-based therapy product, critical quality attributes (CQA) of hCPC should be assessed during manufacturing, to ensure maintenance of product identity, potency, and safety. As mentioned before, several CQA have been associated with these cells, including their capacity to: differentiate into the three main cardiac lineages (CMs, ECs, and SMCs); secrete several key paracrine factors (IGF-1, EGF, VEGF, HGF, and so on); migrate/home to site of injury; proliferate *in vitro*; protect resident myocardial tissue against ischemic injury; and promote neovascularization (Figure 4.1). Therefore, a series of different screening assays are usually performed to evaluate the quality of hCPCs. These screening assays include evaluation of growth kinetics, cell immunophenotype, expression of senescence markers, secretory profile, and angiogenic potential (Nakamura et al. 2016). However, the therapeutic mechanism of action (MoA) for hCPCs is still largely unknown, hampering the design of more precise *in vitro* potency assays for these cells. The development of tools that enable the in-depth characterization of hCPCs is crucial for a deeper understanding on the product quality and respective MoA. Omics tools have contributed to a more comprehensive and defined characterization of these cells, which could have an impact not only on the development of better quality control assays for cell identity and potency but also on bioprocess design and optimization. The depiction of the receptome of hCPCs enables the identification of specific receptors that may be used for the isolation or further characterization of these cells (Gomes-Alves et al. 2015). Also, several studies have been performed to characterize murine CPC whole proteome (Samal et al. 2012), transcriptome (Valiente-Alandi et al. 2016), secretome (Stastna et al. 2010; Park et al. 2016) and, more recently, human CPCs secretome (Sharma et al. 2016). The expression level of several molecules has also already been proposed as markers for higher quality CPCs. For example, podocalyxin (PODXL) has been recently identified as a regulator of hCPC activation, migration and differentiation, and immunoregulatory capacity (Moscoso et al. 2016). Also, high expression levels of the cardiac transcription factor GATA4 in rat CPCs has also been associated with cells with superior growth factor secretion profile and paracrine protection of CMs (Kawaguchi et al. 2010). Recently, a screening assay to characterize CPC immunophenotype by human leukocyte antigen (HLA) profiling was developed, as a novel tool to select allogeneic CPC donors with more suitable immune profiles for each patient (Hocine et al. 2017).

4.4   FUTURE PERSPECTIVES

CPCs have been shown to be able to regenerate myocardium and hold great promise in the clinical field, with several studies supporting them as the best candidate cell type for transplantation-based therapies. However, several challenges still need to be addressed in order to reach the full clinical potential of hCPCs, including: (1) further deciphering the mode of action of hCPCs; (2) improving hCPCs retention in the target tissue; and (3) continuing to invest in the development of efficient and robust controlled bioprocess methods for hCPCs expansion.

The low engraftment level of hCPC supports the notion that the observed functional benefits of hCPC transplantation result from indirect paracrine effects. In fact, hCPCs have shown to secrete factors and chemokines involved in modulation of cell proliferation, angiogenesis, vasculogenesis, and pro-survival of myocardial tissue. However, further understanding of hCPC biology is still needed in order to discover and modulate molecular pathways and mechanisms involved in the regenerative potential of these cells. In recent years, -omics tools have been used to characterize murine (Stastna et al. 2010; Samal et al. 2012; Li et al. 2014b; Valiente-Alandi et al. 2016) and human (Gomes-Alves et al. 2015; Gomes-Alves et al. 2016; Sharma et al. 2016) CPCs molecular landscape. Novel findings will be pivotal for the development of novel clinical strategies, including activation of endogenous hCPCs, preconditioning of cells to be transplanted, and protein-based therapies (for example, combined administration of IGF-1 and HGF has been explored in animal models (Koudstaal et al. 2014; O'Neill et al. 2016). New relevant human cell-based models of cardiac tissue damage should also be pursued to characterize hCPC response to human myocardial injury *in vitro*.

Allogeneic-based approaches constitute one of the most promising clinical solutions as they allow a more widespread off-the-shelf cell availability to the patient while potentially avoiding the variables regarding donor age and co-morbidities that might impact CPCs quality. Clinical trials, such as ALLSTAR and, more recently, CAREMI (Table 4.1) demonstrated the safety of allogeneic CPC transplantation, without raising any immunoresponse issues, while still showing improvements in tissue functionality.

However, enhancement of cell retention at the site of injury still constitutes one of the major challenges inhibiting the full regenerative potential of CPC transplantation-based therapies. Optimization of delivery strategies and imaging technology to guide and improve CPC retention should be pursued, as well as biomaterial-based tissue engineering approaches (such as matrices or patches). Albeit the increased therapy logistics complexity and costs aside, repetitive dosage approaches are also supported by different authors as a novel way to circumvent the low-retention issue (Tokita et al. 2016).

So far, the number of cells transplanted in phase I/II clinical trials ranges from 0.5 to 75 million CPCs per patient. For the large majority of pre-clinical and clinical studies, CPCs are still expanded using planar culture systems (Madonna et al. 2016), which limits the cell numbers available and impairs cell dosage optimization. Also, if considering repetitive dosage and moving forward to large phase III clinical trials with hundreds of patients, novel methods, based on controlled bioreactors, would allow a more efficient, cost-effective, faster, scalable, and controlled expansion of CPCs. Novel bioprocess development should also continue to include the compliancy with GMP guidelines and evaluation of cell's critical quality parameters (identity, purity, and potency). Culture of cells in 3D configurations, such as aggregates or cells adherent to microcarriers or other scaffolds, can also enhance cell quality and cell production yield.

Despite promising results, the full clinical potential of CPC-based therapies is yet to be met. International collaborative consortiums and multicenter studies, such as Translational Alliance for Regenerative Therapies in Cardiovascular Syndromes (TACTIC) (Fernández-Avilés et al. 2017), Consortium for Preclinical Assessment of Cardioprotective Therapies (CAESAR) (Jones et al. 2015), and Cardio Repair European Multidisciplinary Initiative (CARE-MI) (www.cordis. europa.eu), are emerging as important platforms to bring basic researchers and clinicians together to discuss and define common goals and strategies, as well as to standardize protocols and analytical techniques in preclinical and clinical studies.

4.5 CONCLUSION

Although promising, CPC regenerative medicine therapies have yet to prove evident and robust clinical benefit over standard-of-care. Novel directions to potentiate the use of CPCs in myocardial regeneration should include, at the preclinical level, fundamental studies of MoA complemented by the development of novel human cell-based models and -omics tools that will leverage the knowledge on hCPC biology. Although CPC MoA is still not fully understood, therapies employing CPC transplantation and endogenous CPC activation are being pursued, as preclinical and clinical studies continue to yield encouraging results. Bioprocess optimization with scalable bioreactor systems and 3D culture configurations will be a crucial step forward in the successful implementation of large phase III clinical trials that will assess efficacy and allow the implementation of CPC-based therapies for millions of patients worldwide.

LIST OF ABBREVIATIONS

2D	two-dimensional
3D	three-dimensional
AT-MSCs	adipose tissue-derived mesenchymal stem/stromal cells
bFGF	basic fibroblast growth factor
BMMNCs	bone marrow mononuclear cells
BM-MSCs	bone marrow-derived mesenchymal stem/stromal cells
BMP-2	bone morphogenetic protein 2
CABG	coronary artery bypass grafting
CDC	cardiosphere-derived cells
CHF	chronic heart failure
c-kit	tyrosine kinase
CMs	cardiomyocytes
CPC	cardiac stem/progenitor cells
CPP	critical process parameters
CQA	critical quality attributes
CTGF	connective tissue growth factor
EB	embryoid body
ECs	endothelial cells
EGF	epidermal growth factor
ESCs	embryonic stem cells
ET-1	endothelin-1
FACS	fluorescence-activated cell sorting
GMP	good manufacturing practices
hCPCs	human cardiac stem/progenitor cells
hESCs	human embryonic stem cells
HGF	hepatocyte growth factor
HIF-1α	hypoxia inducible factor 1-alpha
hiPSCs	human induced pluripotent stem cells
HLA	human leukocyte antigen

hPSC	human pluripotent stem cells
iCPCs	induced cardiac stem/progenitor cells
IGF-1	insulin-like growth factor 1
IHD	ischemic heart disease
iPSCs	induced pluripotent stem cells
Isl-1	insulin gene enhancer protein
LVEF	left ventricular ejection fraction
MACS	magnetic-activated cell sorting
MESP1	mesoderm posterior protein 1
MI	myocardial infarction
MoA	mechanism of action
O$_2$	oxygen
PDL-1	programmed death ligand 1
PODXL	podocalyxin
PSCs	pluripotent stem cells
RVEF	right ventricular ejection fraction
Sca-1	stem cells antigen-1
SDF-1	stromal cell-derived factor 1
SMCs	smooth muscle cells
SSEA1	stage specific embryonic antigen-1
TGF-β	transforming growth factor beta
VEGF	vascular endothelial growth factor

REFERENCES

Abecasis, B., T. Aguiar, É. Arnault, R. Costa, P. Gomes-Alves, A. Aspegren, M. Serra, and P. M. Alves. 2017. Expansion of 3D human induced pluripotent stem cell aggregates in bioreactors: Bioprocess intensification and scaling-up approaches. *Journal of Biotechnology* 246 (March): 81–93. doi:10.1016/j.jbiotec.2017.01.004.

Abraham, E., B. B. Ahmadian, K. Holderness, Y. Levinson, and E. McAfee. 2017. Platforms for manufacturing allogeneic, autologous and iPSC cell therapy products: An industry perspective. In *Advances in Biochemical Engineering/Biotechnology*. doi:10.1007/10_2017_14.

Aghila, R., K. Gopalakrishnannair, and C. Cheranellore Kartha. 2010. Effects of epidermal growth factor on proliferation and migration of cardiosphere-derived cells expanded from adult human heart. *Growth Factors* 28 (3): 157–165. doi:10.3109/08977190903512628.

Anversa, P., J. Kajstura, A. Leri, and R. Bolli. 2006. Life and death of cardiac stem cells: A paradigm shift in cardiac biology. *Circulation* 113 (11): 1451–1463. doi:10.1161/CIRCULATIONAHA.105.595181.

Assmus, B., D. H. Walter, F. H. Seeger, D. M. Leistner, J. Steiner, I. Ziegler, A. Lutz et al. 2013. Effect of shock wave–facilitated intracoronary cell therapy on LVEF in patients with chronic heart failure. *JAMA* 309 (15): 1622. doi:10.1001/jama.2013.3527.

Badenes, S. M., T. G. Fernandes, C. A. V. Rodrigues, M. Margarida Diogo, and J. M.S. Cabral. 2016. Microcarrier-based platforms for in vitro expansion and differentiation of human pluripotent stem cells in bioreactor culture systems. *Journal of Biotechnology* 234 (September): 71–82. doi:10.1016/j.jbiotec.2016.07.023.

Barile, L., E. Messina, A. Giacomello, and E. Marbán. 2007. Endogenous cardiac stem cells. *Progress in Cardiovascular Diseases* 50 (1): 31–48. doi:10.1016/j.pcad.2007.03.005.

Barile, L., V. Lionetti, E. Cervio, M. Matteucci, M. Gherghiceanu, L. M. Popescu, T. Torre, F. Siclari, T. Moccetti, and G. Vassalli. 2014. Extracellular vesicles from human cardiac progenitor cells inhibit cardiomyocyte apoptosis and improve cardiac function after myocardial infarction. *Cardiovascular Research* 103 (4): 530–541. doi:10.1093/cvr/cvu167.

Bartosh, T. J., Z. Wang, A. A. Rosales, S. D. Dimitrijevich, and R. S. Roque. 2008. 3D-model of adult cardiac stem cells promotes cardiac differentiation and resistance to oxidative stress. *Journal of Cellular Biochemistry* 105 (2): 612–623. doi:10.1002/jcb.21862.

Bauer, M., L. Kang, Y. Qiu, J. Wu, M. Peng, H. H. Chen, G. Camci-Unal et al. 2012. Adult cardiac progenitor cell aggregates exhibit survival benefit both *in vitro* and *in vivo*. *PLoS One* 7 (11): 1–10. doi:10.1371/journal.pone.0050491.

Bearzi, C., M. Rota, T. Hosoda, J. Tillmanns, A. Nascimbene, A. De Angelis, S. Yasuzawa-Amano et al. 2007. Human cardiac stem cells. *Proceedings of the National Academy of Sciences of the United States of America* 104 (35): 14068–14073. doi:10.1073/pnas.0706760104.

Beltrami, A. P., L. Barlucchi, D. Torella, M. Baker, F. Limana, S. Chimenti, H. Kasahara et al. 2003. Adult cardiac stem cells are multipotent and support myocardial regeneration. *Cell* 114 (6): 763–776.

Berlo, J. H. V., O. Kanisicak, M. Maillet, R. J. Vagnozzi, J. Karch, S.-C. J. Lin, R. C. Middleton, E. Marbán, and J. D. Molkentin. 2014. C-Kit+ cells minimally contribute cardiomyocytes to the heart. *Nature* 509 (7500): 337–341. doi:10.1038/nature13309.

Birket, M. J., M. C. Ribeiro, A. O. Verkerk, D. Ward, A. Rita Leitoguinho, S. C. den Hartogh, V. V. Orlova et al. 2015. Expansion and patterning of cardiovascular progenitors derived from human pluripotent stem cells. *Nature Biotechnology* 33 (July): 1–12. doi:10.1038/nbt.3271.

Blin, G., D. Nury, S. Stefanovic, T. Neri, O. Guillevic, B. Brinon, V. Bellamy et al. 2010. A purified population of multipotent cardiovascular progenitors derived from primate pluripotent stem cells engrafts in postmyocardial infarcted nonhuman primates. *Journal of Clinical Investigation* 120 (4): 1125–1139. doi:10.1172/JCI40120.

Bolli, R., A. R. Chugh, D. D'Amario, J. H. Loughran, M. F. Stoddard, S. Ikram, G. M. Beache et al. 2011. Cardiac stem cells in patients with ischaemic cardiomyopathy (SCIPIO): Initial results of a randomised phase 1 trial. *Lancet (London, England)* 378 (9806): 1847–1857. doi:10.1016/S0140-6736(11)61590-0.

Boucek, R. J., J. Steele, J. P. Jacobs, P. Steele, A. Asante-Korang, J. Quintessenza, and A. Steele. 2015. Ex vivo paracrine properties of cardiac tissue: Effects of chronic heart failure. *The Journal of Heart and Lung Transplantation: The Official Publication of the International Society for Heart Transplantation* 34 (6): 839–848. doi:10.1016/j.healun.2014.07.010.

Boukouaci, W., L. Lauden, J. Siewiera, N. Dam, H.-R. Hocine, Z. Khaznadar, R. Tamouza et al. 2014. Natural killer cell crosstalk with allogeneic human cardiac-derived stem/progenitor cells controls persistence. *Cardiovascular Research* 104 (2): 290–302. doi:10.1093/cvr/cvu208.

Bu, L., X. Jiang, S. Martin-Puig, L. Caron, S. Zhu, Y. Shao, D. J. Roberts, P. L. Huang, I. J. Domian, and K. R. Chien. 2009. Human ISL1 heart progenitors generate diverse multipotent cardiovascular cell lineages. *Nature* 460 (7251): 113–117. doi:10.1038/nature08191.

Cao, N., H. Liang, J. Huang, J. Wang, Y. Chen, Z. Chen, and H.-T. Yang. 2013. Highly efficient induction and long-term maintenance of multipotent cardiovascular progenitors from human pluripotent stem cells under defined conditions. *Cell Research* 23 (9). Nature Publishing Group: 1119–1132. doi:10.1038/cr.2013.102.

Ceradini, D. J., A. R. Kulkarni, M. J. Callaghan, O. M. Tepper, N. Bastidas, M. E. Kleinman, J. M. Capla, R. D. Galiano, J. P. Levine, and G. C. Gurtner. 2004. Progenitor cell trafficking is regulated by hypoxic gradients through HIF-1 induction of SDF-1. *Nature Medicine* 10 (8): 858–864. doi:10.1038/nm1075.

Chakravarty, T., R. R. Makkar, D. D. Ascheim, J. H. Traverse, R. Schatz, A. DeMaria, G. S. Francis et al. 2017. ALLogeneic heart STem cells to achieve myocardial regeneration (ALLSTAR) trial: Rationale and design. *Cell Transplantation* 26 (2): 205–214. doi:10.3727/096368916 × 692933.

Chen, I. Y., and J. C. Wu. 2016. Finding expandable induced cardiovascular progenitor cells. *Circulation Research* 119 (1): 16–20. doi:10.1161/CIRCRESAHA.116.308679.

Cheng, K., A. Ibrahim, M. Taylor Hensley, D. Shen, B. Sun, R. Middleton, W. Liu, R. R. Smith, and E. Marbán. 2014. Relative roles of CD90 and c-Kit to the regenerative efficacy of cardiosphere-derived cells in humans and in a mouse model of myocardial infarction. *Journal of the American Heart Association* 3 (5): e001260. doi:10.1161/JAHA.114.001260.

Chimenti, I., R. R. Smith, T.-S. Li, G. Gerstenblith, E. Messina, A. Giacomello, and E. Marbán. 2010. Relative roles of direct regeneration versus paracrine effects of human cardiosphere-derived cells transplanted into infarcted mice. *Circulation Research* 106 (5): 971–980. doi:10.1161/CIRCRESAHA.109.210682.

Chugh, A. R., G. M. Beache, J. H. Loughran, N. Mewton, J. B. Elmore, J. Kajstura, P. Pappas et al. 2012. Administration of cardiac stem cells in patients with ischemic cardiomyopathy: The SCIPIO trial: Surgical aspects and interim analysis of myocardial function and viability by magnetic resonance. *Circulation* 126 (11 Suppl 1): S54–S64. doi:10.1161/CIRCULATIONAHA.112.092627.

Cohen, E. D., Z. Wang, J. J. Lepore, M. M. Lu, M. M. Taketo, D. J. Epstein, and E. E. Morrisey. 2007. Wnt/β-Catenin signaling promotes expansion of Isl-1–positive cardiac progenitor cells through regulation of FGF signaling. *Journal of Clinical Investigation* 117 (7): 1794–1804. doi:10.1172/JCI31731.

Crisostomo, V., C. Baez-Diaz, J. Maestre, M. Garcia-Lindo, F. Sun, J. G. Casado, R. Blazquez et al. 2015a. Delayed administration of allogeneic cardiac stem cell therapy for acute myocardial infarction could ameliorate adverse remodeling: Experimental study in swine. *Journal of Translational Medicine* 13 (1): 156. doi:10.1186/s12967-015-0512-2.

Crisostomo, V., J. G. Casado, C. Baez-Diaz, R. Blazquez, and F. M. Sanchez-Margallo. 2015b. Allogeneic cardiac stem cell administration for acute myocardial infarction. *Expert Review of Cardiovascular Therapy* 13 (3): 285–299. doi:10.1586/14779072.20 15.1011621.

Cunha, B., T. Aguiar, M. M. Silva, R. Silva, M. F. Q. Sousa, E. Pineda, C. Peixoto, M. J. T. Carrondo, M. Serra, and P. M. Alves. 2015. Exploring continuous and integrated strategies for the up- and downstream processing of human mesenchymal stem cells. *Journal of Biotechnology* 213: 97–108. doi:10.1016/j.jbiotec.2015.02.023.

Cunha, B., T. Aguiar, S. B. Carvalho, M. M. Silva, R. A. Gomes, M. J. T. Carrondo, P. Gomes-Alves, C. Peixoto, M. Serra, and P. M. Alves. 2017. Bioprocess integration for human mesenchymal stem cells: From up to downstream processing scale-up to cell proteome characterization. *Journal of Biotechnology* 248 (April): 87–98. doi:10.1016/j.jbiotec.2017.01.014.

Davis, D. R., E. Kizana, J. Terrovitis, A. S. Barth, Y. Zhang, R. Ruckdeschel Smith, J. Miake, and E. Marbán. 2010. Isolation and expansion of functionally-competent cardiac progenitor cells directly from heart biopsies. *Journal of Molecular and Cellular Cardiology* 49 (2): 312–321. doi:10.1016/j.yjmcc.2010.02.019.

Dimmeler, S., and A. Leri. 2008. Aging and disease as modifiers of efficacy of cell therapy. *Circulation Research* 102 (11): 1319–1330. doi:10.1161/CIRCRESAHA.108.175943.

Ellison, G. M., C. D. Waring, C. Vicinanza, and D. Torella. 2012. Physiological cardiac remodelling in response to endurance exercise training: Cellular and molecular mechanisms. *Heart (British Cardiac Society)* 98 (1): 5–10. doi:10.1136/heartjnl-2011-300639.

Ellison, G. M., C. Vicinanza, A. J. Smith, I. Aquila, A. Leone, C. D. Waring, B. J. Henning et al. 2013. Adult c-Kit(pos) cardiac stem cells are necessary and sufficient for functional cardiac regeneration and repair. *Cell* 154 (4): 827–842. doi:10.1016/j.cell.2013.07.039.

Ellison, G. M., D. Torella, S. Dellegrottaglie, C. Perez-Martinez, A. Perez de Prado, C. Vicinanza, S. Purushothaman et al. 2011. Endogenous cardiac stem cell activation by insulin-like growth factor-1/hepatocyte growth factor intracoronary injection fosters survival and regeneration of the infarcted pig heart. *Journal of the American College of Cardiology* 58 (9): 977–986. doi:10.1016/j.jacc.2011.05.013.

Fernandes, S., J. J. H. Chong, S. L. Paige, M. Iwata, B. Torok-Storb, G. Keller, H. Reinecke, and C. E. Murry. 2015. Comparison of human embryonic stem cell-derived cardiomyocytes, cardiovascular progenitors, and bone marrow mononuclear cells for cardiac repair. *Stem Cell Reports* 5 (5): 753–762. doi:10.1016/j.stemcr.2015.09.011.

Fernández-Avilés, F., R. Sanz-Ruiz, A. M. Climent, L. Badimon, R. Bolli, D. Charron, V. Fuster et al. 2017. Global position paper on cardiovascular regenerative medicine: Scientific statement of the transnational alliance for regenerative therapies in cardiovascular syndromes (TACTICS) international group for the comprehensive cardiovascular application of regenerative medicinal products. *European Heart Journal*, May. doi:10.1093/eurheartj/ehx248.

Forte, G., S. Pietronave, G. Nardone, A. Zamperone, E. Magnani, S. Pagliari, F. Pagliari et al. 2011. Human cardiac progenitor cell grafts as unrestricted source of supernumerary cardiac cells in healthy murine hearts. *Stem Cells (Dayton, Ohio)* 29 (12): 2051–2061. doi:10.1002/stem.763.

Gaetani, R., D. A. M. Feyen, V. Verhage, R. Slaats, E. Messina, K. L. Christman, A. Giacomello, P. A. F. M. Doevendans, and J. P. G. Sluijter. 2015. Epicardial application of cardiac progenitor cells in a 3D-printed gelatin/hyaluronic acid patch preserves cardiac function after myocardial infarction. *Biomaterials* 61 (August): 339–348. doi:10.1016/j.biomaterials.2015.05.005.

Gharibi, B., and F. J. Hughes. 2012. Effects of medium supplements on proliferation, differentiation potential, and in vitro expansion of mesenchymal stem cells. *Stem Cells Translational Medicine* 1 (11): 771–782. doi:10.5966/sctm.2010-0031.

Goichberg, P., J. Chang, R. Liao, and A. Leri. 2014. Cardiac stem cells: Biology and clinical applications. *Antioxid Redox Signal* 21 (14): 2002–2017. doi:10.1089/ars.2014.5875.

Gomes-Alves, P., M. Serra, C. Brito, C. Pinto Ricardo, R. Cunha, M. Sousa, B. Sanchez et al. 2016. In vitro expansion of human cardiac progenitor cells: Exploring'omics tools for characterization of cell-based allogeneic products. *Translational Research* 171 (May): 96–110.e1–e3. doi:10.1016/j.trsl.2016.02.001.

Gomes-Alves, P., M. Serra, C. Brito, L. R. Borlado, J. A. Opez, J. U. Azquez, M. J. T. Carrondo, A. Onio Bernad, and P. M. Alves. 2015. Exploring analytical proteomics platforms toward the definition of human cardiac stem cells receptome. *Proteomics* 15: 1332–1337. doi:10.1002/pmic.201400318.

Hocine, H. R., H. E. L. Costa, N. Dam, J. Giustiniani, I. Palacios, P. Loiseau, A. Benssusan et al. 2017. Minimizing the risk of allo-sensitization to optimize the benefit of allogeneic cardiac-derived stem/progenitor cells. *Scientific Reports* 7 (January): 41125. doi:10.1038/srep41125.

Hong, K. U., and R. Bolli. 2014. Cardiac stem cell therapy for cardiac repair. *Current Treatment Options in Cardiovascular Medicine* 16 (7): 324. doi:10.1007/s11936-014-0324-3.

Hosoyama, T., M. Samura, T. Kudo, A. Nishimoto, K. Ueno, T. Murata, T. Ohama et al. 2015. Cardiosphere-derived cell sheet primed with hypoxia improves left ventricular function of chronically infarcted heart. *American Journal of Translational Research* 7 (12): 2738–2751.

Hosseinkhani, H., M. Hosseinkhani, S. Hattori, R. Matsuoka, and N. Kawaguchi. 2010. Micro and nano-scale in vitro 3D culture system for cardiac stem cells. *Journal of Biomedical Materials Research. Part A* 94 (1): 1–8. doi:10.1002/jbm.a.32676.

Ibrahim, A. G.-E., K. Cheng, and E. Marbán. 2014. Exosomes as critical agents of cardiac regeneration triggered by cell therapy. *Stem Cell Reports* 2 (5): 606–619. doi:10.1016/j.stemcr.2014.04.006.

Ishigami, S., S. Ohtsuki, S. Tarui, D. Ousaka, T. Eitoku, M. Kondo, M. Okuyama et al. 2015. Intracoronary autologous cardiac progenitor cell transfer in patients with hypoplastic left heart syndrome: The TICAP prospective phase 1 controlled trial. *Circulation Research* 116 (4): 653–664. doi:10.1161/CIRCRESAHA.116.304671.

Ishigami, S., S. Ohtsuki, T. Eitoku, D. Ousaka, M. Kondo, Y. Kurita, K. Hirai et al. 2017. Intracoronary cardiac progenitor cells in single ventricle physiology: The PERSEUS (Cardiac Progenitor Cell Infusion to Treat Univentricular Heart Disease) randomized phase 2 trial. *Circulation Research* 120 (7): 1162–1173. doi:10.1161/CIRCRESAHA.116.310253.

Jones, S. P., X.-L. Tang, Y. Guo, C. Steenbergen, D. J. Lefer, R. C. Kukreja, M. Kong et al. 2015. The NHLBI-sponsored consortium for preclinicAl assESsment of cARdioprotective therapies (CAESAR): A new paradigm for rigorous, accurate, and reproducible evaluation of putative infarct-sparing interventions in mice, rabbits, and pigs. *Circulation Research* 116 (4): 572–586. doi:10.1161/CIRCRESAHA.116.305462.

Kanazawa, H., E. Tseliou, J. F. Dawkins, G. De Couto, R. Gallet, K. Malliaras, K. Yee et al. 2016. Durable benefits of cellular postconditioning: Long-term Effects of allogeneic cardiosphere-derived cells infused after reperfusion in pigs with acute myocardial infarction. *Journal of the American Heart Association* 5 (2): e002796. doi:10.1161/JAHA.115.002796.

Kapelios, C. J., J. N. Nanas, and K. Malliaras. 2016. Allogeneic cardiosphere-derived cells for myocardial regeneration: Current progress and recent results. *Future Cardiology* 12 (1): 87–100. doi:10.2217/fca.15.72.

Kawaguchi, N., A. J. Smith, C. D. Waring, M. D. Kamrul Hasan, S. Miyamoto, R. Matsuoka, and G. M. Ellison. 2010. C-Kitpos GATA-4 high rat cardiac stem cells foster adult cardiomyocyte survival through IGF-1 paracrine signalling. *PLoS One* 5 (12): e14297. doi:10.1371/journal.pone.0014297.

Koudstaal, S., M. M. C. Bastings, D. A. M. Feyen, C. D. Waring, F. J. van Slochteren, P. Y. W. Dankers, D. Torella et al. 2014. Sustained delivery of insulin-like growth factor-1/hepatocyte growth factor stimulates endogenous cardiac repair in the chronic infarcted pig heart. *Journal of Cardiovascular Translational Research* 7 (2): 232–241. doi:10.1007/s12265-013-9518-4.

Krinner, A., M. Zscharnack, A. Bader, D. Drasdo, and J. Galle. 2009. Impact of oxygen environment on mesenchymal stem cell expansion and chondrogenic differentiation. *Cell Proliferation* 42 (4): 471–484. doi:10.1111/j.1365-2184.2009.00621.x.

Kryukov, O., E. Ruvinov, and S. Cohen. 2014. Three-dimensional perfusion cultivation of human cardiac-derived progenitors facilitates their expansion while maintaining progenitor state. *Tissue Engineering. Part C* 00 (00): 1–9. doi:10.1089/ten.tec.2013.0528.

Kwon, C., J. Arnold, E. C. Hsiao, M. M. Taketo, B. R. Conklin, and D. Srivastava. 2007. Canonical Wnt signaling is a positive regulator of mammalian cardiac progenitors. *Proceedings of the National Academy of Sciences* 104 (26): 10894–10899. doi:10.1073/pnas.0704044104.

Lalit, P. A., M. R. Salick, D. O. Nelson, J. M. Squirrell, C. M. Shafer, N. G. Patel, I. Saeed et al. 2016. Lineage reprogramming of fibroblasts into proliferative induced cardiac progenitor cells by defined factors. *Cell Stem Cell* 18 (3): 354–367. doi:10.1016/j.stem.2015.12.001.

Lauden, L., W. Boukouaci, L. R. Borlado, I. P. López, P. Sepúlveda, R. Tamouza, D. Charron, and R. Al-Daccak. 2013. Allogenicity of human cardiac stem/progenitor cells orchestrated by programmed death ligand 1. *Circulation Research* 112 (3): 451–464. doi:10.1161/CIRCRESAHA.112.276501.

Leri, A., M. Rota, T. Hosoda, P. Goichberg, and P. Anversa. 2014. Cardiac stem cell niches. *Stem Cell Research*. doi:10.1016/j.scr.2014.09.001.

Li, N. A., C. Wang, L. I. X. Jia, and J. Du. 2014a. Heart regeneration, stem cells, and cytokines. *Regenerative Medicine Research* 2 (1): 6. doi:10.1186/2050-490X-2-6.

Li, T.-S., K. Cheng, K. Malliaras, N. Matsushita, B. Sun, L. Marbán, Y. Zhang, and E. Marbán. 2011. Expansion of human cardiac stem cells in physiological oxygen improves cell production efficiency and potency for myocardial repair. *Cardiovascular Research* 89 (1): 157–165. doi:10.1093/cvr/cvq251.

Li, T.-S., K. Cheng, K. Malliaras, R. Ruckdeschel Smith, Y. Zhang, B. Sun, N. Matsushita et al. 2012. Direct comparison of different stem cell types and subpopulations reveals superior paracrine potency and myocardial repair efficacy with cardiosphere-derived cells. *Journal of the American College of Cardiology* 59 (10): 942–953. doi:10.1016/j.jacc.2011.11.029.

Li, X., Y. Ren, V. Sorokin, K. Keong Poh, H. Hwa Ho, C. Neng Lee, D. de Kleijn, S. Kiang Lim, J. P. Tam, and S. Kwan Sze. 2014b. Quantitative profiling of the rat heart myoblast secretome reveals differential responses to hypoxia and re-oxygenation stress. *Journal of Proteomics* 98 (February): 138–149. doi:10.1016/j.jprot.2013.12.025.

Linke, A., P. Müller, D. Nurzynska, C. Casarsa, D. Torella, A. Nascimbene, C. Castaldo et al. 2005. Stem cells in the dog heart are self-renewing, clonogenic, and multipotent and regenerate infarcted myocardium, improving cardiac function. *Proceedings of the National Academy of Sciences of the United States of America* 102 (25): 8966–8971. doi:10.1073/pnas0502678102.

Machida, M., Y. Takagaki, R. Matsuoka, and N. Kawaguchi. 2011. Proteomic comparison of spherical aggregates and adherent cells of cardiac stem cells. *International Journal of Cardiology* 153 (3): 296–305. doi:10.1016/j.ijcard.2010.08.049.

Madonna, R., L. W. Van Laake, S. M. Davidson, F. B. Engel, D. J. Hausenloy, S. Lecour, J. Leor et al. 2016. Position paper of the European society of cardiology working group cellular biology of the heart: Cell-based therapies for myocardial repair and regeneration in ischemic heart disease and heart failure. *European Heart Journal*, ehw113. doi:10.1093/eurheartj/ehw113.

Makkar, R. R., R. R. Smith, K. Cheng, K. Malliaras, L. E. J. Thomson, D. Berman, L. S. C. Czer et al. 2012. Intracoronary cardiosphere-derived cells for heart regeneration after myocardial infarction (CADUCEUS): A prospective, randomised phase 1 trial. *Lancet (London, England)* 379 (9819): 895–904. doi:10.1016/S0140-6736(12)60195-0.

Marbán, E. 2014. Breakthroughs in cell therapy for heart disease: Focus on cardiosphere-derived cells. *Mayo Clinic Proceedings* 89 (6): 850–858. doi:10.1016/j.mayocp.2014.02.014.

Mathur, A., F. Fern Andez-Avilés, S. Dimmeler, C. Hauskeller, S. Janssens, P. Menasche, W. Wojakowski, J. F. Martin, A. Zeiher, and B. Investigators. 2017. The consensus of the task force of the European society of cardiology concerning the clinical investigation of the use of autologous adult stem cells for the treatment of acute myocardial infarction and heart failure: Update 2016. *European Heart Journal* 0: 1–6. doi:10.1093/eurheartj/ehw640.

Ménard, C., A. A. Hagège, O. Agbulut, M. Barro, M. C. Morichetti, C. Brasselet, A. Bel et al. 2005. Transplantation of cardiac-committed mouse embryonic stem cells to infarcted sheep myocardium: A preclinical study. *The Lancet* 366 (9490): 1005–1012. doi:10.1016/S0140-6736(05)67380-1.

Menasché, P., V. Vanneaux, A. Hagège, A. Bel, B. Cholley, I. Cacciapuoti, A. Parouchev et al. 2015a. Human embryonic stem cell-derived cardiac progenitors for severe heart failure treatment: First clinical case report. *European Heart Journal* 36 (30): 2011–2017. doi:10.1093/eurheartj/ehv189.

Menasché, P., V. Vanneaux, J.-R. Fabreguettes, A. Bel, L. Tosca, S. Garcia, V. Bellamy et al. 2015b. Towards a clinical use of human embryonic stem cell-derived cardiac progenitors: A translational experience. *European Heart Journal* 36 (12): 743–750. doi:10.1093/eurheartj/ehu192.

Messina, E., L. De Angelis, G. Frati, S. Morrone, S. Chimenti, F. Fiordaliso, M. Salio et al. 2004. Isolation and Expansion of Adult Cardiac Stem Cells from Human and Murine Heart. *Circulation Research* 95 (9): 911–921. doi:10.1161/01.RES.0000147315.71699.51.

Miyamoto, S., N. Kawaguchi, G. M. Ellison, R. Matsuoka, T. Shin'oka, and H. Kurosawa. 2010. Characterization of long-term cultured c-Kit+ cardiac stem cells derived from adult rat hearts. *Stem Cells and Development* 19 (1): 105–116. doi:10.1089/scd.2009.0041.

Moretti, A., L. Caron, A. Nakano, J. T. Lam, A. Bernshausen, Y. Chen, Y. Qyang et al. 2006. Multipotent embryonic Isl1+ progenitor xells lead to cardiac, smooth muscle, and endothelial cell diversification. *Cell* 127 (6): 1151–1165. doi:10.1016/j.cell.2006.10.029.

Moscoso, Isabel, Naiara Tejados, Olga Barreiro, Pilar Sepúlveda, Alberto Izarra, Enrique Calvo, Akaitz Dorronsoro, et al. 2016. Podocalyxin-like protein 1 is a relevant marker for human c-Kit[pos] cardiac stem cells. *Journal of Tissue Engineering and Regenerative Medicine* 10 (7): 580–590. doi:10.1002/term.1795.

Nadal-Ginard, B., G. M. Ellison, and D. Torella. 2014. The absence of evidence is not evidence of absence: The pitfalls of cre knock-ins in the C-Kit locus. *Circulation Research*, 115 (May): e21–e23. doi:10.1161/CIRCRESAHA.114.304676.

Nakamura, T., T. Hosoyama, D. Kawamura, Y. Takeuchi, Y. Tanaka, M. Samura, K. Ueno et al. 2016. Influence of aging on the quantity and quality of human cardiac stem cells. *Scientific Reports* 6 (1): 22781. doi:10.1038/srep22781.

Nienow, A. W., Q. A. Rafiq, K. Coopman, and C. J. Hewitt. 2014. A potentially scalable method for the harvesting of hMSCs from microcarriers. *Biochemical Engineering Journal* 85 (April): 79–88. doi:10.1016/j.bej.2014.02.005.

Nigro, P., G. L. Perrucci, A. Gowran, M. Zanobini, M. C. Capogrossi, and G. Pompilio. 2015. C-Kit+ cells: The tell-tale heart of cardiac regeneration? *Cellular and Molecular Life Sciences* 72 (9): 1725–1740. doi:10.1007/s00018-014-1832-8.

O'Neill, H. S., J. O'Sullivan, N. Porteous, E. Ruiz-Hernandez, H. M. Kelly, F. J. O'Brien, and G. P. Duffy. 2016. A collagen cardiac patch incorporating alginate microparticles permits the controlled release of hepatocyte growth factor and insulin-like growth factor-1 to enhance cardiac stem cell migration and proliferation. *Journal of Tissue Engineering and Regenerative Medicine*, December. doi:10.1002/term.2392.

Park, C. Y., S. C. Choi, J. Ho Kim, J. Hyun Choi, H. Joon Joo, S. Jun Hong, and D. Sun Lim. 2016. Cardiac stem cell secretome protects cardiomyocytes from hypoxic injury partly via monocyte chemotactic protein-1-dependent mechanism. *International Journal of Molecular Sciences* 17 (6): 6–7. doi:10.3390/ijms17060800.

Rajabi-Zeleti, S., S. Jalili-Firoozinezhad, M. Azarnia, F. Khayyatan, S. Vahdat, S. Nikeghbalian, A. Khademhosseini, H. Baharvand, and N. Aghdami. 2014. The behavior of cardiac progenitor cells on macroporous pericardium-derived scaffolds. *Biomaterials* 35 (3) 970–982. doi:10.1016/j.biomaterials.2013.10.045.

Rota, M., M. Elena Padin-Iruegas, Y. Misao, A. De Angelis, S. Maestroni, J. Ferreira-Martins, E. Fiumana et al. 2008. Local activation or implantation of cardiac progenitor cells rescues scarred infarcted myocardium improving cardiac function. *Circulation Research* 103 (1): 107–116. doi:10.1161/CIRCRESAHA.108.178525.

Ryzhov, S., A. E. Goldstein, S. V. Novitskiy, M. R. Blackburn, I. Biaggioni, and I. Feoktistov. 2012. Role of A2B adenosine receptors in regulation of paracrine functions of stem cell antigen 1-positive cardiac stromal cells. *The Journal of Pharmacology and Experimental Therapeutics* 341 (3): 764–774. doi:10.1124/jpet.111.190835.

Samal, R., S. Ameling, K. Wenzel, V. Dhople, U. Völker, S. B. Felix, S. Könemann, and E. Hammer. 2012. OMICS-based exploration of the molecular phenotype of resident cardiac progenitor cells from adult murine heart. *Journal of Proteomics* 75 (17): 5304–5315. doi:10.1016/j.jprot.2012.06.010.

Santini, M. P., E. Forte, R. P. Harvey, and J. C. Kovacic. 2016. Developmental origin and lineage plasticity of endogenous cardiac stem cells. *Development* 143: 1242–1258. doi:10.1242/dev.111591.

Serra, M., C. Brito, C. Correia, and P. M. Alves. 2012. Process engineering of human pluripotent stem cells for clinical application. *Trends in Biotechnology* 30 (6): 350–359. doi:10.1016/j.tibtech.2012.03.003.

Serra, M., C. Brito, M. F. Q. Sousa, J. Jensen, R. Tostões, J. Clemente, R. Strehl, J. Hyllner, M. J. T. Carrondo, and P. M. Alves. 2010. Improving expansion of pluripotent human embryonic stem cells in perfused bioreactors through oxygen control. *Journal of Biotechnology* 148 (4): 208–215. doi:10.1016/j.jbiotec.2010.06.015.

Sharma, S., R. Mishra, G. E. Bigham, B. Wehman, M. M. Khan, H. Xu, P. Saha et al. 2016. A Deep proteome analysis identifies the complete secretome as the functional unit of human cardiac progenitor cells. *Circulation Research.* doi:10.1161/CIRCRESAHA.116.309782.

Smith, R. R., L. Barile, H. Cheol Cho, M. K. Leppo, J. M. Hare, E. Messina, A. Giacomello, M. R. Abraham, and E. Marbán. 2007. Regenerative potential of cardiosphere-derived cells expanded from percutaneous endomyocardial biopsy specimens. *Circulation* 115 (7): 896–908.

Soh, B.-S., S.-Y. Ng, H. Wu, K. Buac, J.-H. C. Park, X. Lian, J. Xu et al. 2016. Endothelin-1 supports clonal derivation and expansion of cardiovascular progenitors derived from human embryonic stem cells. *Nature Communications* 7 (March): 10774. doi:10.1038/ncomms10774.

Stastna, M., and J. E. Van Eyk. 2012. Investigating the secretome: Lessons about the cells that comprise the heart. *Circulation. Cardiovascular Genetics* 5 (1): o8–o18. doi:10.1161/CIRCGENETICS.111.960187.

Stastna, M., I. Chimenti, E. Marbán, and J. E. Van Eyk. 2010. Identification and functionality of proteomes secreted by rat cardiac stem cells and neonatal cardiomyocytes. *Proteomics* 10 (2): 245–253. doi:10.1002/pmic.200900515.

Synnergren, J., L. Drowley, A. T. Plowright, G. Brolén, M.-J. Goumans, A. C. Gittenberger-de Groot, P. Sartipy, and Q.-D. Wang. 2016. Comparative transcriptomic analysis identifies genes differentially expressed in human epicardial progenitors and hiPSC-derived cardiac progenitors. *Physiological Genomics*, no. 25: physiolgenomics.00064.2016. doi:10.1152/physiolgenomics.00064.2016.

Takehara, N., M. Nakata, T. Ogata, T. Nakamura, S. Matoba, S. Gojo, T. Sawada, H. Yaku, and H. Matsubara. 2012. The ALCADIA (AutoLogous Human CArdiac-Derived Stem Cell To Treat Ischemic cArdiomyopathy) trial. *Circulation* 126: 2776–2799.

Tang, X.-L., G. Rokosh, S. K. Sanganalmath, F. Yuan, H. Sato, J. Mu, S. Dai et al. 2010. Intracoronary administration of cardiac progenitor cells alleviates left ventricular dysfunction in rats with a 30-Day-Old infarction. *Circulation* 121 (2): 293–305. doi:10.1161/CIRCULATIONAHA.109.871905.

Tiwari, A., C. S. Wong, L. P. Nekkanti, J. A. Deane, C. McDonald, G. Jenkin, and M. A. Kirkland. 2016. Impact of oxygen levels on human hematopoietic stem and progenitor cell expansion. *Stem Cells and Development* 25 (20): 1604–1613. doi:10.1089/scd.2016.0153.

Tokita, Y., X.-L. Tang, Q. Li, M. Wysoczynski, K. U. Hong, S. Nakamura, W.-J. Wu et al. 2016. Repeated administrations of cardiac progenitor cells are markedly more effective than a single administration: A new paradigm in cell therapy. *Circulation Research* 119 (5): 635–651. doi:10.1161/CIRCRESAHA.116.308937.

Tomescot, A., J. Leschik, V. Bellamy, G. Dubois, E. Messas, P. Bruneval, M. Desnos et al. 2007. Differentiation *in vivo* of cardiac committed human embryonic stem cells in postmyocardial infarcted rats. *STEM CELLS* 25 (9): 2200–2205. doi:10.1634/stemcells.2007-0133.

Torella, D., G. M. Ellison, I. Karakikes, and B. Nadal-Ginard. 2007. Growth-factor-mediated cardiac stem cell activation in myocardial regeneration. *Nature Clinical Practice. Cardiovascular Medicine* 4 Suppl 1 (February): S46–S51. doi:10.1038/ncpcardio0772.

Tropepe, V., M. Sibilia, B. G. Ciruna, J. Rossant, E. F. Wagner, and D. van der Kooy. 1999. Distinct neural stem cells proliferate in response to EGF and FGF in the developing mouse telencephalon. *Developmental Biology* 208 (1): 166–188. doi:10.1006/dbio.1998.9192.

Urbanek, K., D. Torella, F. Sheikh, A. De Angelis, D. Nurzynska, F. Silvestri, C. Alberto Beltrami et al. 2005a. Myocardial regeneration by activation of multipotent cardiac stem cells in ischemic heart failure. *Proceedings of the National Academy of Sciences of the United States of America* 102 (24): 8692–8697. doi:10.1073/pnas.0500169102.

Urbanek, K., M. Rota, S. Cascapera, C. Bearzi, A. Nascimbene, A. De Angelis, T. Hosoda et al. 2005b. Cardiac stem cells possess growth factor-receptor systems that after activation regenerate the infarcted myocardium, improving ventricular function and long-term survival. *Circulation Research* 97 (7): 663–673. doi:10.1161/01.RES.0000183733.53101.11.

Valiente-Alandi, I., C. Albo-Castellanos, D. Herrero, I. Sanchez, and A. Bernad. 2016. Bmi1 + cardiac progenitor cells contribute to myocardial repair following acute injury. *Stem Cell Research & Therapy* 7 (1): 100. doi:10.1186/s13287-016-0355-7.

Wang, L., H. Gu, M. Turrentine, and M. Wang. 2014. Estradiol treatment promotes cardiac stem cell (CSC)-derived growth factors, thus improving CSC-mediated cardioprotection after acute ischemia/reperfusion. *Surgery* 156 (2): 243–252. doi:10.1016/j.surg.2014.04.002.

Waring, C. D., C. Vicinanza, A. Papalamprou, A. J. Smith, S. Purushothaman, D. F. Goldspink, B. Nadal-Ginard, D. Torella, and G. M. Ellison. 2014. The adult heart responds to increased workload with physiologic hypertrophy, cardiac stem cell activation, and new myocyte formation. *European Heart Journal* 35 (39): 2722–2731. doi:10.1093/eurheartj/ehs338.

Weinberger, F., D. Mehrkens, F. W. Friedrich, M. Stubbendorff, X. Hua, J. Christina Müller, S. Schrepfer, S. M. Evans, L. Carrier, and T. Eschenhagen. 2012. Localization of Islet-1-Positive cells in the healthy and infarcted adult murine heart. *Circulation Research* 110 (10): 1303–1310. doi:10.1161/CIRCRESAHA.111.259630.

WHO|Cardiovascular Diseases Fact Sheet. 2016. *World Health Organization.* World Health Organization. http://www.who.int/mediacentre/factsheets/fs317/en/ Accessed June 2017.

Wu, Q., J. Zhan, S. Pu, L. Qin, Y. Li, and Z. Zhou. 2016. Influence of aging on the activity of mice sca-1+CD31- Cardiac stem cells. *Oncotarget* 8 (1): 29–41. doi:10.18632/oncotarget.13930.

Yusen, R. D., L. B. Edwards, A. Y. Kucheryavaya, C. Benden, A. I. Dipchand, S. B. Goldfarb, B. J. Levvey et al. 2015. The registry of the international society for heart and lung transplantation: Thirty-second official adult lung and heart-lung transplantation report—2015; focus theme: Early graft failure. *Journal of Heart and Lung Transplantation* 34 (10): 1264–1277. doi:10.1016/j.healun.2015.08.014.

Zakharova, L., H. Nural-Guvener, J. Nimlos, S. Popovic, and M. A. Gaballa. 2013. Chronic heart failure is associated with transforming growth factor beta-dependent yield and functional decline in atrial explant-derived c-Kit+ cells. *Journal of the American Heart Association* 2 (5): e000317. doi:10.1161/JAHA.113.000317.

Zhang, Y., N. Cao, Y. Huang, C. Ian Spencer, J.-D. Fu, C. Yu, K. Liu et al. 2016. Expandable cardiovascular progenitor cells reprogrammed from fibroblasts. *Cell Stem Cell* 18 (3): 368–381. doi:10.1016/j.stem.2016.02.001.

Zwetsloot, P. P., A. M. D. Végh, S. J. J. Lorkeers, G. P. J. Van Hout, G. L. Currie, E. S. Sena, H. Gremmels et al. 2016. Cardiac stem cell treatment in myocardial infarction: A systematic review and meta-analysis of preclinical studies. *Circulation Research* 118 (8): 1223–1232. doi:10.1161/CIRCRESAHA.115.307676.

5 Bioreactor Protocols for the Expansion and Differentiation of Human Neural Precursor Cells in Targeting the Treatment of Neurodegenerative Disorders

Arindom Sen, Behnam A. Baghbaderani,
Michael S. Kallos, Ivar Mendez, and Leo A. Behie

CONTENTS

5.1   INTRODUCTION

Central nervous system (CNS) disorders, such as Parkinson's disease (PD), Huntington's disease (HD), and spinal cord injuries are devastating diseases that currently afflict more than 30 million individuals in North America. These disorders, which can cause both motor and cognitive impairment, can significantly reduce quality of life for both patients and their families, and place them under psychological and financial strain. Treating the huge number of individuals suffering from these disorders places an enormous cost on health care systems. In North America alone, the annual cost of treating the over 1 million people diagnosed with PD is over $25 billion (Parkinson's Foundation, 2017; Parkinson Canada, 2017). For Huntington's disease, there are more than 30,000 patients in North America, and more than 200,000 people are genetically at risk. Total annual cost of treatment for just those patients with late stage HD is estimated to be over $2.5 billion in the North America (Huntington's Disease Society of America, 2017; Huntington Society of Canada, 2017).

5.1.1   CELL REPLACEMENT THERAPY

Many disorders of the CNS are caused by the loss of functional cells. For example, PD and HD are neurodegenerative disorders characterized by the selective and gradual loss in the brain of dopaminergic neurons and GABAergic neurons, respectively, resulting in progressive motor and cognitive impairment. Injuries to the spinal cord can also result in significant cell loss, thereby compromising neural system function. Drug therapy, which is the conventional treatment for CNS disorders, has mainly focused on treating the symptoms associated with these medical conditions; however, this does not address the root causes of cell loss, and has not yet resulted in the regeneration of cells. In addition, drugs become less effective over time due to habituation, and can lead to a series of debilitating side effects. A potential treatment alternative that has recently gained more attention is cell replacement therapy. In this approach, the aim is to transplant viable cells into an affected region, so they can integrate into the brain or spinal cord, thereby replacing the dead or non-functional cells, and restoring the functionality that existed prior to disease or injury. There is mounting evidence to suggest that this may be an effective treatment pathway.

Cell replacement therapy in which primary human fetal CNS tissue is processed and then directly implanted into patients has shown encouraging results in clinical trials of medical conditions such as PD (Mendez et al., 2008), HD (Bachoud-Levi et al., 2006), and demyelination disorders (Gupta et al., 2012). However, the use of fetal tissue is mired in ethical controversy. In addition, there are issues related to tissue source variability, quality, and sterility, and the pooling of heterogeneous tissues procured from several fetuses to provide sufficient material for a single transplantation procedure introduces major logistical problems (Martín-Ibáñeza et al., 2017). Moreover, there is certainly not a sufficient supply of primary human tissue to treat the millions of individuals who could benefit from this potential treatment approach. For this reason alone, it is now clear that the use of primary human tissue in cell replacement therapies will not be of widespread clinical utility.

For cell replacement strategies to become a routine therapeutic option for the treatment of CNS disorders, an alternate approach that has emerged is to procure a small amount of primary neural tissue, and subject the precursor cells contained therein to reliable and standardized methods that support their proliferation in culture, thereby providing a means of generating clinically relevant numbers of therapeutically useful cells. This chapter will review and discuss the robust methods that have already been developed for the rapid expansion and differentiation of human neural precursor cell populations. It should be noted that, although the focus will be on neural precursor cells, the methods described here have the potential to serve as a model system for other cell types with potential clinical relevance, such as human induced pluripotent stem cells (hiPSCs) (Pagliuca et al., 2014).

5.1.2 NEURAL PRECURSOR CELLS

Human neural stem cells and progenitor cells, which are collectively termed human neural precursor cells (hNPCs), have received considerable attention as an alternative to the direct use of primary human tissue in clinical cell replacement strategies for the treatment of neurodegenerative disorders. Methods now exist to isolate hNPCs from the adult or developing human CNS, to expand their numbers in culture using mitogenic factors, and to direct their fate towards the major functional cell types of the CNS. Moreover, experimental studies have demonstrated the safety and feasibility of transplanting *in vitro* expanded hNPCs and their differentiated derivatives in animal models of neurodegenerative disorders and central nervous system injury (Svendsen et al., 1996; Svendsen et al., 1997; Vescovi et al., 1999; Ostenfeld et al., 2000; Kelly et al., 2011; Uchida et al., 2012; van Gorp et al., 2013; Zhou et al., 2015). In order for these expanded cell populations to be approved for use in clinical settings, they must be generated in a reproducible manner under controlled, standard conditions. Well-mixed suspension bioreactors are scalable vessels that have been shown to support the efficient proliferation of hNPCs isolated from a single source, thereby providing a reliable means to convert a very small number of isolated cells to large quantities that would be sufficient to treat hundreds of patients.

5.2   COMPUTER CONTROLLED BIOREACTORS FOR HUMAN NEURAL PRECURSOR CELL PRODUCTION

5.2.1   INITIAL HUMAN NEURAL PRECURSOR CELL EXPANSION STUDIES

Initial work around NPC-expansion technology utilized murine NPCs as a model cell type due to ease of availability, and the similarity of their characteristics to their human counterparts. Several years of focused research have now resulted in the development of robust expansion protocols for human NPCs in 125 mL stirred suspension bioreactors (Corning® vessels with a paddle impeller) (Sen and Behie, 1999; Kallos et al., 1999; Sen et al., 2001, 2002a, 2002b; Kallos et al., 2003). Subsequently, human telencephalon-derived hNPC production was successfully scaled-up from 125 to 250 mL suspension bioreactors. Appropriate operating conditions in the 250 mL bioreactors were chosen based on previous studies performed in standard 125 mL bioreactors, as well as oxygen transfer and hydrodynamics studies in the larger bioreactors. Given the importance of controlling agitation-induced shear stresses in stem cell cultures, constant peak shear stress at the impeller tip served as the basis for scale-up. Human NPCs exhibited a doubling time of 69 hours, reached a cell-fold expansion of 44 after 16 days, and maintained a viability of approximately 90% over that same time period in the 250 mL suspension bioreactors. Controlling culture hydrodynamics by keeping the agitation rate at 100 rpm successfully maintained the mean aggregate diameter of the tissue-like cell aggregates below 600 μm, which was found to be necessary to achieve high cell densities (approximately 4×10^6 cells/mL) and viabilities more than 90%. Cell culture medium analyses revealed that profiles of nutrients (glucose and glutamine), by-products (lactate and ammonia), and amino acids in the 250 mL bioreactors were very comparable to those in the standard small-scale bioreactors, suggesting no significant change in cell metabolism with culture scale.

5.2.2   SCALE-UP TO 500 ML BIOREACTOR

Once the scale-up of hNPC expansion from standard 125 mL small scale bioreactors to larger 250 mL bioreactors was achieved, studies were performed to further scale-up the expansion of hNPCs to 500 mL computer-controlled bioreactors. Wheaton 500 mL stirred suspension bioreactors (Wheaton Science Products) with a four-blade impeller (Wheaton Science Products) and measuring probes (dissolved oxygen (DO) sensor, Broadley-James Corporation; pH sensor, Mettler-Toledo; temperature

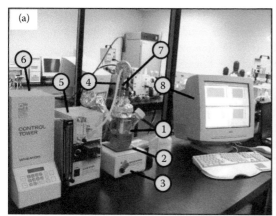

FIGURE 5.1 Photographs of 500 mL computer-controlled suspension bioreactor. (a) Experimental set-up for the 500 mL suspension bioreactor showing: (1) 500 mL suspension bioreactor; (2) heating pad; (3) stir plate; (4) gas inlets to the bioreactor connected to O_2, CO_2, N_2, and air gas cylinders via Wheaton pump; (5) Wheaton pump; (6) the control tower (Wheaton) connected to the dissolved oxygen (DO), pH, and temperature probes to monitor the level of each parameter; (7) DO, pH, and temperature probes; and (8) data acquisition computer. (b) A closer view of the 500 mL suspension bioreactor and measuring probes.

sensor, MINCO) were used to develop protocols to expand human telencephalon-derived populations under computer-controlled conditions. The 500 mL bioreactors (Figure 5.1) had two side-arms, which enabled sampling and the addition of growth medium to the bioreactor and provided the option of adding filtered aeration and venting tubes. Prior to use, the inner surface of the bioreactor vessel and outer surface of the impeller were siliconized using a 1:9 ratio of Sigmacote® in hexane. The bioreactor was then sterilized and filled with 500 mL of PPRF.h2 growth medium. The measuring probes were calibrated per the instructions provided by Wheaton. The temperature of the bioreactor was maintained at 37°C by controlling a heating pad located directly beneath the vessel. The agitation rate of the culture was controlled using a Micro Stir Model II Single Place stir plate (Wheaton).

To develop a successful scale-up process, a few important issues were taken into consideration. First, the vessel geometry and impeller design for the 500 mL bioreactor (four-blade impeller/Wheaton) differed significantly from the 125 mL and 250 mL Corning® spinner flasks (each of these had a standard paddle impeller) in which expansion protocols had previously been established. Moreover, due to the size of the vessel it was now possible to incorporate measuring probes (absent at smaller scales) into the process to measure the temperature, pH, and dissolved oxygen levels in culture. These modifications dramatically affect both the mass transfer characteristics and hydrodynamics in suspension culture, which could significantly impact cell growth and viability in the large-scale bioreactor. Therefore, studies were conducted in these larger vessels to evaluate the oxygen transfer profile and culture hydrodynamics to identify appropriate operating conditions that could support expansion of hNPCs. Standard process control techniques were used to control temperature, dissolved oxygen level, and pH.

5.2.2.1 Oxygen Transfer in 500 mL Bioreactors

In small bioreactors, gas transfer from the headspace through the culture medium surface can typically satisfy the oxygen demand of the cells. This has previously proven to be a satisfactory approach for murine NPCs (Kallos and Behie, 1999; Kallos et al., 1999; Sen et al., 2001; Sen et al., 2002a, 2002b; Kallos et al., 2003; Gilbertson et al., 2006), mammary epithelial stem cells (MESCs) (Youn et al., 2005), and breast cancer stem cells (BrCSCs) (Youn et al., 2006) in small bioreactors where headspace O_2, CO_2, and N_2 levels were controlled. Moreover, it was also shown to be the case for hNPC expansion in both the 125 mL and 250 mL spinner bioreactors. However, since the surface area to volume ratio (SA/V) decreases with increasing culture volume, initially it was unknown if surface aeration alone would be enough to satisfy the overall oxygen demand of hNPCs in larger 500 mL bioreactors. If not, then direct sparging would be necessary. However, sparging is not desirable when working with shear sensitive mammalian cells as these cells can become entrained in the liquid boundary layer around a bubble, and subsequently be damaged or killed when that bubble bursts at the culture surface. Thus, a series of preliminary studies were conducted to evaluate whether the oxygen transfer rates (OTR) across the liquid medium surface would be sufficient to meet the oxygen uptake rate of the cells. We found that headspace aeration alone could meet the oxygen demand of the hNPCs in 500 mL vessels by controlling the gas mixture (O_2, CO_2, and N_2) composition in the headspace. A headspace DO set point value of 70% air saturation was maintained as it was found to be sufficient for the cells thereby providing a simple oxygen control strategy. Moreover, it could be assumed that the headspace oxygen concentration in the 500 mL bioreactor (C^*_{O2}) was at 95% air saturation. As a result, the minimum volumetric mass transfer coefficient (k_La, for oxygen transfer) required for cell growth in the 500 mL bioreactor was different compared to smaller bioreactors.

The actual k_La values in the 500 mL bioreactors were measured at various agitation rates (i.e., 70, 85, 100, 130 rpm) using the unsteady-state method (Shuler and Kargi, 2002). A k_La value of 1.9 h^{-1} was measured in the 500 mL computer-controlled bioreactor containing 500 mL of cytokine-free growth medium at an agitation rate of 85 rpm. The measured value was comparable to the k_La value of 1.79 h^{-1} calculated using the correlation developed by Aunins and co-workers (Aunins et al., 1989) at 85 rpm in the 500 mL bioreactor with additional probes.

It is important to note that growth of many mammalian cell lines, in general, might be optimal at a DO level below the maximum oxygen solubility at 100% air saturation (Butler, 2004). Several studies have shown that lowered oxygen concentration may not alter proliferation activity of mammalian forebrain derived NPCs in culture (Studer et al., 2000; Storch et al., 2001). Indeed, the actual physiological level of oxygen in the human brain (1%–3% O_2) is much lower than the oxygen concentration typically present in the gas phase of many cultures (i.e., atmospheric) (Zhu et al., 2005). This literature data along, with the results of the oxygen transfer calculations, suggested that expansion of telencephalon derived hNPCs in 500 mL computer-controlled bioreactors should not face oxygen limitations even at lowered oxygen concentrations and at high cell densities.

5.2.2.2   Process Control of Bioreactors

To ensure the consistency of the cell population characteristics between production runs, critical process parameters must be controlled and maintained at their optimal levels. Temperature, oxygen concentration, and pH are among the most important process variables (PV) that can be regulated in mammalian cell cultures using standard process control techniques (Shuler and Kargi, 2002; Butler, 2004). Computerized control systems receive input from a variety of sensors designed to measure key components within the culture environment, including dissolved gas levels, pH, nutrients and temperature. These inputs are compared to set point values, which are predetermined to maintain the culture under optimum conditions. Depending on the magnitude of the deviation, the controller may initiate corrective action to restore that parameter in the culture environment back to the set point value. Controller targets include bioreactor heating systems, which can be turned on and off to maintain temperature, valves that can be opened and closed to control the levels of gases such as O_2, CO_2, and N_2, and pumps that can be activated to add nutrients when they are depleted or to restore pH by adding alkali. The control parameters and types of controllers used to regulate the culture parameters are usually recommended by the bioreactor manufacturer. Control strategies include proportional, integral and derivative control, or a combination of these approaches (Svrcek et al., 2000).

5.2.2.1.1   Temperature Control

A temperature of 37°C and pH of 7.2–7.4 are usually considered optimal culture conditions for mammalian cells. For NPCs, a culture temperature between 36.5°C and 38°C has been recommended to allow optimal cell proliferation and viability (Sen, 2003). Exposure to a temperature lower than the physiological temperature of 37°C may have an adverse effect on cell growth and proliferation rates, although the cells themselves should not be damaged. However, continuous exposure to above-physiological temperatures (i.e., 39–40°C) may destroy cells (Butler, 2004). Considering the sensitivity of the mammalian cells to temperature fluctuations, the culture temperature must be carefully controlled. It has been found that the temperature within a 500 mL Wheaton bioreactor can be maintained at 37°C by using a PID controller with a controller gain (K_C) value of 50, an integral time constant (t_I) of 0.03 and a derivative time constant (t_d) of 1.5 (as recommended by the bioreactor manufacturer).

5.2.2.1.2   pH Control

To maintain hNPCs in the physiological pH range of 7.2–7.4, sodium bicarbonate and HEPES ([4-(2-hydroxyethyl]-1-piperazineethanesulfonic acid) were included in the cell culture medium to provide a buffer system that works in conjunction with CO_2 gas added to the head space. For small-scale bioreactors maintained in humidified incubators (adjusted at 37°C and 5% CO_2 and 95% air saturation), it has been demonstrated that PPRF-h2 medium, which contains 20 mM sodium bicarbonate and 4.9 mM HEPES, can support the proliferation of hNPCs (Baghbaderani et al., 2008). However, even with these buffers in place, the pH levels in culture tends to decrease over time due to the build-up of metabolic by-products such as lactate, which in turn can adversely affect the proliferation activity of the cells (Sen and Behie, 1999; Butler, 2004). To minimize this pH drop in 500 mL hNPC cultures, a proportional controller can be employed to adjust the headspace CO_2 gas concentration. Moreover, a proportional controller can be used to control the dissolved oxygen at levels that minimize the production of lactic acid by promoting aerobic cellular respiration.

5.2.2.3 Hydrodynamics Within 500 mL Bioreactors

Agitation rate is known to affect oxygen transfer, hydrodynamic shear, and mixing times in stirred suspension cultures. Amongst these parameters, hydrodynamic shear has been suggested as a basis for scale-up of mammalian cells that propagate as tissue-like cell aggregates (Youn et al., 2005). It has previously been shown for murine NPCs that aggregate size can be controlled by altering shear stresses (Kallos et al., 1999; Sen et al., 2001), which is an important consideration, as cells within large aggregates may die due to diffusional limitations in oxygen and nutrients, and a build-up of metabolic by-products. However, excessive hydrodynamic shear can also have a serious deleterious effect on cell viability, growth, and stem cell characteristics. These effects can be mitigated by monitoring aggregate size distribution, viabilities, and viable cell densities for the tissue aggregates. Since the greatest shear experienced by the cells occurs at the tip of the impeller, impeller speed was used as the primary basis for scale-up of hNPCs. It should be noted that mass transfer rates and culture hydrodynamics in suspension bioreactors are also influenced by fluid properties, vessel geometry, impeller design, and the presence of measuring probes. Therefore, it is crucial to take these factors into consideration when scaling-up a cell production process.

Previous studies in the literature have shown that scale-up from 125 mL bioreactors (standard paddle impeller/Corning) to 500 mL bioreactors (four-bladed impeller/Wheaton) can be achieved for the production of murine neural precursor cells (mNPCs) (Gilbertson et al., 2006) and BrCSCs (Youn et al., 2006). Gilbertson et al. (2006) studied the effect of vessel geometry, impeller design, and the addition of measuring probes on large-scale expansion of mNPCs. The addition of measuring probes was found to significantly change the hydrodynamics in large-scale computer-controlled bioreactors. To avoid detrimental hydrodynamic effects on mNPC expansion, Gilbertson et al. (2006) reduced the agitation rate in the 500 mL bioreactor from 100 rpm (that was previously shown to be favorable for the smaller bioreactors) to 60 rpm. On the other hand, BrCSCs were found to expand well in 500 mL computer-controlled bioreactor at an agitation rate of 52 rpm, which was similar to the optimal rate of 50 rpm used in 125 mL bioreactors (Youn et al., 2006). Based on the correlation by Nagata (Nagata, 1975) and the equation by Cherry and Kwon (Cherry and Kwon, 1990), the maximum shear stress, occurring at the tip of the impeller, to which the cells would be exposed in a 500 mL bioreactor operated at 100 rpm, is 0.6 Pa. As previously shown by the hydrodynamics study in 125 mL bioreactors, this shear level is within the acceptable range for expansion of hNPCs in suspension culture.

Considering the change in the hydrodynamics of large-scale computer-controlled bioreactors due to the presence of measuring probes, a preliminary experiment was conducted for a period of 10 days to examine the feasibility of hNPC expansion in 500 mL computer-controlled bioreactors operated at 100 rpm. In this experiment, human telencephalon derived NPC aggregates (passage 8 [P8]) obtained from 125 mL suspension bioreactors were enzymatically dissociated into single cells and inoculated at 100,000 cells/mL into a computer-controlled suspension bioreactor containing 500 mL of PPRF-h2 medium. The cells were also inoculated into two 125 mL Corning spinner flasks (each containing 100 mL PPRF-h2 medium) for comparison. An agitation rate of 100 rpm was used at both scales (125 mL and 500 mL bioreactors) to maintain the cells in suspension. Standard process control techniques were used to control the temperature at 37°C, pH at 7.2 and dissolved oxygen concentration at 70% air saturation and monitored using a computer. The 125 mL bioreactors were housed in a humidified incubator maintained at 37°C, 95% air saturation (20% O_2), and 5% CO_2. All bioreactors were fed every five days by replacing 40% of the spent medium with fresh medium, while maintaining the cells within the bioreactor. To perform cell counts, samples were procured from each bioreactor every other day using standard protocols and analyzed.

Comparison of the growth kinetics and aggregate size distribution in small-scale and large-scale bioreactors showed that the hydrodynamics within the 500 mL bioreactor at an agitation rate of 100 rpm were not favorable for the expansion of hNPC populations. Compared to the 125 mL bioreactors, the cells exhibited dramatically lower proliferation activity in the 500 mL bioreactor. After 10 days in culture, a viable cell density of $3.86 \pm 0.4 \times 10^5$ cells/mL was obtained in the 500 mL bioreactor at 100 rpm as opposed to $14.8 \pm 2.0 \times 10^5$ cells/mL in 125 mL bioreactors operated at the same agitation rate. Moreover, whereas several hNPC aggregates with an average diameter of about 500 μm could be observed on day 10 in the 125 mL bioreactors, aggregates larger than 300–400 μm were rarely observed over the same period in the 500 mL bioreactor. Aggregate size is an important consideration; the diffusional limitations that can form with growing aggregate size can limit the penetration of gases and nutrients, leading to cell death and necrosis within the core of larger aggregates, thereby making smaller aggregates sizes in culture more desirable (Baghbaderani et al., 2008).

The results of preliminary experiments might imply that the correlation by (Nagata, 1975) and/or the equation by (Cherry and Kwon, 1990) underestimates the maximum level of shear in a 500 mL bioreactor with measuring probes. In fact, Aunins et al. (1989), and (Moreira et al., 1995) reported that using the Nagata correlation may result in an underestimation of the true power dissipation in large vessels, which would mean that the shear stress actually experienced by cells in the 500 mL vessel at 100 rpm would be greater than calculated. To eliminate this concern, the agitation rate in 500 mL computer-controlled bioreactors was reduced to 85 rpm to avoid the detrimental effects of high agitation rate (100 rpm) that were previously observed.

5.3 EXPERIMENTAL RESULTS FOR HUMAN NEURAL PRECURSOR CELLS IN 500 mL BIOREACTORS

After identifying the appropriate oxygen and hydrodynamic parameters and validating the process control methods used to maintain a stable culture environment, a study was performed to investigate hNPC growth kinetics, aggregate size distribution, and basic cell metabolism in the 500 mL computer-controlled bioreactors.

An expanded human telencephalon derived NPC (P9) population was obtained from one 125 mL suspension bioreactor, enzymatically dissociated into single cells and inoculated at 100,000 cells/mL into a 500 mL computer-controlled suspension bioreactor containing 500 mL of PPRF-h2 medium. For comparison, cells were also inoculated into two 125 mL Corning spinner flasks with paddle impellers (each containing 100 mL PPRF-h2 medium). The 125 mL bioreactors were placed in a humidified incubator to maintain the culture conditions at 37°C, 5% CO_2 and 95% air (20% O_2). The 500 mL bioreactor was placed on a heating pad to maintain a temperature of 37°C. A stir plate was used to control the agitation rate at 85 rpm. Filtered oxygen, CO_2, N_2, and air were supplied to the bioreactor via tubing connected to gas cylinders. Process control techniques were used to control temperature, dissolved oxygen level, and pH in the 500 mL bioreactor (Figure 5.2). Temperature control (37°C) was accomplished using a proportional-integral-derivative (PID) controller with a proportional gain of 50 and integral and derivative times of 0.03 and 1.5, respectively (Figure 5.2a). Dissolved oxygen level was controlled at 14.7% (equivalent to 70% air saturation) using a proportional controller with a gain of 45 (Figure 5.2b). pH control (7.3) was accomplished using a proportional controller with a gain of 30 (Figure 5.2c). It should be noted that the recommended control parameters provided a good first-estimation to control the hNPC expansion process. However, the actual value for each control parameter was subsequently adjusted using a standard trial and error approach using the cell culture medium alone prior to inoculating the cells into the bioreactor.

The growth kinetics and aggregate size distributions of hNPCs in the large bioreactor were measured over the course of 20 days. All cultures were fed every five days by replacing 40% of the spent medium with fresh medium. Samples were taken every other day from all vessels to perform cell counts. To maintain a constant culture volume over the course of the study, a volume of fresh PPRF.h2 medium equivalent to the sample volume was added back to each bioreactor after each sampling event. Following hNPC harvest, supernatant was stored at −20°C for future cell culture medium analysis. The aggregates were enzymatically dissociated, and cell viability was evaluated using the trypan blue exclusion method.

5.3.1 Cell Growth Kinetics

Figure 5.3 shows the growth kinetics for human telencephalon-derived hNPCs grown in 500 mL and 125 mL bioreactors. hNPCs in the 500 mL bioreactor achieved a maximum viable cell density of $3.7 \pm 0.5 \times 10^6$ cells/mL and a doubling time of 84 h, whereas the cells in the 125 mL bioreactors reached a maximum viable cell density of $3.2 \pm 0.4 \times 10^6$ cells/mL and a doubling time of 81 h. These results show that the growth characteristics of the cells grown in the large-scale bioreactor were comparable to the growth kinetics observed in the small bioreactors ($p = 0.16$, ANOVA on maximum cell densities). Noticeably, the growth curve for the cells in the small bioreactors was very similar to the growth curve previously shown for the cells at an agitation rate of 100 rpm at the same passage level, demonstrating the reproducibility

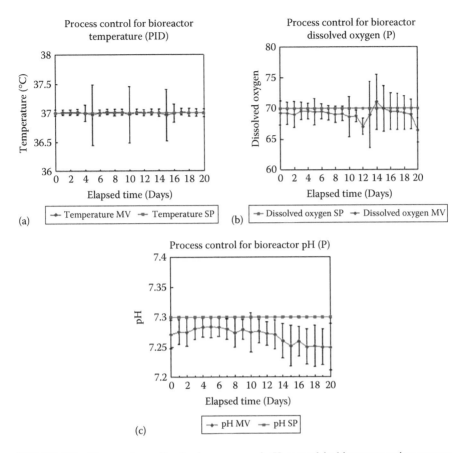

FIGURE 5.2 Temperature, dissolved oxygen, and pH control in bioreactor using proportional control (P), and proportional-integral-differential (PID) control methods.

Shown are set point (SP) and measured manipulated variable (MV) data for temperature (a), dissolved oxygen (b), and pH (c) level control at 37°C, 70% air saturation, and 7.3, respectively, over a period of 20 days in a 500 mL bioreactor. Temperature control (37°C) was accomplished using a PID controller with a proportional gain of 50 and integral and derivative times of 0.03 and 1.5, respectively. pH control (7.3) was accomplished using a proportional controller with a gain of 30. DO level was controlled at 14.7% of dissolved oxygen (70% air saturation) using a proportional controller with a gain of 45. Error bars for each parameter represent average of 144 measurements per day (one measurement every 10 minutes).

of the results obtained using the described expansion protocols. The viability of the cells grown in the 500 mL bioreactor and 125 mL bioreactors remained above 90% for the first 14 days of the study. Although the viability gradually decreased after day 14 at both scales, it remained acceptable (above 85%) up to day 18 in both cases.

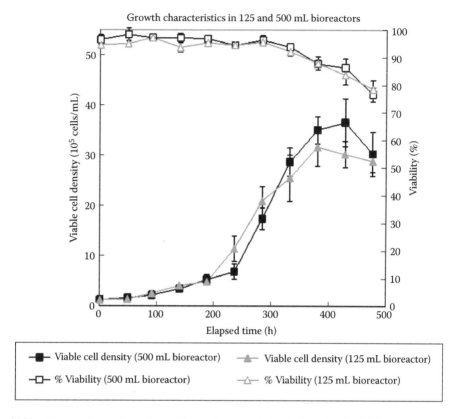

FIGURE 5.3   Comparison of hNPC growth curves in 125 mL and 500 mL bioreactors.

Human telencephalon-derived hNPCS (P9) obtained from one 125 mL suspension bioreactor were enzymatically dissociated into a single cell suspension and inoculated at 100,000 cells/mL into one computer-controlled bioreactor containing 500 mL PPRF-h2 medium and two 125 mL bioreactors each containing 100 mL of fresh PPRF-h2 medium. The cells were cultured over the course of 20 days during which time they were fed every five days by replacing 40% of the spent medium with fresh medium. All data points represent the average of duplicate sampling each counted twice. Error bars demonstrate the standard deviation for each measurement.

5.3.2   TISSUE AGGREGATE SIZE DISTRIBUTIONS

Photomicrographs of hNPC cell aggregates sampled from the 500 mL computer-controlled bioreactor showed that their morphology was similar to the aggregates formed in 125 mL bioreactors (Figure 5.4). The average aggregate size increased over time in the 500 mL bioreactor and aggregates with diameters of approximately 500 μm could be observed by the end of day 18 (432 h) (Figure 5.4d), at which time the maximum viable cell density was obtained.

Aggregate size distributions for the 500 mL bioreactor at an agitation rate of 85 rpm are shown in Figure 5.5. Compared to the aggregate diameters for 125 mL and 250 mL bioreactors, the aggregates formed in 500 mL bioreactors were smaller

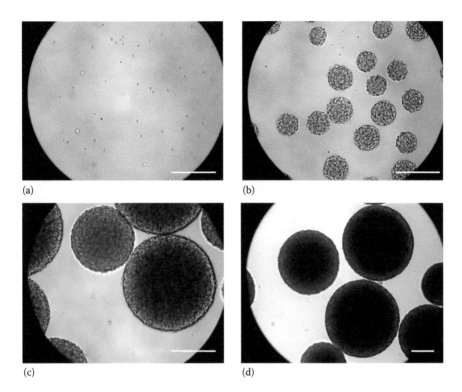

(a)   (b)

(c)   (d)

FIGURE 5.4 Photomicrographs of the bioreactor-expanded hNPCs grown in a 500 mL computer-controlled bioreactor.

Shown are the cells immediately following inoculation (a), four days post-inoculation (b), 10 days post-inoculation (c), and 18 days post-inoculation (d). Scale bars represent 200 μm.

in size. For instance, two days post-inoculation, more than 80% of the aggregates formed in the 500 mL bioreactor were less than 100 μm in diameter, whereas most of the aggregates formed in the 125 mL bioreactor were between 100 and 300 μm in diameter. The smaller aggregate size observed in the 500 mL bioreactor could be due to the energy dissipation and shear stress in the 500 mL bioreactor at an agitation rate of 85 rpm, which was still higher than the shear experienced by the cells in the smaller bioreactors at an agitation rate of 100 rpm. The increased shear, if present, may be attributed to the different impeller design and the presence of measuring probes in the 500 mL computer-controlled bioreactor. Nevertheless, it was not detrimental to the growth kinetics and cell viability. Moreover, more than 75% of the aggregates remained under 700 μm by day 18, when the maximum viable cell density as measured, and viability of the cells remained well above 85%. In agreement with the results observed in the smaller bioreactors (125 mL and 250 mL), as well as the results reported in the literature (Svendsen et al., 1998; Wright et al., 2003; Suzuki et al., 2004), it is evident that hNPCs can generate aggregates larger than 500 μm without compromising cell viability in a large-scale suspension bioreactor.

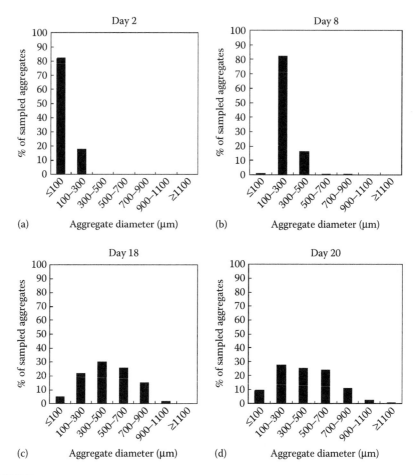

FIGURE 5.5 Tissue aggregate size distribution (hNPCs) in a 500 mL computer-controlled bioreactor.

An agitation rate of 85 rpm was used to maintain the cells in suspension. Shown are aggregate size distributions two days (48 hours) post-inoculation (a), eight days (192 h) post-inoculation (b), 16 days (384 h) post-inoculation (c), and 20 days (480 h) post-inoculation (d) in 500 mL suspension bioreactors. Human NPC aggregate diameters were determined by taking photomicrographs of bioreactor samples and measuring two perpendicular diameters for each aggregate present in the sample. The average diameter for each aggregate was determined by averaging the two measured perpendicular diameters. All data points represent the average of duplicate bioreactor runs.

5.3.3 CULTURE MEDIUM ANALYSES

Following the evaluation of the growth kinetics and aggregate size distribution, cell culture medium analyses were performed to determine nutrient (glucose and glutamine), metabolite (lactate and ammonia), and amino acid profiles in the 500 mL bioreactor. Sample preparation and medium analyses were performed in the same manner as previously described for hNPCs expanded in the 125 mL and 250 mL bioreactors (Baghbaderani et al., 2008).

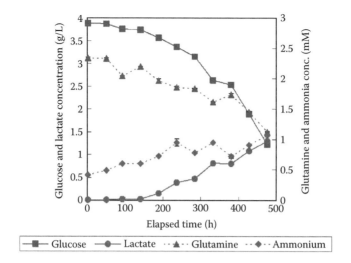

FIGURE 5.6   Concentrations of glucose, glutamine, lactate, and ammonia in the 500 mL suspension bioreactors.

Human telencephalon-derived hNPCs were inoculated at 100,000 cells/mL into one computer-controlled bioreactor containing 500 mL of PPRF-h2 medium. The cells were cultured over the course of 20 days during which time they were fed every five days by replacing 40% of the spent medium with fresh medium. Two samples (5 mL) were taken from the 500 mL bioreactor to perform cell counts. After harvesting the cells, the supernatant was stored at −20°C for culture medium analysis. Prior to analysis, each sample was thawed at 4°C overnight and incubated at room temperature for one hour. Concentrations of glucose, glutamine, lactate, and ammonia were measured using a Nova Biomedical Bioprofile 100 Analyzer.

Figure 5.6 shows the profiles of essential nutrients (glucose and glutamine) and metabolites (lactate and ammonium) determined from growth medium analysis over a course of 20 days in the 500 mL suspension bioreactors. The profiles of nutrients and metabolites in the 500 mL bioreactor were comparable to those in smaller bioreactors. Glucose concentration steadily dropped from 3.88 to 1.23 g/L (day 20) in the 500 mL bioreactor (versus 3.81 to 1.94 g/L in the 125 mL bioreactors). Over that same period, glutamine concentration decreased from 2.34 to 1.13 mM in the 500 mL bioreactor (versus 2.22 to 1.21 mM in the 125 mL bioreactors). In contrast, lactate concentration increased from 0.01 to 1.3 g/L (day 20) in the 500 mL bioreactor (versus 0.0 to 0.81 g/L in the 125 mL bioreactors). Ammonia concentration also increased from 0.41 to 1.07 mM (day 20) in the 500 mL bioreactor (versus 0.35 mM to 1.0 mM in the 125 mL bioreactors). These results indicated that hNPCs grown at both scales had comparable cell metabolism. Once again, it was shown that semi-fed batch culture was able to maintain lactate levels in large-scale bioreactors below a 2.0 g/L, which is known to be detrimental to mammalian cells (Ozturk et al., 1997).

Figure 5.7 shows concentrations of 15 amino acids in the 500 mL suspension bioreactors over time. Amino acid profiles in both small-scale and large-scale of hNPCs expansion were very similar. Moreover, amino acid concentrations were maintained at acceptable levels using a semi-fed batch approach. Altogether, the results of the

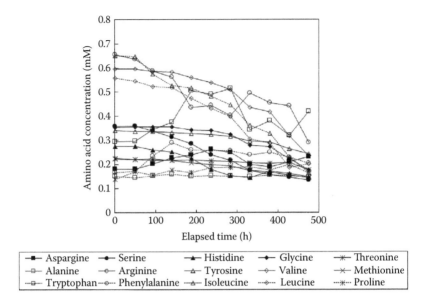

FIGURE 5.7 Changes in amino acid concentrations for human NPCs in the 500 mL computer-controlled bioreactor.

Human telencephalon-derived hNPCS were inoculated at 100,000 cells/mL into one computer-controlled bioreactor containing 500 mL PPRF-h2 medium. The cells were cultured over the course of 20 days during which time they were fed every five days by replacing 40% of the spent medium with fresh medium. Two samples (5 mL) were taken from the 500 mL bioreactor to perform cell counts. After harvesting the cells, the supernatant was stored at −20°C for culture medium analysis. Prior to analysis, each sample was thawed at 4°C overnight and incubated at room temperature for one hour. Concentrations of 15 amino acids were measured using a reverse phase high performance liquid chromatography (HPLC) column.

growth kinetics, aggregate size distribution, and medium analyses suggest that production of hNPCs can be scaled-up successfully from small-scale bioreactors to large-scale computer-controlled bioreactors without adversely impacting cell proliferation and metabolism.

5.3.4 Immunocytochemical Analyses of Bioreactor-Expanded Human Neural Precursor Cells

When developing a hNPC production system, it is important to ensure that the bioreactor-expanded cells retain their neural precursor cell markers, and exhibit multilineage potential after differentiation *in vitro*. Nestin, a well-known marker for neural stem and progenitor cells (Buc-Caron, 1995; Chalmers-Redman et al., 1997; Svendsen et al., 1998; Carpenter et al., 1999; Suzuki et al., 2004) was used to determine if the expanded hNPC populations remained undifferentiated after expansion. After being exposed to a differentiation protocol, antibodies against β-tubulin-III (TUJ1), Glial Fibrillary Acidic Protein (GFAP), and O4 were used to detect neurons, astrocytes, and oligodendrocytes, respectively to determine if the expanded cells retained their multilineage differentiation ability. In addition, efforts were undertaken to detect the presence of tyrosine hydroxylase (TH) and gamma-aminobutyric acid (GABA) positive cells to determine whether or not the bioreactor-expanded hNPCs could differentiate towards DAergic and GABAergic neuronal phenotypes, respectively. This was particularly important as the ability to generate these particular cell types are believed to be pivotal for the success of cell restoration strategies aimed at treating specific neurodegenerative disorders.

5.3.4.1 Immunocytochemical Analyses for Bioreactor-Expanded Human Neural Precursor Cell Aggregates and Differentiated Cells

Bioreactor-expanded hNPC aggregates (i.e., undifferentiated) were taken from the 500 mL suspension bioreactors on day 10 and added to poly-D-lysine/laminin coated wells containing PPRF-h2 medium. About 10–15 aggregates were added to each well of a chamber slide. Upon plating, the aggregates were stained for nestin. Figure 5.8a shows a brightfield image of a typical hNPC tissue aggregate after plating on a poly-D-lysine/laminin coated surface, and Figure 5.8b shows a magnification of the same aggregate positively stained for nestin suggesting that hNPCs expanded in large-scale computer-controlled bioreactors retained the expression of nestin, a marker for neural stem and progenitor cells.

5.3.4.2 Characterization and *In Vitro* Differentiation of Human Neural Precursor Cells

To evaluate the multilineage potential of differentiated bioreactor-expanded hNPCs, aggregates were taken on day 10 from the 500 mL bioreactor, dissociated into single cells, and plated at 20,000 cells/well on poly-D-lysine/laminin coated wells containing cytokine-free PPRF-h2 medium. The cells were differentiated for a period of 10 days in the cytokine-free medium. Upon plating, the cells adhered to the surface of each well and over the next two days developed processes. After 10 days, a majority of the cells had formed distinct interconnected processes (Figure 5.8c). Following immunocytochemical analyses of the differentiated cells, a majority of the cells (64.5% ± 8%) positively stained for the neuronal marker β-tubulin-III. Figure 5.8d–f illustrate immunostaining of the differentiated cells shown in Figure 5.8c. Cell nuclei staining using 4′,6-Diamidino-2-Phenylindole, Dihydrochloride (DAPI) was used to identify the total number of cells. All of the differentiated cells that stained positively for β-tubulin-III also co-labeled for GABA. There were no statistically significant differences between the percentage of cells positively stained for both markers (p-value = 0.81, ANOVA) (Figure 5.8g–i). This is not surprising since many published studies suggest that forebrain derived NPCs mostly differentiate into a GABAergic phenotype *in vitro* (Storch and Schwarz, 2002; Lindvall et al., 2004; Storch et al., 2004). Furthermore, about 22% ± 3% of the differentiated bioreactor-expanded human NPCs were positively stained for the astrocytic marker GFAP (Figure 5.8j–l). No O4-IR (oligodendrocyte cell marker), or TH-IR (DAergic neuronal marker) were found amongst the differentiated bioreactor-expanded hNPCs. Approximately 13.5% ± 7% of the cells did not stain positive for any of the analyzed cell markers. It is not clear why these cells did not show immunoreactivity for any of the neuronal or astroglial

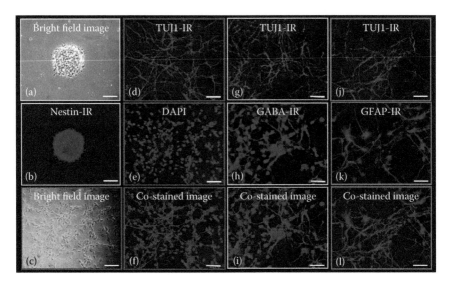

FIGURE 5.8 Immunocytochemical analyses of (1) *Undifferentiated* bioreactor-expanded hNPCs grown in a 500 mL computer-controlled bioreactor and (2) extent of *Multi-Lineage Differentiation* into neuronal and astroglial phenotypes (e.g., GABA neurons).

Shown are (a) brightfield image of one hNPC aggregate taken from a 500 mL suspension bioreactor (day 10), and (b) the same aggregate positively stained for nestin (a neural stem and progenitor cell marker). Photomicrograph (c) shows a brightfield image of differentiated hNPCs taken from a 500 mL suspension bioreactor on day 10 and stained for (d) neuronal marker TUJ1 (red), (e) cell nuclei marker (DAPI; blue), and (f) co-stained for cell nuclei marker (DAPI; blue) and neuronal marker TUJ1 (red). Photomicrographs (g–i) display differentiated hNPCs taken from 500 mL bioreactor and stained for (g) neuronal marker TUJ1 (red), (h) GABA producing neurons marker (GABA; blue), and (i) co-labeled for both neuronal marker (TUJ1; red) and GABA. Photomicrographs (j–l) show differentiated hNPCs taken from a 500 mL bioreactor and stained for (j) neuronal marker TUJ1 (red), (k) astrocyte marker (GFAP; blue), and (l) co-stained for neuronal marker (TUJ1; red) and astrocyte marker (GFAP; blue). Scale bars in (a) and (b) are 200 μm. Scale bars in (c) to (l) are 50 μm.

markers, but Carpenter et al. (1999) previously observed the same results after differentiation of human forebrain derived NPCs grown in the presence of EGF (20 μg/L), bFGF (20 μg/L), and LIF (10 μg/L). They suggested non-labelled cells might be a nestin-negative population of cells that did not undergo differentiation.

The results of immunocytochemical analysis for hNPCs grown in the 500 mL computer-controlled bioreactors were comparable to those observed in small scale bioreactors. These results showed that the bioreactor-expanded telencephalon-derived hNPCs might be bi-potent precursor cells with the capacity to produce neuronal and astroglial progeny *in vitro*. Moreover, these initial results demonstrated that it is possible to generate large populations of cells with a neuronal phenotype. This is important because a large population of hNPCs expanded under controlled-conditions can be differentiated into neuronal phenotypes, and therefore might be used as a reliable source of cells in treatment of neurodegenerative disorders using cell replacement strategy.

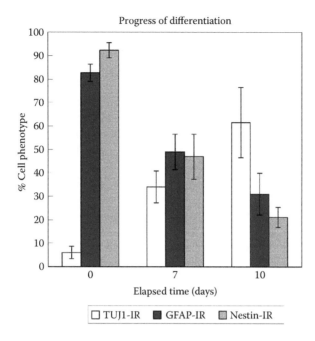

FIGURE 5.9   Progress of differentiation of hNPCs in cytokine-free PPRF-h2.

Human telencephalon-derived hNPC tissue aggregates were taken on day 10 of passage level 12 from one 125 mL suspension bioreactor, enzymatically dissociated into single cells, and plated at 20,000 cells/well on poly-D-Lysine/laminin pre-coated surfaces of 16 well-plate Lab-Tek chamber slides. Each well contained 200 μL of cytokine-free PPRF-h2 medium. The experiment was performed for a period of 10 days, and the number of nestin-IR (neural stem or progenitor cells), TUJ1-IR (neuronal phonotype), O4-IR (oligodendrocyte phenotype) and GFAP-IR cells (astrocyte phenotype) were quantified on day 0, day 7, and day 10 of the study (one slide per each day). Co-labelled cells for TUJ1 and DAPI were used to quantify TUJ1-IR cells (neurons). Quantifications were performed in three separate wells of each chamber slide by scanning six random fields of the differentiated cells using a Zeiss Axiovert 200M microscope with a 40× objective. In our studies, we were finally able to achieve an in vitro differentiation of bioreactor expanded hNPCs of a highly-enriched population of GABAergic neurons that expressed β-III-Tubulin, TUJ1 (64.19% ± 2.93%), GABA (99.10% ± 0.19%).

Considering the eventual goal of hNPC population expansion is to generate large numbers of clinically relevant specialized cells for the treatment of different neurological disorders, a study was undertaken to more rigorously evaluate the capacity of the bioreactor generated cells to differentiate into different specialized neural cell types. Figure 5.9 provides a comparison between the populations of TUJ1-IR cells (neuronal phenotype), GFAP-IR (astrocytic phenotype), and nestin-IR cells (neural stem or progenitor cells) generated over a period of 10 days in cytokine-free medium on poly-D-lysine/laminin coated surface. These results show that the number of nestin-IR cells and GFAP-IR cells decreased over time whilst the number of TUJ1-IR cells increased over that same period. On day 0 of differentiation (about 30 minutes after inoculation of the cells), 92.3% ± 3.6% of the cells were nestin-IR cells, which significantly decreased to 46.9% ± 7.7% on day 7 (p < 0.00001) and 21.9% ± 8.9 % on day 10 (p < 0.00001). The difference between the number of nestin-IR cells on day 7 and day 10 was also statistically significant (p < 0.01). Regarding the population of GFAP-IR cells, 82.7% ± 18.9% of the cells were GFAP-IR cells on day 0, which significantly decreased to 49.0% ± 2.6% on day 7 (p < 0.01) and 30.9% ± 2.3% on day 10 (p < 0.00001). The difference between the number of GFAP-IR cells on day 7 and day 10 was also statistically significant (p < 0.001). In comparison, 5.9% ± 3.8% of the cells were TUJ1-IR cells on day 0 of the differentiation. The number of TUJ1-IR cells significantly increased to 34.0% ± 3.1% on day 7 (p < 0.0001) and 61.9% ± 9.2% on day 7 (p < 0.0001). The difference between the number of TUJ1-IR cells on day 7 and day 10 was also statistically significant (p < 0.001). No O4-IR cells were observed over the entire course of the experiment. Also, less than 10% of the quantified cells were equally found to be non-labelled on day 0, day 7, and day 10 of the study (p = 0.8).

Figure 5.10 shows the progress of differentiation of hNPCs in cytokine-free PPRF-h2 medium. Photomicrographs of the cells stained for nestin (neural stem/progenitor cells marker), TUJ1 (neuronal marker), and GFAP (astrocytic marker).

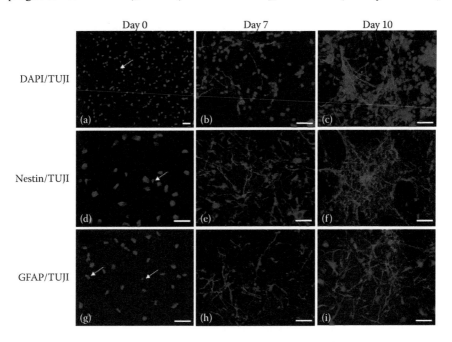

FIGURE 5.10 Progress of differentiation of hNPCs in the cytokine-free PPRF-h2 medium (Immunocytochemistry).

Shown in Row 1 are immunocytochemical analyses of the bioreactor-expanded hNPCs grown in a 125 mL bioreactor and differentiated towards neuronal phenotype TUJ1 (red) and co-labelled for DAPI (cell nuclei stain, blue) on day 0 (a), day 7 (b), and day 10 (c). Also, shown in Row 2 are the cells labelled for neuronal marker TUJ1 (red) and nestin (blue) on day 0 (d), day 7 (e), and day 10 (f). Row 3 shows cells labelled for neuronal marker TUJ1 (red) and astrocyte marker (GFAP; blue) on day 0 (g), day 7 (h), and day 10 (i). Scale bars represent 50 μm.

DAPI has been used to stain cell nuclei. The increase in the number of neuro-nal phenotype agreed with the decrease in the number of nestin-IR and GFAP-IR cells. Except for a few cells that double-labelled for both TUJ1 and nestin on day 0 (Figure 5.10d), no double-labelled cells for TUJ1 and nestin or TUJ1 and GFAP were found over the course of the study. These double-labeled cells for nestin and TUJ1 observed at the beginning of the study could be early committed neuronal progenitor cells.

The results of our study agreed with the results of differentiation studies per-formed by Kallur et al. (2006). They found that upon differentiation of human forebrain derived NPCs in the cytokine-free medium supplemented with 1% fetal bovine serum (FBS) on poly-L-lysine coated surface, the number of nestin-IR cells decreased from an average of 70% on day 1 to about 20% on day 28 of differentiation. While less than 5% of the cells positively stained for neuronal marker β-tubulin III on day 1 of differentiation, the number of GFAP-IR cells was about 50% on the same day. The number of neurons increased to about an aver-age of 20% on day 28. In contrast, the number of GFAP-IR cells declined to less than 10% on day 28 of differentiation (Kallur et al., 2006). In agreement with the results of the current study, Carpenter et al. (1999) also showed that when human cortex derived NPCs were differentiated in the cytokine-free medium supple-mented with 1% FBS on poly-ornithine coated surface, the percentage of nestin-IR cells declined from 75% on day 1 to about 5% on day 7 of differentiation. Moreover, the percentage of TUJ1-IR cells increased from about 5% on day 1 to about 25% on day 7 but the percentage of TUJ1-IR reached a plateau after seven days. The change in the population of GFAP-IR cells contrasted with the results of the current study, showing an increase in the percentage of GFAP-IR cells from day 1 (about 10%) to day 7 (about 30%). The percentage of GFAP-IR cells also reached a plateau after day 7 (Carpenter et al., 1999). Duittoz and Hevor (2001) also found a comparable trend regarding the change in population of nestin-IR and TUJ1-IR generated after differentiation of the ovine forebrain derived NPCs in the cytokine-free medium over a period of 15 days. About 90% of the cells were found to be nestin-IR cells at the beginning of the study, which declined over time in the cytokine-free medium. The population of TUJ1-IR cells also increased over time. However, the number of TUJ1-IR cells (about 10% on day 15 of differentiation) was significantly lower than the population of TUJ1-IR cells observed in the cur-rent study. In contrast to the results of the current study, the number of GFAP-IR cells peaked on day 5 of differentiation but the GFAP-IR cells decreased from day 5 to day 15. Interestingly, the study found that about 50% of the cells did not positively stained for any of the markers, suggesting these cells were nestin-negative cells that had undergone programmed cell death or had not fully differentiated (Duittoz and Hevor, 2001).

Regarding the population of GFAP-IR and nestin-IR cells, Wright et al., (2003) found that more than 92% of the proliferating human forebrain-derived hNPCs grown in the growth medium supplemented with EGF (20 μg/L), bFGF (20 μg/L) and LIF (10 μg/L) were nestin-IR. The current study showed that 90% of these nestin-IR cells were also GFAP-IR. It was suggested that these nestin/GFAP-IR cells were actively dividing NPCs capable of growing for an extended period of time in culture (Wright et al., 2003).

Overall, these studies showed that the majority of telencephalon-derived cells grown in culture were nestin-IR neural precursor cells. However, small populations of TUJ1-IR cells (neurons or early neural progenitor cells) or non-labelled cells were also found among the expanded cells, highlighting the heterogeneity of human neural precursor cell populations. Moreover, it was shown that the increase in the number of neurons was associated with decrease in the number of nestin-IR cells and GFAP-IR cells, indicating that some of the neural precursor cells had undergone differentiation over time, preferentially towards neuronal phenotype rather than glial phenotype.

5.3.5 BIOREACTOR SCALE-UP CONCLUSIONS

Scale-up of hNPC expansion was successfully performed from 125 mL suspension bioreactors to 250 mL suspension bioreactors as an important link to continue the production of human NPCs in 500 mL computer-controlled bioreactors. Human NPCs reached a cell-fold expansion of 44 (a doubling time of 69 hours) after 16 days in the 250 mL suspension bioreactors. Aggregate size analysis and growth medium analysis in the 250 mL bioreactors showed comparable aggregate size distribution and cell metabolism to the 125 mL bioreactors, suggesting that scale-up of hNPCs expansion from standard small-scale bioreactors to larger bioreactors is feasible.

Expansion of hNPCs was further scaled-up to the 500 mL computer-controlled bioreactors. Despite the changes in the hydrodynamics of the large-scale bioreactor, expansion of hNPCs was not compromised in the 500 mL bioreactors. hNPC populations reached a cell-fold expansion of approximately 37 (a doubling time of 84 h) after 18 days in the 500 mL bioreactors. The average hNPC aggregate diameter in the 500 mL bioreactors was maintained below a target value of 500 μm by controlling the shear field. Moreover, the cells grown in the large-scale bioreactor had a similar metabolic profile to those grown in the small-scale bioreactors. Immunocytochemical analysis revealed that the bioreactor-expanded telencephalon-derived hNPCs retained the ability to stain for nestin and give rise to neuronal and glial phenotypes after differentiation *in vitro*. Moreover, an enriched population of neurons, including GABAergic phenotype was generated upon differentiation of the bioreactor-expanded hNPCs in culture. This study showed that large quantities of hNPCs could be successfully and reproducibly generated under standardized conditions in computer-controlled suspension bioreactors, and that scale-up did not compromise important characteristics of the cells, including their proliferation activity and multilineage differentiation potential.

Our results also showed that withdrawal of cytokines from the growth medium and plating the cells on a pre-coated surface promoted neuronal cell generation. The simultaneous decrease in the number of nestin-IR cells over time confirmed that the neurons observed on day 10 of differentiation were derived from neuronal precursor cells after induction of in vitro differentiation. While a small number of neurons were observed at the beginning of the differentiation period, a majority of the cells were GFAP-IR. Because a similar secondary antibody was used for nestin-IR and GFAP-IR cells, it was not possible to co-label the cells positively stained for these two markers. However, considering that the number of nestin-IR cells and GFAP-IR cells at each analyzed time point were comparable, it is likely that the nestin-IR cell population were composed primarily of GFAP-IR cells.

5.4 TRANSPLANTATION STUDIES IN ANIMAL MODELS OF NEURODEGENERATIVE DISORDERS

5.4.1 LITERATURE REVIEW

Over the last two decades, the development of cell-based therapies for the treatment of CNS disorders has depended upon studies performed in animal models. There are validated animal models for neurodegenerative disorders such as PD, HD, and neuropathic pain. Inducing a pattern of cell loss in the striatum, substantia nigra, or nigrostiatal pathway, which is compatible with the pathophysiology of PD or HD using neurotoxins such as 6-hydroxydopamine (6-OHDA) and quinolinic acid (QA) are well established models of these conditions (Dunnett and Rosser, 2004; Melrose et al., 2006). For allodynia, a neuropathic pain model has been developed by ligation of the L_5 spinal nerve alone or both the L_5 and L_6 spinal nerves of one side of rat (Kim and Chung, 1992). The pain behavior expressed following nerve ligation is associated with the appearance of abnormal sensory function including allodynia that results from dysfunction in the spinal cord (Bridges et al., 2001). Although these animal models do not demonstrate all features of the human neurodegenerative disorders, they do mimic the deficits observed in the disease, are highly reproducible, and thus are widely used to experimentally study potential therapies (Mendez et al., 2000a; Bridges et al., 2001; Dunnett and Rosser, 2004, 2007; McLeod et al., 2006b; Mukhida et al., 2001, 2006, 2007).

Moreover, cell transplantation into animal models of neurodegenerative disorders has been widely investigated in the literature (Svendsen et al., 1996, 1997; Vescovi et al., 1999; Hagan et al., 2003; McBride et al., 2004; Ryu et al., 2004; Yasuhara et al., 2006; Anderson et al., 2007; Anderson and Caldwell, 2007). Behavioral assessments have been used in these studies to investigate functional benefits resulting from transplantation of hNPCs. Further, histochemical analyses have been performed to study the survival, differentiation, and innervation of *in vitro* expanded hNPCs into the host tissue. The results of these studies have demonstrated that transplantation of hNPCs in animal models is feasible, and that transplantation of tissue-specific stem cells will not result in tumor formation, which mitigates a huge concern of safety, as it would otherwise invalidate the clinical translation of this approach. Some of these studies suggested that, in order for transplantation studies to result in functional outcomes, the engrafted cells should be differentiated into the specific neuronal phenotype that has been lost upon lesioning of the brain (for instance, differentiation of hNPCs into DAergic neurons in parkinsonian rats) (Svendsen et al., 1996, 1997). Other studies suggested that even in the absence of the cells that can replace the lost cells, the transplanted hNPCs may result in functional benefits by releasing neurotrophic factors into the host tissue and inducing survival of the host neurons (Hagan et al., 2003; Ryu et al., 2004; Yasuhara et al., 2006; Anderson and Caldwell, 2007). Despite the encouraging results observed in these studies, one major issue associated with studies was that transplanted hNPCs exclusively expanded in static culture were problematic due to the uncontrolled culture conditions. Expansion of hNPCs in standard suspension bioreactors offers the possibility of producing large quantities of clinical grade hNPCs in a standardized method, which would enable widespread therapeutic application of these cells. Specifically, McLeod et al. (2006b) reported transplantation of murine NPCs grown in suspension bioreactors in an animal model of PD. The results of this study did not show any functional recovery due to the inability of the cells to produce DAergic neurons. However, positive results achieved from this study, including long-term graft survival, no intended migration of the cells to other areas of the CNS and lack of tumor formation, revealed that the use of bioreactor-expanded neural stem cells for transplantation studies could be a viable option in treatment of neurodegenerative disorders. This lays the foundation for further animal studies using bioreactor expanded hNPC populations.

5.4.2 OUR TRANSPLANTATION STUDIES

Previous sections of this chapter have shown that large populations of human neural precursor cells can be expanded under controlled and standardized operating conditions in suspension bioreactors. In fact, bioreactor expanded hNPCs retained important cellular characteristics including multilineage differentiation capacity and production of large population of neurons *in vitro*. However, in order to show the viability of using bioreactor-expanded hNPCs for the treatment of neurodegenerative disorders, it was necessary to perform transplantation studies in animal models to evaluate the safety, feasibility, and therapeutic efficacy of transplanting bioreactor-expanded hNPCs for therapeutic effect.

Reported herein are our results from studies that evaluated the transplantation of bioreactor-expanded human telencephalon derived NPCs into animal models of HD and neuropathic pain. The overall goal of our studies was to test the functional and anatomical characteristics of transplanted pre-differentiated and undifferentiated bioreactor-expanded hNPCs in animal models of neurodegenerative disease and injury. In particular, these studies aimed at; (1) examining the anatomical reconstruction and function recovery of the striatum following transplantation of bioreactor-expanded hNPCs in the QA rat model of HD, (2) determining the functional effects of transplanting bioreactor-expanded hNPCs in a nerve ligation model of neuropathic pain in the rat, and (3) determining if the cells formed tumors or exhibited inappropriate migration from the transplantation target areas that may preclude their clinical utility.

5.4.2.1 Overview of Transplantation Studies

Bioreactor expanded hNPCs in Calgary were cryopreserved in PPRF.h2 with 10% dimethyl sulfoxide (DMSO) and shipped frozen to the Cell Restoration Laboratory (CRL) in Halifax, Canada to perform transplantation studies in animal models of neurodegenerative disease. hNPCs were transplanted as undifferentiated or differentiated cells into the central nervous system of each animal. We present here the phenotypic characteristics and ability of bioreactor-expanded hNPCs to improve functional behavior in animal models of HD (McLeod et al., 2013) and neuropathic pain (mechanical allodynia) (Mukhida et al., 2007).

5.4.2.2 Huntington's Disease Model—Transplantation Experiments (McLeod et al., 2013)

5.4.2.2.1 Differentiation of Bioreactor Expanded Human Neural Stem Cells and Progenitor Cells

Prior to transplantation, the ability of bioreactor-expanded hNPCs to differentiate into neuronal and glial phenotypes were studied in a valproic acid and BDNF supplemented differentiation medium. The study was performed for a period of seven days on a poly-L-lysine coated surface. The intermediate filament nestin was used as a marker for undifferentiated neural stem or progenitor cells. The cell-cycle-associated protein, Ki-67 was used to label the nuclei of actively proliferating cells during late G1, S, G2 and M phases of the cell cycle. The cells that had differentiated towards neuronal phenotypes were labeled with the neuronal filament TUJ1 (β-Tubulin-III). Those cells, which had specifically differentiated into GABA producing neurons, were detected using three markers: (1) GABA as a neurotransmitter produced by GABAergic neurons, (2) GAD as GABA synthesizing enzyme, and (3) the GABA$_A$ receptor β-subunit. Moreover, the intracellular calcium binding protein calretinin was also used as a marker that co-labels with more than 70% of GABA-immunoreactive cells in the human striatum. The opioid peptide leu-enkephalin was used as a marker to further identify the subpopulation of the GABAergic neurons following differentiation. Leu-enkephalin is a pentapeptide enkephalin subtype. Enkephalin is one of the two natural pain-killers of the human body, believed to be found in spiny projection neurons in the striatum.

The differentiation results were compared with undifferentiated cells using the same cell markers. Figures 5.11 and 5.12 give the results of immunostaining the undifferentiated and differentiated human bioreactor-derived hNPCs, respectively. Quantitative analysis of these results revealed a dramatic decrease in the population of nestin-positive cells following differentiation. While nearly the entire population (100.00% $\pm$ 1.81%) of undifferentiated human NPCs were nestin-immunoreactive (Figure 5.11a), the majority (11.56% $\pm$ 1.81%, p<0.001) of the cells did not stain for nestin following differentiation (Figure 5.12a). This was in agreement with the significant decrease observed in the population of actively proliferating Ki-67-IR cells. The population of Ki-67-IR cells decreased from 37.04% $\pm$ 1.51% among the undifferentiated cell population (Figure 5.11f) to 4.66% $\pm$ 0.67% (p<0.001) after differentiation (Figure 5.12m). On the other hand, immunocytochemical analysis showed that the population of the neuronal phenotypes significantly increased following seven days of exposure to differentiation medium. While only 5.61% $\pm$ 1.08% of undifferentiated human NPCs (Figure 5.11c) were TUJ1-IR, 64.19% $\pm$ 2.93% (p<0.001) of the cells became TUJ1-IR after differentiation (Figure 5.12c). Regarding the population of GABAergic neurons, 1.32% $\pm$ 0.68%, 0.27% $\pm$ 0.2%, 0%, and 0% of undifferentiated hNPCs positively stained for the GABAergic neurotransmitter GABA (Figure 5.11d), the GABA synthesizing enzyme GAD, the GABA$_A$ receptor β-subunit, and the intracellular calcium binding protein calretinin, respectively. This was significantly lower than 99.10% $\pm$ 0.19% (p<0.001), 99.00% $\pm$ 0.35% (p<0.001), 73.90% $\pm$ 2.03% (p<0.001), and 58.82% $\pm$ 2.43% (p<0.0) of differentiated

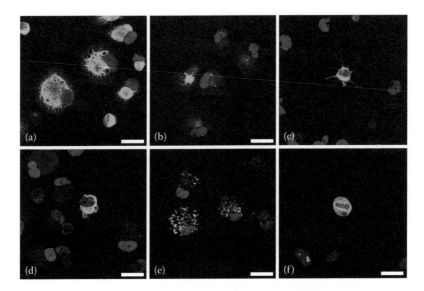

FIGURE 5.11 Immunocytochemical analysis of the undifferentiated bioreactor-expanded hNPCs *in vitro*.

Shown are undifferentiated bioreactor-expanded hNPCs expressing human specific nuclear antigen (red) and co-labelled with (a) nestin, (b) GFAP, (c) TUJ1, (d) GABA, and (e) leu-enkephalin (green). Also, shown is (f) the cell-cycle-associated protein Ki-67 (green) co-labelled with the DNA bound nuclear counterstain TO-PRO-3 (blue). Scale bar represents 20 µm.

hNPCs that positively stained for GABA (Figure 5.12d), the GABA synthesizing enzyme GAD (Figure 5.12j), the $GABA_A$ receptor β-subunit (Figure 5.12h), and the intracellular calcium binding protein calretinin (Figure 5.12k), respectively. It was interesting to note that the population of neuronal phenotypes positively stained for the opioid peptide leu-enkephalin significantly decreased from 43.90% $\pm$ 3.38% among the undifferentiated hNPCs (Figure 5.11e) to 7.28% $\pm$ 1.74% ($p<0.001$) among the differentiated hNPCs (Figure 5.12k). Following seven days of differentiation, 64.19% $\pm$ 2.93% of the bioreactor-expanded hNPCs were co-labelled for TUJ1 and GABA (Figure 5.12d–f), and 73.90% $\pm$ 2.03% of the cells co-labelled for GABA and the $GABA_A$ receptor β-subunit (Figure 5.12g–i). In contrast to the neuronal cell population, the population of undifferentiated hNPCs positively stained for astrocytic cell marker, GFAP, (1.03% $\pm$ 0.38%) (Figure 5.11b) did not altered after differentiation (1.30% $\pm$ 0.31%, $p>0.05$) (Figure 5.12). Both undifferentiated and differentiated bioreactor-expanded hNPCs did not express the phenotypic oligodendrocyte marker O4, the intracellular calcium binding protein calbindin-(Calb) D-28-K, the opioid peptide met-enkephalin, somatostatin, parvalbumin, substance P, the post-synaptic D2 dopamine receptor, and the post-synaptic glutamate receptor (data not shown).

The results of *in vitro* differentiation suggested that the differentiation medium induced phenotypic maturation of the bioreactor-expanded human NPCs towards

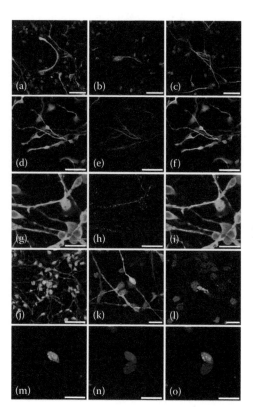

FIGURE 5.12 Immunocytochemical analysis of the differentiated bioreactor-expanded human NPCs *in vitro*.

Shown are bioreactor-expanded hNPCs following seven days of differentiation and expressing (a) nestin, (b) GFAP and (c) TUJ1 (green) that also co-labelled with human specific nuclear antigen (red) (scale bar is 50 µm). The differentiated cells co-expressed (d) GABA (green) and (e) TUJ1 (red); (f) merged image (scale bar is 50 µm). Also, shown are the differentiated cells co-labelled with (g) GABA (green), (h) GABA$_A$ receptor β-subunit (red), and (i) merged image (scale bar is 20 µm). The differentiated bioreactor-expanded human NPCs also expressed (j) GABA synthesizing enzyme GAD (green) (scale bar is 50 µm), (k) calretinin (green) (scale bar is 20 µm), and (l) leu-enkephalin (green) (scale bar is 20 µm) that co-labelled with human specific nuclear antigen (red). Also, shown are the cells positively stained for (m) cell-cycle-associated protein Ki-67 (green) and (n) DNA bound nuclear counterstain TO-PRO-3 (blue), and co-labeled for both (o) Ki-67 and TO-PRO-3 (scale bar is 20 µm).

a highly-enriched population of striatal GABAergic cells, positively staining for GABA, GAD, the GABA$_A$ receptor β-subunit, calretinin and TUJ-1.

5.4.2.3 Huntington's Disease Model—Behavioral Study

In order to evaluate the behavioral changes that occurred prior to and after transplantation of the cells, the QA-lesioned animals were equally divided into three experimental groups: *Group 1-black*, considered as control group, had animals that did not receive a transplant, whereas the animals in *Group 2-red* received 800,000 undifferentiated bioreactor-expanded hNPCs, and the animals in *Group 3-blue* received 800,000 cells, which were obtained by differentiating the bioreactor-expanded hNPCs. Three tests were performed to evaluate the behavioral analysis as follows.

5.4.2.3.1 Huntington's Disease Model–Behavioral Animal Tests

Figure 5.13 summarizes the results of the amphetamine induced rotations test in all three groups of QA-lesioned rats at 0, 2, 4, 6, 8, and 10 weeks post-transplantation. The number of amphetamine-induced rotations significantly decreased in the QA-lesioned animals transplanted with differentiated hNPCs (Group 3) at 2 (3.93 ± 1.06 vs 12.71 ± 1.41 rotations/min, $p<0.001$), 4 (3.12 ± 1.12 vs 13.09 ± 1.99 rotations/min, $p<0.001$), 6 (3.97 ± 1.47 vs 13.12 ± 1.89 rotations/min, $p<0.01$), 8 (2.91 ± 1.49 vs 13.77 ± 1.82 rotations/minute, $p<0.01$), and 10 weeks post-transplantation (3.81 ± 1.86 vs 12.62 ± 1.33 rotations/minute, $p<0.01$) when compared to the control animals. Moreover, the rotational behavior significantly declined in animals transplanted with differentiated hNPCs (Group 3) at 2 (3.93 ± 1.06 vs 14.04 ± 1.35 rotations/min, $p<0.001$), 4 (3.12 ± 1.12 vs 13.91 ± 2.00 rotations/min, $p<0.001$), 6 (3.97 ± 1.47 vs 14.12 ± 2.42 rotations/min, $p<0.01$), 8 (2.91 ± 1.49 vs 14.47 ± 2.38 rotations/min, $p<0.001$), and 10 weeks post-transplantation

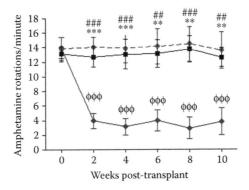

FIGURE 5.13 Asymmetrical rotational behaviour analysis following an amphetamine challenge.

Number of unilateral rotations of the animals was studied after QA lesioning and transplantation of the bioreactor-expanded hNPCs into the striatum at 2, 4, 6, 8, and 10 weeks post-transplantation. The animals were equally divided into three study groups. There was a significant decrease in asymmetrical rotational behaviour following transplantation of differentiated bioreactor-expanded hNPCs (Group 3-blue) at 2, 4, 6, 8, and 10 weeks post-transplantation when compared to control animals (Group 1-black) (**: $p<0.01$, ***: $p<0.001$), as well as animals transplanted with undifferentiated bioreactor-expanded hNPCs (Group 2-red) (##: $p<0.01$, ###: $p<0.001$) (mean $\pm$ S.E.M.). There was a significant decrease in rotational behaviour following transplantation of differentiated cells (Group 3-blue) at 2, 4, 6, 8, and 10 weeks when compared to baseline pre-transplantation rotational behaviour ($\phi\phi\phi$: $p<0.001$) (mean $\pm$ S.E.M.).

(3.81 ± 1.86 vs 13.56 ± 2.51 rotations/min, p<0.01) when compared to those transplanted with undifferentiated hNPCs (Group 2). There was no significant difference in rotational behavior between Group 1 and Group 2 at all time points (p>0.05). In comparison with the rotational behavior of the base-line post-lesion animals (13.96 ± 1.46 rotations/min), rotational behavior of the animals transplanted with the differentiated human NPCs (Group 3) significantly decreased at 2 (3.93 ± 1.06 rotations/min, p<0.001), 4 (3.12 ± 1.12 rotations/min, p<0.001), 6 (3.97 ± 1.47 rotations/min, p<0.001), 8 (2.91 ± 1.49 rotations/min, p<0.001), and 10 weeks post-transplantation (3.81 ± 1.86 rotations/min, p<0.001). In contrast, there was no significant change in rotational behavior in Group 1 and Group 2 when compared to post-lesion baseline rotational behavior at all time points (p>0.05).

This study showed that *transplantation of differentiated bioreactor-expanded human NPCs into the QA-lesioned animals resulted in substantial improvement in the amphetamine-induced rotational asymmetry by an average of 74.59% ± 1.58%.* A viable explanation for this improvement could be reinnervation of the grafted differentiated hNPCs into the host nigrostriatal dopaminergic system and re-construction of the functional synaptic contacts. This explanation was supported by the presence of synaptophysin and dopamine D2 expression in the host striatum of the animals transplanted with differentiated hNPCs.

Similarly, in the cylinder test, which is used to assess forelimb asymmetry, right forelimb asymmetry of animals engrafted with differentiated hNPCs significantly decreased at 4, 6, 8, and 10 weeks post-transplantation (results not shown). This behavioral improvement was not observed in animals transplanted with undifferentiated cells. Lastly, in the T-Maze test, which is used to assess spatial working memory, significant improvement in animal alternation behavior was observed in lesioned animals transplanted with hNPCs, but not in animals receiving transplants with undifferentiated cells (results not shown), suggesting that cells cultured in the bioreactor can be used to decrease the extent of impairment associated with working memory in these animals. Overall the results of these behavioral studies suggest that transplantation of cells, especially those which have been differentiated prior to implanted into the central nervous system, can make a positive impact on the amelioration of symptoms associated with HD.

5.4.2.3.2 Huntington's Disease Model—Histochemical Analyses

Histochemical analyses were conducted as a complementary study to the behavioral analysis to investigate cytoarchitecture of the lesioned area in the absence and presence of transplanted cells, as well as phenotypical analysis of the engrafted area. In particular, the study focused on evaluation of cell survival, innervation of the cells into the transplantation site, and tumor formation after transplantation. Furthermore, different neuronal and glial markers were used to monitor neuronal or glial phenotypes survived or differentiated after transplantation. This study was conducted at the end of the behavioral study at 10 weeks post-transplantation.

Figure 5.14 illustrates graft morphology and cytoarchitecture of the engrafted site for all three different groups of animals. Human neuron specific enolase (hNSE) was

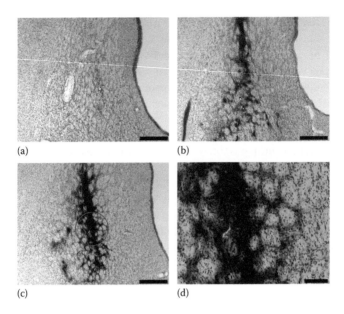

(a)   (b)

(c)   (d)

FIGURE 5.14   Graft morphology of the grafted striatum 10 weeks post-transplantation. Shown are coronal sections of (a) control (Group 1), (b) undifferentiated bioreactor-expanded hNPCs (Group 2), and (c) differentiated bioreactor-expanded hNPCs (Group 3) grafts stained with hNSE and cresyl violet following QA lesioning of the rodent striatum at 10 weeks post-transplantation. (d) demonstrates high power image of differentiated bioreactor-expanded hNPCs graft in (c) (shown as a red circle) with hNSE positive loop structures. Scale bars in (a), (b), and (c) are 500 μm. Scale bar in (d) is 100 μm.

used to stain cell bodies and fibers of the differentiated or undifferentiated hNPCs in the transplantation area. Crystal violet was used to counter-stain neuronal and glial cells of the host tissue as well as transplanted human cells. As shown in Figure 5.14, robust cell survival could be observed for both bioreactor-expanded undifferentiated hNPCs (*Group 2*) (Figure 5.14b) and differentiated hNPCs (*Group 3*) (Figure 5.14c and d) with extensive staining for hNSE in cell bodies and fibers. However, there were significant differences in the distribution and histological morphology of the hNSE positive cells for each study group. Undifferentiated hNPCs that positively stained for hNSE were widely distributed throughout the rostrocaudal axis of the striatum, and even migrated to adjacent regions such as corpus callosum (Figure 5.14b). In contrast, differentiated hNPCs did not migrate or widely distribute themselves throughout the rostrocaudal axis of the striatum (Figure 5.14c). Moreover, differentiated hNPCs formed dense clusters of cells at the primary transplantation site (Figure 5.14c). It is important to note that differentiated hNPCs were able to form distinct loop structures similar to the architecture arrangement of the normal structure of the non-lesioned animals (Figure 5.14d). Stereological analysis revealed that the graft volume, defined as the portion of the striatum containing hNSE positive bodies and fibers, was significantly larger in undifferentiated Group 2 (3.01 ± 0.18 mm^3), when compared to the graft volume measured for differentiated Group 3 (1.94 ± 0.19 mm^3, $p<0.001$).

5.4.2.3.3   Huntington's Disease Model—Conclusions

In the QA rat model of Huntington's disease, bioreactor-expanded hNPCs were transplanted into the lesioned striatum as undifferentiated bioreactor-expanded hNPCs or pre-differentiated bioreactor-expanded hNPCs that were immunoreactive for gamma amino butyric acid (GABAergic phenotype). Pre-differentiation studies *in vitro* suggested that the maturation of hNPCs towards a GABAergic phenotype in a highly-enriched manner was dependent upon the differentiation medium. Following implantation of pre-differentiated cells, results from the rotational behavior test, cylinder test and T-Maze test showed marked improvements in behavior and functionality. Immunohistochemical analysis of the engrafted tissue demonstrated robust cell survival for both undifferentiated hNPCs and differentiated hNPCs. Moreover, a GABAergic phenotype previously acquired by *in vitro* differentiation was maintained following transplantation of differentiated hNPCs into the lesioned animals. Differentiated bioreactor-expanded hNPCs did not undergo cell proliferation or tumor formation following transplantation and did not express astrocytic marker; however, a sparse population of nestin-positive cells was found in the engrafted tissues. In comparison, undifferentiated hNPCs did not show immunoreactivity for neuronal markers and underwent differentiation towards an astrocytic phenotype, although they also did not exhibit evidence of cell proliferation or tumor formation following transplantation. This study suggested that the *in vitro* differentiation of bioreactor-expanded hNPCs may be a safe and reliable source of GABAergic neuronal phenotype that can result in functional recovery following transplantation into the animal model of HD.

5.4.3   Neuropathic Pain (Allodynia) Transplantation Studies

5.4.3.1   Differentiation of Bioreactor Expanded Human Neural Precursor Cells

Prior to transplantation, bioreactor-expanded hNPCs were differentiated for a period of seven days in a differentiation medium containing valproic acid and BDNF. For comparison, the cells were also differentiated in the same medium excluding valproic acid and BDNF over the same period of time. Following differentiation, the cells were stained for neuronal, astrocytic, or oligodendroglial markers. Bioreactor-expanded hNPCs were also stained for nestin prior to the differentiation. Figure 5.15 displays the results of immunocytochemical analysis performed after differentiation of hNPCs. Bioreactor-expanded hNPCs exhibited immunoreactivity for nestin, suggesting the presence of a population of proliferative neural precursor cells in the human neurospheres. Following differentiation in the absence of valproic acid and BDNF, some of the cells showed immunoreactivity for nestin. About 16.26% ± 2.12% GFAP-IR, 30.74% ± 1.40% TUJ1-IR, and MBP-IR cells were detected after differentiation in the valproic acid- and BDNF-free medium. These results demonstrated that the bioreactor-expanded hNPCs were able to give rise to neuronal, astrocytic, and oligodendroglial phenotypes in this medium, suggesting the multipotentiality of bioreactor-derived hNPCs. Moreover, a small population of the cells (13.14% ± 2.57%) positively stained for GABA. In the presence of valproic acid and BDNF, bioreactor-expanded hNPCs primarily differentiated into a GABAergic

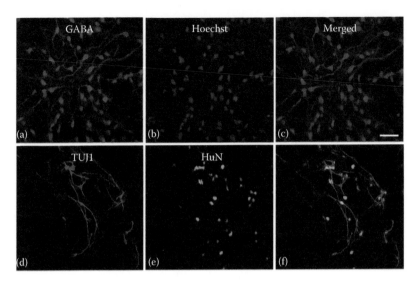

FIGURE 5.15   Bioreactor-derived hNPCs primarily differentiated into GABAergic phenotype in the presence of valproic acid- and BDNF.

Shown are differentiated bioreactor-expanded hNPCs in the differentiation medium supplemented with valproic acid and BDNF positively stained for (a) GABA, (b) cell nuclei staining Hoechst, (c) both GABA and Hoechst, (d) TUJ1, (e) human cell nuclei staining HuN,
and (f) both TUJ1 and HuN. Scale bar represents 50 µm. (From Mukhida, K. et al.: Spinal
GABAergic transplants attenuate mechanical allodynia in a rat model of neuropathic pain.
Stem Cells. 2007. 25(11). 2874–2885 Copyright Wiley-VCH Verlag GmbH & Co. KGaA.
Reproduced with permission.)

phenotype (93.09% ± 3.70%). The majority of cells (74.59% ± 6.34%) were also positively stained for the neuronal marker, TUJ1 (Figure 5.15). These results showed that
bioreactor-expanded hNPCs could be induced to differentiate into a large population
of cells with a GABAergic phenotype *in vitro*.

5.4.3.2   Neuropathic Pain Model—Transplantation Results

In the L5-L6 spinal nerve ligation model of neuropathic pain, undifferentiated
bioreactor-expanded hNPCs or GABAergic pre-differentiated bioreactor-expanded
hNPCs were transplanted into the spinal cord above the ligation level. In a separate
group of animals, GABAergic cells of the human striatal primordia were transplanted
into the spinal cord for comparison. Behavioral assessments of animals receiving
transplants was conducted before and after nerve root ligation, and at 1, 2, 4, and
6 weeks after transplantation using von Frey filaments applied to the plantar aspect
of the hind paw to assess paw withdrawal threshold levels. The behavioral analysis
revealed that animals receiving transplants of GABAergic pre-differentiated hNPCs,
or human striatal primordial cells, demonstrated a significant attenuation in pain
threshold scores. In contrast, transplantation of undifferentiated hNPCs did not
attenuate pain threshold scores. Immunohistochemical analysis of the spinal cord
transplants of GABAergic pre-differentiated hNPCs demonstrated robust graft

survival and migration of the cells for bioreactor-expanded differentiated hNPCs. The majority of transplanted undifferentiated hNPCs were immunoreactive for nestin and appeared to differentiate into astrocytes post-transplantation. In both undifferentiated and pre-differentiated hNPC grafts, a few cells were immunoreactive for human proliferating cells marker Ki-67, however no tumor formation was observed. Overall, this study suggested that the production of hNPCs in standard suspension bioreactors may represent a viable option to produce a reliable source of GABAergic neurons as a potential therapeutic option in the treatment of neuropathic pain.

5.4.3.3 Neuropathic Pain Model—Animal Behavioral Study

An animal model of neuropathic pain was developed by unilateral ligation of L5-L6 spinal nerve root. The animals were transplanted with fetal green fluorescent protein (GFP) mouse striatal primordial cells, differentiated bioreactor-expanded hNPCs, or undifferentiated bioreactor-expanded hNPCs. To serve as control, some of the ligated animals were also transplanted with only cell culture medium according to the same procedure used for cell transplantation. Post-ligation and post-transplantation behavioral analyses were conducted using von Frey filaments applied to the plantar aspect of hind paw. Figure 5.16 demonstrate the results of behavioral assessments conducted after transplantation into the spinal cord of ligated animals, illustrating the results as the normalized left hind paw withdrawal thresholds level with respect to the right hind paw. One-week post-transplantation, paw withdrawal thresholds of the animals that received human striatal primordial cells or differentiated bioreactor-expanded hNPCs (GABAergic phenotype) significantly increased. This increase in the paw withdrawal thresholds was maintained throughout the whole period of the study (up to six weeks post-transplantation). As shown in Figure 5.16, prior to the lesioning the left hind paw withdrawal thresholds level of the animals transplanted with human striatal primordial cells or predifferentiated bioreactor-expanded hNPCs were 0.98 ± 0.024 or 1 ± 0, respectively. Following lesioning, the left hind paw withdrawal thresholds level of the animals transplanted with fetal striatal primordial cells or predifferentiated bioreactor-expanded hNPCs decreased to 0.12 ± 0.08 or 0.14 ± 0.014 ($p<0.001$), respectively. One week following transplantation of the human striatal primordial cells or predifferentiated bioreactor-expanded hNPCs, the left hind paw withdrawal thresholds level increased to 0.48 ± 0.07 or 0.31 ± 0.022 ($p<0.001$), respectively. Although maximum improvement in the behavioral assessment was observed at four weeks post-transplantation, the level of paw withdrawal thresholds in the animals transplanted with fetal striatal primordial cells or pre-differentiated bioreactor-expanded hNPCs remained as high as 0.71 ± 0.07 or 0.55 ± 0.02 ($p<0.001$), respectively, at six weeks post-transplantation. On the contrary, paw withdrawal thresholds of the animals that received undifferentiated bioreactor-expanded hNPCs or only cell culture medium did not significantly change ($p>0.05$) over the course of study, suggesting lack of any behavioral benefit from the transplantation procedure.

The results of behavioral assessment showed that transplantation of GABAergic neurons obtained from either differentiated bioreactor-expanded hNPCs significantly attenuated low threshold neuropathic pain. There could be several explanations for the observed behavioral benefit. It is possible that intraspinal transplanted GABAergic neurons have replaced those GABAergic inhibitory interneurons that have been shown

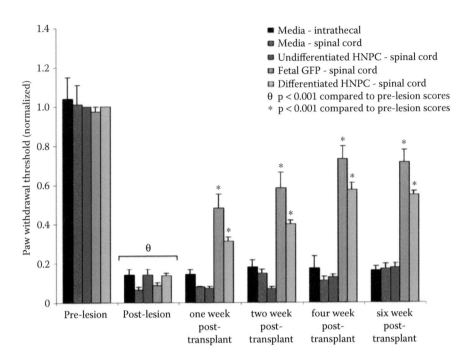

FIGURE 5.16 Behavioral improvement following intraspinal transplantation of predifferentiated bioreactor-expanded hNPCs or GFP mouse striatal primordia cells (Normalized Left Hind Paw Withdrawal Thresholds Level).

Behavioral assessment was conducted before and after lesioning of the spinal nerve of rats as wells as following transplantation of (1) fetal GFP mouse striatal primordia cells, (2) predifferentiated bioreactor-expanded hNPCs, (3) undifferentiated bioreactor-expanded hNPCs, and (4) only cell culture medium. Behavioral assessment was conducted using von Ferry filaments applied to the plantar aspect of both left and right hind paw. Shown are the normalized threshold levels of left hind paw compared to the thresholds level of right hind paw. * represents a statistically significant difference at p<0.001 compared to post-lesion paw withdrawal scores, which are expressed as the mean score for the left hind paw normalized for each animal's right hind paw. θ represents a statistically significant difference at p<0.001 compared to pre-lesion paw withdrawal scores. (From Mukhida, K. et al.: Spinal GABAergic transplants attenuate mechanical allodynia in a rat model of neuropathic pain. *Stem Cells.* 2007. 25(11). 2874–2885 Copyright Wiley-VCH Verlag GmbH & Co. KGaA. Reproduced with permission.)

to be permanently lost in the ipsilateral spinal dorsal horn laminae I to III after spinal nerve injury. These GABAergic cells have been shown to play a significant role in reducing the hyperexcitability developed in dorsal horn projection neurons following injury to the peripheral or spinal nerves. Therefore, transplantation of these GABAergic phenotypes may have restored the function of lost GABAergic neurons after injury.

5.4.3.4   Neuropathic Pain Model—Conclusions

Transplantation studies in an animal model of neuropathic pain showed that intraspinal transplantation of GABAergic neurons derived from *in vitro* differentiation of bioreactor-expanded hNPCs or fetal GFP mouse could give rise to significant behavioral improvement. It was suggested that maintaining GABAergic phenotype is essential to reduce mechanical allodynia. The study also showed that GABAergic phenotype of the predifferentiated hNPCs was maintained *in vivo*. No evidence of tumor formation was observed following transplantation of differentiated or undifferentiated hNPCs into the spinal cord of ligated animals. Together, this study showed that production of hNPCs in standard suspension bioreactors, and their *in vitro* differentiation into GABAergic neurons could be a viable option to produce a reliable source of clinical grade cells. These results have a high potential of translation into the clinical realm as the initial aim would be to control severe intractable pain syndromes such as cancer pain that can produce unbearable suffering.

SUMMARY

The transplantation of *in vitro* expanded human neural precursor cells (hNPCs) and their differentiated progeny (e.g. neurons) represents a potential new treatment alternative for individuals suffering from incurable disorders of the central nervous system (CNS) such as Parkinson's disease, Huntington's disease and chronic neuropathic pain. However, for CNS-targeted cell restorative therapies to have widespread therapeutic significance, it will be necessary to generate large quantities of clinical grade hNPCs and their differentiated progeny using standardized manufacturing processes. This chapter focuses on the development of robust protocols for the large-scale expansion and differentiation of therapeutically effective hNPC populations for the treatment of neurodegenerative disorders. It describes proper hNPC handling and characterization methods, the large-scale expansion of hNPC populations under controlled conditions in suspension bioreactors, and animal model studies where differentiated cells derived from bioreactor-generated hNPCs were transplanted to effectively treat Huntington's disease and spinal cord pain (allodynia).

REFERENCES

Anderson L, Burnstein RM, He X, Luce R, Furlong R, Foltynie T, Sykacek P, Menon DK, Caldwell MA. 2007. Gene expression changes in long term expanded human neural progenitor cells passaged by chopping lead to loss of neurogenic potential *in vivo*. *Exp Neurol* 204(2):512–524.

Anderson L, Caldwell MA. 2007. Human neural progenitor cell transplants into the subthalamic nucleus lead to functional recovery in a rat model of Parkinson's disease. *Neurobiol Dis* 27(2):133–140.

Aunins JG, Woodson BA, Hale LC, Wang DIC. 1989. Effects of paddle impeller geometry on power input and mass transfer in small-scale animal cell culture vessels. *Biotechnol Bioeng* 43:1127–1132.

Bachoud-Levi AC, Gaura V, Brugieres P, Lefaucheur JP, Boisse MF, Maison P et al. 2006. Effect of fetal neural transplants in patients with Huntington's disease 6 years after surgery: A long-term follow-up study. *Lancet Neurol* 5(4):303–309.

Baghbaderani BA, Mukhida K, Sen A, Hong M, Mendez I, Behie LA. 2008. Expansion of human neural precursor cells in large-scale bioreactors for the treatment of neurodegenerative disorders. *Biotechnol Prog* 24(4):859–870.

Bridges D, Thompson SWN, Rice ASC. 2001. Mechanisms of neuropathic pain. *British Journal of Anaesthesia* 87(1):12–26.

Buc-Caron M-H. 1995. Neuroepithelial progenitor cells explanted from human fetal brain proliferate and differentiatein vitro. *Neurobiol Dis* 2(1):37–47.

Butler M. 2004. *Animal Cell Culture & Technology*. BIOS Scientific Publishers, London, UK, 243 p.

Carpenter MK, Cui X, Hu Z-Y, Jackson J, Sherman S, Seiger A, Wahlberg LU. 1999. *In vitro* expansion of a multipotent population of human neural progenitor cells. *Exp Neurol* 158(2):265–278.

Cherry RS, Kwon K-Y. 1990. Transient shear stresses on a suspension cell in turbulence. *Biotechnol Bioeng* 36(6):563–571.

Duittoz AH, Hevor T. 2001. Primary culture of neural precursors from the ovine central nervous system (CNS). *Journal of Neuroscience Methods* 107(1–2):131–140.

Dunnett SB, Rosser AE. 2004. Cell therapy in Huntington's disease. *NeuroRx* 1(4):394–405.

Dunnett SB, Rosser AE. 2007. Cell transplantation for Huntington's disease: Should we continue? *Brain Res Bull* 72(2–3):132–147.

Gilbertson JA, Sen A, Behie LA, Kallos MS. 2006. Scaled-up production of mammalian neural precursor cell aggregates in computer-controlled suspension bioreactors. *Biotechnol Bioeng* 94(4):783–792.

Gupta N, Henry RG, Strober J, Kang SM, Lim DA, Bucci M et al. 2012. Neural stem cell engraftment and myelination in the human brain. *Sci Transl Med* 4:155ra137.

Hagan M, Wennersten A, Meijer X, Holmin S, Wahlberg L, Mathiesen T. 2003. Neuroprotection by human neural progenitor cells after experimental contusion in rats. *Neuroscience Letters* 351(3):149–152.

Huntington's Disease Society of America. 2017. Huntington's disease in the United States. http://www.hdac.org/. New York.

Huntington Society of Canada. 2017. Huntington's disease in Canada. http://www.parkinson.org. Kitchener, ON.

Kallos MS, Behie LA. 1999. Inoculation and growth conditions for high-cell-density expansion of mammalian neural stem cells in suspension bioreactors. *Biotechnol Bioeng* 63(4):473–483.

Kallos MS, Behie LA, Vescovi AL. 1999. Extended serial passaging of mammalian neural stem cells in suspension bioreactors. *Biotechnol Bioeng* 65(5):589–599.

Kallos MS, Sen A, Behie LA. 2003. Large-scale expansion of mammalian neural stem cells: A review. *Med Biol Eng Comput* 41(3):271–282.

Kallur T, Darsalia V, Lindvall O, Kokaia Z. 2006. Human fetal cortical and striatal neural stem cells generate region-specific neurons *in vitro* and differentiate extensively to neurons after intrastriatal transplantation in neonatal rats. *J Neurosci Res* 84(8):1630–1644.

Kelly CM, Precious SV, Torres EM, Harrison AW, Williams D, Scherf C et al. 2011. Medical terminations of pregnancy: A viable source of tissue for cell replacement therapy for neurodegenerative disorders. *Cell Trans* 20:503–513.

Kim SH, Chung JM. 1992. An experimental model for peripheral neuropathy produced by segmental spinal nerve ligation in the rat. *Pain* 50(3):355–363.

Lindvall O, Kokaia Z, Martinez-Serrano A. 2004. Stem cell therapy for human neurodegenerative disorders-how to make it work. *Nat Med* 10 Suppl:S42–S50.

Martín-Ibáñeza R, Guardia I, Pardo M, Herranz C, Zietlowe R, Vinhe N, Rossere A, Canals JM. 2017. Insights in spatio-temporal characterization of human fetal neural stem cells. *Exp Neurol* 291:20–35.

McBride JL, Behrstock SP, Chen EY, Jakel RJ, Siegel I, Svendsen CN, Kordower JH, 2004. Human neural stem cell transplants improve motor function in a rat model of Huntington's disease. *J Comp Neurol* 475(2):211–219.

McLeod MC, Kobayashi NR, Sen A, Baghbaderani BA, Sadi D, Ulalia R, Behie LA, Mendez I. 2013. Transplantation of GABAergic cells derived from bioreactor-expanded neural precursor cells restores motor and cognitive behavioral deficits in a rodent model of Huntington's disease. *Cell Trans* 22:2237–2256.

Melrose HL, Lincoln SJ, Tyndall GM, Farrer MJ. 2006. Parkinson's disease: A rethink of rodent models. *Exp Brain Res* 173(2):196–204.

Mendez I, Baker KA, Hong M. 2000a. Simultaneous intrastriatal and intranigral grafting (double grafts) in the rat model of Parkinson's disease. *Brain Res Rev* 32(1):328–339.

Mendez I, Vinuela A, Astradsson A, Mukhida K, Hallett P, Robertson H et al. 2008. Dopamine neurons implanted into people with Parkinson's disease survive without pathology for 14 years. *Nat Med* 14(5):507.

Moreira JL, Aunins JG, Carrondo MJ. 1995. Hydrodynamics effects on BHK cells grown as suspended natural aggregates *Biotechnol Bioeng* 46:351–360.

Mukhida K, Baker KA, Sadi D, Mendez I. 2001. Enhancement of sensorimotor behavioral recovery in hemiparkinsonian rats with intrastriatal, intranigral, and intrasubthalamic nucleus dopaminergic transplants. *J Neurosci* 21(10):3521–3530.

Mukhida K, Hong M, Behie LA, Mendez I. 2006. Co-grafting with bioreactor-expanded human neural precursor cells enhances survival of fetal dopaminergic transplants in Hemiparkinsonian rodents. *Exp Neurol* 198(2):582.

Mukhida K, Mendez I, McLeod M, Kobayashi N, Haughn C, Milne B, Baghbaderani B, Sen A, Behie LA, Hong M. 2007. Spinal GABAergic transplants attenuate mechanical allodynia in a rat model of neuropathic pain. *Stem Cells* 25(11):2874–2885.

Nagata S. 1975. *Mixing: Principles and Applications.* New York: Wiley, 458 p.

Parkinson's Foundation. 2017. Parkinson's disease in the United States of America. http://www.parkinson.org. Miami, FL.

Ostenfeld T, Caldwell MA, Prowse KR, Linskens MH, Jauniaux E, Svendsen CN. 2000. Human neural precursor cells express low levels of telomerase *in vitro* and show diminishing cell proliferation with extensive axonal outgrowth following transplantation. *Exp Neurol* 164(1):215–226.

Ozturk SS, Thrift JC, Blackie JD, Naveh D. 1997. Real-time monitoring and control of glucose and lactate concentrations in a mammalian cell perfusion reactor *Biotechnol Bioeng* 53:372–378.

Pagliuca FW, Millman JR, Gürtler M, Segel M, Van Dervort A, Ryu JH, Peterson QP, Greiner D, Melton DA. (2014) Generation of functional human pancreatic β cells in vitro. *Cell* 159(2):428–439.

Ryu JK, Kim J, Cho SJ, Hatori K, Nagai A, Choi HB, Lee MC, McLarnon JG, Kim SU. 2004. Proactive transplantation of human neural stem cells prevents degeneration of striatal neurons in a rat model of Huntington disease. *Neurobiol Dis* 16(1):68–77.

Sen A. 2003. *Bioreactor Protocols for Neural Stem Cells [PhD].* University of Calgary, Calgary, Canada, 613 p.

Sen A, Behie LA. 1999. The development of a medium for the *in vitro* expansion of mammalian neural stem cells. *Can J Chem Eng* 77(5):963–972.

Sen A, Kallos MS, Behie LA. 2001. Effects of hydrodynamic on extended cultures of mammalian neural stem cell aggregates in suspension culture. *Ind Eng Chem Res* 40:5350–5357.

Sen A, Kallos MS, Behie LA. 2002a. Expansion of mammalian neural stem cells in bioreactors: Effect of power input and medium viscosity. *Brain Res Dev Brain Res* 134(1–2):103–113.

Sen A, Kallos MS, Behie LA. 2002b. Passaging protocols for mammalian neural stem cells in suspension bioreactors. *Biotechnol Prog* 18(2):337–345.

Shuler ML, Kargi F. 2002. *Bioprocess Engineering Basic Concepts.* Prentice Hall PTR, Upper Saddle River, NJ, 553 p.

Storch A, Paul G, Csete M, Boehm BO, Carvey PM, Kupsch A, Schwarz J. 2001. Long-term proliferation and dopaminergic differentiation of human mesencephalic neural precursor cells. *Exp Neurol* 170(2):317–325.

Storch A, Sabolek M, Milosevic J, Schwarz SC, Schwarz J. 2004. Midbrain-derived neural stem cells: From basic science to therapeutic approaches. *Cell Tissue Res* 318(1):15–22.

Storch A, Schwarz J. 2002. Neural stem cells and neurodegeneration. *Curr Opin Investig Drugs* 3(5):774–781.

Studer L, Csete M, Lee SH, Kabbani N, Walikonis J, Wold B, McKay R. 2000. Enhanced proliferation, survival, and dopaminergic differentiation of CNS precursors in lowered oxygen. *J Neurosci* 20(19):7377–7383.

Suzuki M, Wright LS, Marwah P, Lardy HA, Svendsen CN. 2004. Mitotic and neurogenic effects of dehydroepiandrosterone (DHEA) on human neural stem cell cultures derived from the fetal cortex. *PNAS* 101(9):3202–3207.

Svendsen CN, Caldwell MA, Shen J, ter Borg MG, Rosser AE, Tyers P, Karmiol S, Dunnett SB. 1997. Long-term survival of human central nervous system progenitor cells transplanted into a rat model of Parkinson's disease. *Exp Neurol* 148(1):135–146.

Svendsen CN, Clarke DJ, Rosser AE, Dunnett SB. 1996. Survival and differentiation of rat and human epidermal growth factor-responsive precursor cells following grafting into the lesioned adult central nervous system. *Exp Neurol* 137(2):376–388.

Svendsen CN, ter Borg MG, Armstrong RJE, Rosser AE, Chandran S, Ostenfeld T, Caldwell MA. 1998. A new method for the rapid and long term growth of human neural precursor cells. *J Neurosci Methods* 85(2):141–152.

Svrcek WY, Mahoney DP, Young BR. 2000. *A Real-Time Approach to Process Control*. John Wiley & Sons, Chichester, UK, 307 p.

Uchida N, Chen K, Dohse M, Hansen KD, Dean J, Buser JR et al. 2012 Human neural stem cells induce functional myelination in mice with severe bysmyelination. *Sci Trans Med* 4(155):155ra136.

Van Gorp S, Leerink M, Kakinohana O, Platoshyn O, Santucci C, Galik J et al. 2013. Amelioration of motor/sensory dysfunction and spasticity in a rat model of acute lumbar spinal cord injury by human neural stem cell transplantation. *Stem Cell Res Ther* 4:57.

Vescovi AL, Parati EA, Gritti A, Poulin P, Ferrario M, Wanke E et al. 1999. Isolation and cloning of multipotential stem cells from the embryonic human CNS and establishment of transplantable human neural stem cell lines by epigenetic stimulation. *Exp Neurol* 156(1):71–83.

Wright LS, Li J, Caldwell MA, Wallace K, Johnson JA, Svendsen CN. 2003. Gene expression in human neural stem cells: Effects of leukemia inhibitory factor. *J Neurochem* 86(1):179–195.

Yasuhara T, Matsukawa N, Hara K, Yu G, Xu L, Maki M, Kim SU, Borlongan CV. 2006. Transplantation of human neural stem cells exerts neuroprotection in a rat model of Parkinson's Disease. *J Neurosci* 26(48):12497–12511.

Youn BS, Sen A, Behie LA, Girgis-Gabardo A, Hassell JA. 2006. Scale-up of breast cancer stem cell aggregate cultures to suspension bioreactors. *Biotechnol Prog* 22(3):801–810.

Youn BS, Sen A, Kallos MS, Behie LA, Girgis-Gabardo A, Kurpios N, Barcelon M, Hassell JA. 2005. Large-scale expansion of mammary epithelial stem cell aggregates in suspension bioreactors. *Biotechnol Prog* 21(3):984–993.

Zhou F-W, Fortin JM, Chen H-X, Martinez-Diaz H, Chang L-J, Reynolds BA et al. 2015. Functional integration of human neural precursor cells in mouse cortex. *PLoS One* 10(3):e0120281.

Zhu LL, Wu LY, Yew DT, Fan M. 2005. Effects of hypoxia on the proliferation and differentiation of NSCs. *Mol Neurobiol* 31(1–3):231–242.

6 Bioprocessing of Human Stem Cells for Therapeutic Use through Single-Use Bioreactors

Aletta C. Schnitzler, Mark Lalli, Manjula Aysola, Janmeet Anant, and Julie Murrell

CONTENTS

6.1 RISE OF CELL THERAPIES

Treatments for disease-states varying from acute infections to chronic autoimmune disorders and cancer incorporate a correspondingly varied arsenal of therapeutic approaches. These include chemical entities (e.g., small molecules and nucleic acids), larger molecules (e.g., peptides and proteins), organisms (e.g., bacteria and viruses), and cells or tissues such as mesenchymal stem/stromal cells and bone marrow transplants. There are also combination treatments that vary from administration of multiple entities to a patient to true combination therapies, such as antibody drug conjugates and gene modified T cells. The vast majority of approved therapeutics are chemical compounds and recombinant proteins. However, cellular therapies have emerged as a new front runner in the race to remedy disease, even chronic illnesses.

Cell-based therapies are an entirely different class compared to traditional biopharmaceuticals because the cell is the therapeutic product rather than simply how a recombinant protein is produced. Stem cells exert their influence once inside the patient through engraftment for the repair or replacement of diseased tissue, or through secretion of molecules that drive an endogenous response in the patient. The administration of this complex, heterogeneous therapeutic can be influenced by the patient's inherent biology or niche (Augello et al. 2010, Chamberlain et al. 2007).

Cellular therapies can be classified by the relationship of the donor to the patient: autologous or allogeneic. An autologous donor is also the recipient of the treatment, termed a "one-to-one" patient specific therapy. An allogeneic donor provides the tissue from which cells are isolated and typically used to produce the large number of therapeutic cells that will be administered to multiple recipients; thus, considered a "one-to-many" therapy (Figure 6.1). Although "one-to-one" allogeneic approaches are used in organ transplantation, this is less common in cell therapies. The criteria

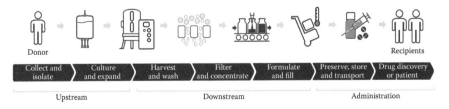

FIGURE 6.1 Summary of manufacturing steps for mesenchymal stem/stromal cells as an example of an allogeneic cell therapy.

that must be considered to meet the needs of manufacturers and patients are distinct for the two classes of cell therapies. Autologous therapies may require expansion of cells prior to administration; thus, the manufacturing process for many patients in parallel is considered "scale-out." The key needs for this approach include automated and closed manufacturing platforms, as well as faster quality control testing for lot release since these cells are often administered to the patient as a fresh substance (i.e., non-cryopreserved). Allogeneic therapies typically require the generation of large numbers of high quality cells in order to prepare ready-to-use doses (often under cryopreservation) for multiple patients, and thus these larger volume lot sizes are considered "scale-up." Vessels that are scalable from approximately 50 L up to 1000 L cell culture volume remain a key need for the expansion of allogeneic cells (Rowley et al. 2012, Simaria et al. 2014). In both cases, clinical grade raw materials, including serum replacement reagents and, in the future, chemically-defined formulations are important for reduction of safety risks associated with cell therapies that undergo minimal downstream processing before introduction into patients.

6.1.1 Mesenchymal Stem/Stromal Cell Focus

A vast array of cell types is under clinical investigation for use as cell therapies, as is apparent by searching the clinicaltrials.gov database that is maintained by the National Library of Medicine at the National Institutes of Health (Health 2017). This database tracks trial design, enrollment, and results from clinical studies conducted in all 50 US states and over 200 countries. Stem cell therapies are a subset of cell therapies and can range from minimally manipulated adipose stromal vascular fraction to scalably-produced induced pluripotent stem cells (iPSC). Embryonic stem cells and iPSCs have the potential to develop into the three germ layers (endoderm, ectoderm, and mesoderm) whereas multipotent stem cells, such as mesenchymal stem/stromal cells (MSCs), have the potential to develop into only a limited number of cell types. MSCs have been the focus of intense study and development over the past decades due to their demonstrated ability to ameliorate disease states that have an inflammation component. The clinical and commercial activities in the cell therapy space include hundreds of global trials and several marketed products for stem cells (Health 2017).

6.1.2 Cell Therapy Manufacturing

Traditional cell culture methods are still commonly utilized to produce stem cells for use in clinical trials. This can include tissue culture flasks and multilayer culture vessels. However, these vessels were not designed to be high-throughput or closed, and thus operate in a very manual-based environment. An operator must take each culture vessel in and out of a biosafety cabinet to perform sampling and medium exchanges, the original cell seeding, the final cell detachment (in the case of anchorage-dependent cells as MSCs), and harvest. Moreover, planar culture vessels limit the ability to monitor cell characteristics during the expansion process. High frequency manipulation and lack of in-process monitoring options contribute to a high-risk process concerning safety. Additionally, it requires a highly skilled workforce. Although these processes are cumbersome and lack scalability, they are still manageable in early clinical phases given the low numbers of patients enrolled in the trials and the high hurdle to change manufacturing processes.

The generation of cells for use in late-stage clinical trials and for approved therapies will require manufacturing platforms that ensure the safety of the cells, improve operator ease of use, and increase lot-to-lot reproducibility in the final product. In the industry, there is a move away from planar culture toward stirred-tank bioreactors (Simaria et al. 2014) that can achieve large scale cell expansion in a single unit operation. Although there is a focus on increasing yields to drive down costs (Rowley et al. 2012), investigations into increasing the potency of the cells produced may be a path to achieving the same result. An ideal treatment would require only the cells with the specific quality attributes identified by a true understanding of the mechanism of action. Implementation of these therapies would depend on the ability to produce cells with these specific quality attributes. The result would be lower batch size requirements and a lower cost of goods. Development of scalable manufacturing systems and identification of the most potent cells should be pursued in parallel to increase the likelihood of cell therapy successes, in terms of both *in vivo* efficacy and economic viability. In this quickly evolving field, manufacturers have the opportunity to develop regulatory-compliant, cost-effective processes for the industrialization of stem cell production.

6.1.2.1 Single-Use Systems

Whereas the end-product of traditional bioreactor systems may be a pharmacological compound, such as a protein or virus that undergoes significant downstream processing, cells generated for stem cell therapies undergo very little downstream processing. Biopharmaceutical downstream processing has for many years served a dual role: purification of proteins and clearance of adventitious agents. With cell therapies gaining a path toward approval, there exists a need to replace the clearance function present in downstream processing of biopharmaceuticals with a safety mechanism that will work with cell therapies. As a result, single-use technologies have been identified as not only a way to increase operator ease of use, but also to replace the adventitious agent clearance functionality. In addition to having lower start-up costs than their stainless-steel counterparts, single-use technologies and products help reduce the likelihood of contamination and the resultant loss of product (Simaria et al. 2014, Lopez et al. 2010). Another advantage of single-use systems compared to traditional reactors is the relative ease of customization. This adaptability allows a wide range of process parameters to be met without requiring a large investment in custom, permanent reactor systems.

Single-use systems for expansion with working volumes as large as 2000 L have been developed and are available for implementation. Single-use technologies have been developed and are in use for medium preparation, cell separation, sterile-to-sterile connections, storage, and transport for traditional biologic manufacturing and can be applied to therapeutic cell production as well. In short, single-use technologies help address the need for closed systems and high-quality materials needed for therapeutic cell production. Nonetheless, there are challenges that remain with a fully single-use approach—for example, ensuring the material components are low in extractables and maintain high physical integrity, avoiding materials derived from animal sources, and meeting cell yield and performance requirements.

6.2 CELL EXPANSION IN SINGLE-USE BIOREACTORS

The scalability of cell expansion for stem cell-based therapy has been a factor limiting its practicality and economics. In traditional reactor engineering, the evolution from a batch-operated reactor to a stirred-tank reactor run continuously has several advantages including homogeneity of reagent distribution, increased reaction rate, and greater ease of sampling. In stem cell manufacturing, several vessels have been utilized (Table 6.1) for the expansion of adherent cells on suspended microcarriers rather than on a planar substrate, which offers the previously mentioned advantages. Stirred systems such as spinner flasks and stirred-tank bioreactors, as well as non-stirred systems, such as oscillating bioreactors, allow expansion of MSCs on microcarriers. The Mobius® 3-L bioreactor, the first single-use stirred-tank reactor reported to expand human MSCs on microcarriers, demonstrated yields of up to 600 million cells with a working volume of 2.4 L (Jing et al. 2013). Collagen-coated microcarriers were used and the cells attached to them were found to remain as undifferentiated stem cells throughout the expansion process.

Timmins et al. expanded the operational space of MSC manufacturing by demonstrating the expansion of placental-origin MSCs in the Wave bioreactor™ on macroporous gelatin microcarriers (Timmins et al. 2012). Such oscillating bioreactors utilize a rocking stage to circulate medium and encourage gas transfer. These wave-mixed systems tune the maximum angle of the rocking stage and the oscillation rate (reviewed in [Eibl and Eibl 2009]), whereas a stirred-tank system modulates the rotational speed of the impeller to control agitation. Hybrid systems, such as the Integrity™ Xpansion™ Multiplate Bioreactor (Pall Corporation), allow adhesion of cells onto the same surface that they would be exposed to within a standard planar flask, but circulate culture medium via forced pumping and allow aeration via diffusion through a central gas exchanger. Although ease of cell sampling is sacrificed in this system, it offers the advantage of a maximum shear stress of 10 mPa which makes it a favorable system for cells, which are sensitive to or intolerant of large hydrodynamic forces (Chisti 2001).

6.2.1 STIRRED-TANK BIOREACTORS FOR ADHERENT CELL CULTURE

In traditional chemical reactors and bioreactors, scale-up is preferred over scale-out in production lines since the reactor volume is proportional to the cubic length of the reactor whereas the reactor footprint scales only with the square length. However, for cells that require adhesion to a surface, a method of suspension is necessary in order to utilize the increased volume present in a bioreactor compared to a planar substrate. Microcarriers, particles of various materials, which may be coated with adhesion assisting proteins, provide a solid interface for cell adhesion while also

TABLE 6.1

Summary of Available Single-Use Cellular Expansion Technologies. Approximate Surface Areas for Suspension Systems Are Based on 15 g/L of a 360 cm²/g Microcarrier

Technology	Volume	Surface Area (cm²)	Example Technologies	Features
Multilayer stacks (10–40 stack layers)	1.4–5.5 L	6,360–25,440	CellSTACK® Cell Culture Chambers (Corning), Nunc Cell Factory™ Systems (Thermo Scientific)	• Single-batch expansion • Small scale • High volume:area ratio • No inline control or monitoring • Cumbersome at larger scale, may require robotic handling
Closed system, multilayer stacks (12–120 layers)	1.3–13 L	6,000–60,000	HYPERFlask® Cell Culture Vessels (Corning)	• Single-batch expansion • Condensed, gas-permeable culture layers • Closed venting/gassing • Lower volume:area ratio than traditional stacks • No inline control or monitoring • Cumbersome at larger scale, may require robotic handling
Spinner flasks	12 mL–3 L	675–2,700	Corning® Disposable Spinner Flasks (Corning)	• Suspension systems offer lower volume:area ratio than planar culture • Suitable for small-scale process development and expansion • No inline control or monitoring

(Continued)

TABLE 6.1 (*Continued*)
Summary of Available Single-Use Cellular Expansion Technologies. Approximate Surface Areas for Suspension Systems Are Based on 15 g/L of a 360 cm²/g Microcarrier

Technology	Volume	Surface Area (cm²)	Example Technologies	Features
Mini-reactor systems	3–250 mL	16–1,350	DASbox®Mini Bioreactor System (Eppendorf), BioLevitator™ 3D Cell Culture System (Hamilton), TAP ambr™ microbioreactor (Sartorius), Micro-24 MicroReactor System (Pall Corporation)	• Suitable for small-scale process development with up to 24 parallel reactors • Controlled gassing • Inline monitoring • BioLevitator™ exclusively used with GEM™ magnetic microcarriers
Benchtop stirred reactors	1–5 L	5,400–27,000	Mobius® (EMD Millipore), CelliGen® BLU (Eppendorf), UniVessel® SU (Sartorius)	• Suitable for small-scale process development and expansion • Controlled gassing • Inline monitoring
Pilot scale, stirred reactors	50–300 L	2.7×10^5–1.62×10^6	Mobius® (EMD Millipore), CelliGen® BLU (Eppendorf), BIOSTAT® STR (Sartorius), Xcellerex™ XDR (GE Healthcare), HyPerforma™ Single-use Bioreactor (Thermo Scientific), Nucleo™ Single-use Bioreactor (Pall Corporation), Allegro™ STR 200 (Pall Corporation)	• Suitable for pilot scale and clinical scale manufacturing • Many systems scalable from lower volume offerings • Controlled gassing • Inline monitoring
Production scale, stirred reactors	500–2000 L	2.7×10^6–1.08×10^7	Mobius® (EMD Millipore), BIOSTAT® STR (Sartorius), Xcellerex™ XDR (GE Healthcare), HyPerforma™ Single-use Bioreactor (Thermo Scientific), Nucleo™ Single-use Bioreactor (Pall Corporation)	• Suitable for clinical scale manufacturing • Many systems scalable from lower volume offerings • Installation of large scale single-use bags is cumbersome • Controlled gassing • Inline monitoring

(*Continued*)

TABLE 6.1 (Continued)

Summary of Available Single-Use Cellular Expansion Technologies. Approximate Surface Areas for Suspension Systems Are Based on 15 g/L of a 360 cm²/g Microcarrier

Technology	Volume	Surface Area (cm²)	Example Technologies	Features
Oscillating motion reactors, surface aeration only	300 mL–500 L	1,620– 2.7×10^6	WAVE bioreactor™ system (GE Healthcare), BIOSTAT® RM (Sartorius), SmartBag™ containers (Finesse), Appliflex™ systems (Applicon), CELL-tainer® (CELLution Biotech/Lonza), XRS-20 Bioreactor System (Pall Life Sciences)	• Suitable for development to clinical scale manufacturing • Many systems scalable from lower volume offerings • No sparging • Inline monitoring
Oscillating motion reactors, sparging	30–1000 L	162,000– 5.4×10^6	BaySHAKE® (Bayer)	• Suitable for development to clinical scale manufacturing • Many systems scalable from lower volume offerings • Gas transfer via sparging • Inline monitoring
Vertical wheel/bubble column	50 mL–500 L	270– 2.7×10^6	Vertical-Wheel™ reactor (PBS), CellMaker PLUS™ system (Cellexus)	• Suitable for development to clinical scale manufacturing • Gas transfer via sparging • Mixing by air- or magnetic-driven wheel • Inline monitoring
Pilot scale, static	19.8 L	1.15×10^5	Integrity™ Xpansion™ Multiplate (Pall Corporation)	• Suitable for development to small scale manufacturing • Growth on planar surface • Gentle media circulation and gas transfer • Requires large incubator for temperature control • Inline monitoring

Source: Schmitzler, A. C. et al., *Biochem. Eng. J.*, 108, 3–13, 2016.

dispersing into the culture medium via mechanical agitation. In this way, the homo-geneity of the reactor working volume is maintained so concentration gradients in nutrients, cell waste, and dissolved gas do not form (Sun et al. 2011). Additionally, controlled parameters such as temperature, pH, and dissolved oxygen (DO) can be measured accurately.

6.2.1.1 Microcarrier Materials and Surface Chemistries

Whereas many microcarriers are spherical, alternative geometries such as cylin-ders (DE53, Whatman) and disks (Fibra-Cel®, Eppendorf) also exist. Manufacturers utilize a variety of unreactive, compatible materials in the production of microcarri-ers including glass, polystyrene, alginate, cellulose, dextran, and gelatin. Even mag-netic microcarriers, such as GEM™, which is designed for use with the Hamilton small-scale BioLevitator™ system, have been developed that offer a novel, low-shear suspension method and potentially more efficient cell-microcarrier separation.

The surface properties of the microcarriers are considered critical since the solid interface contributes to cell adhesion, proliferation, and plays a role in differentia-tion of stem cells. Many of the materials used in microcarrier manufacturing can be further tailored to the unique needs of specific cells via functionalization to present electrostatic charges, short-chain peptides or extracellular matrix (ECM) proteins on their surface. ECM proteins, such as fibronectin, laminin, and collagen, can be coated to the microcarrier surface to facilitate cell attachment. Collagen is a major ECM protein secreted by MSCs themselves and is often utilized in microcarrier systems, though most commercially-available collagen products remain of animal origin. In order to reduce batch-to-batch variability, animal-free components are preferred in stem cell culture medium. Some substitutes, which are sourced animal-free include laminin, poly-d-lysine, fibronectin (Qian and Saltzman 2004), and derivatives such as superfibronectin (Morla et al. 1994).

Media are often supplemented with serum, particularly fetal bovine serum, which includes an undefined cocktail of various proteins. Some of these, especially ECM proteins, can nonspecifically bind to the surface of the microcarriers and encourage cell adhesion. However, certain proteins preferentially bind with different surfaces. Microcarriers carrying a positive charge attract albumin, and microcarriers that are coated with gelatin will bind predominantly with fibronectin (Mukhopadhyay et al. Talwar 1993). In this way, specific pairings can be selected during process optimiza-tion. Although culture medium optimization is cell type-dependent, media should also be matched properly with tissue culture-treated surfaces to prevent any loss in differentiation potential relative to standard collagen-coated surfaces (Mauney et al. 2004) (Mauney et al. 2005).

Recent advances in material science and engineering have resulted in novel, syn-thetic materials, which support expansion of MSCs even in the absence of serum (Dolley-Sonneville et al. 2013). These studies demonstrate that surface properties including stiffness, curvature, and nanoscale topography contribute to the compat-ibility of the surface for successful cell proliferation as well as induction/prevention of differentiation (Mauney et al. 2005, Soininen et al. 2014, Zhao et al. 2014). Stiff gels have been shown to support differentiation of MSCs into an osteoblastic pheno-type, whereas soft gels preferentially lead to differentiation into neuronal-like cells

(Engler et al. 2006). Likewise, the nanoscale topography of a surface has been shown to influence culture and maintenance of MSCs. McMurray et al. demonstrated that pits 120 nm in size either prevent differentiation or induce osteogenesis depending on the regularity of their orientation (McMurray et al. 2011). So far, only the surface properties of planar substrates have been correlated with the growth and differentiation profiles of MSCs, but microcarrier systems are likely to share some similarities in the relationship between surface properties and culture of MSCs. Therefore, investigating and tailoring the surface properties, both chemical and mechanical, of microcarriers is a potentially exciting avenue for optimizing reactor-based human MSC culture.

6.2.2 FLUID DYNAMICS IN MICROCARRIER-BASED SYSTEMS

In standard bioreactor systems, the pH, DO, temperature, and agitation rate are monitored and controlled in order to externally maintain homeostasis for the expanding cells. In traditional chemical stirred-tank reactor systems, the agitation rate is sufficient to maintain homogeneity of the vessel contents. However, where microcarriers are involved, the vessel contains a solid-liquid dispersion rather than a chemical solution. As a result, the minimum agitation rate is governed by the sedimentation rate of the microcarriers. This operational space is also limited on the upper end by the limit of shear tolerance on cultured cells. Hydrodynamic shear has been observed to induce cell lysis, decrease biopharmaceutical production (Chisti 2001), and trigger MSC differentiation (Schätti et al. 2011). Moreover, the shear sensitivity of cells attached to microcarriers may be increased (Cherry and Papoutsakis 1988, Chen et al. 2013).

Several empirical studies as well as theoretical extrapolations have been completed to define the bounds on successful agitation strategies. To prevent hydrodynamic shear, the minimum agitation rate necessary to maintain homogenous dispersion of microcarriers is favored. However, the agitation rate also affects bulk mixing in solution as well as oxygen dissolution and carbon dioxide evolution, with higher rates favorable. The correct balance between maintaining vessel homogeneity and protecting the cultured cells against shear often proves challenging. Figure 6.2 summarizes factors that influence the hydrodynamic environment of microcarrier-attached cultured cells (Schnitzler et al. 2016).

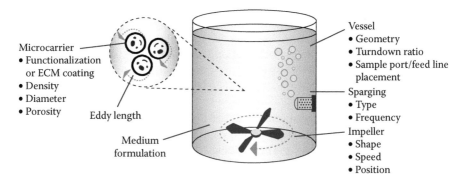

FIGURE 6.2 Considerations for microcarrier-based cell expansion in stirred-tank bioreactors.

6.2.2.1 Defining Minimal Agitation Requirements

Determination of the appropriate agitation requirement for microcarrier suspension in human MSC culture is critical. Cell growth has been demonstrably poor under conditions with non-homogenous microcarrier suspensions (Clark and Hirtenstein 1981). The fluid mechanics of stirred-tank reactors, including mixing, have been studied quite extensively. However, the mechanics of dispersion, rather than dissolution, of solid particles has been studied less extensively than liquid-liquid or gas-liquid interactions (Chapman et al. 1983). In 1958, Zwietering first developed an empirical correlation of microcarrier suspension (Zwietering 1958) and, despite the passage of nearly 60 years, is still the most widely-used correlation between reactor conditions and microcarrier suspension (Chapman et al. 1983) (Ibrahim and Nienow 2004). The Zwietering correlation is used to calculate the minimum agitation rate sufficient to maintain suspension of microcarriers such that no microcarrier remains out of solution for more than one to two seconds (Zwietering 1958). This term, N_{js}, is regarded as the lower limit for bioreactor operation with microcarriers, despite non-homogenous dispersal of microcarriers in solution. Indeed, at N_{js}, microcarrier distribution is a function of height and approaches zero at the liquid surface. The distribution of microcarriers only approaches homogeneity at impeller speeds greater than N_{js} and, at agitation speeds lower than N_{js}, microcarriers sediment from the suspension. Sedimentation not only blocks portions of the surface from interacting with the culture medium, but also allows cells to grow across multiple microcarriers and form bridges (Clark and Hirtenstein 1981). Since cell growth increases the density of microcarriers throughout culture, sedimentation and bridging become more prevalent toward the end of culture (Clark and Hirtenstein 1981).

The correlation provided by Zwietering is dependent on factors related to vessel geometry, medium properties, and the material properties and amount of microcarriers used, shown here in Equation (6.1) (Zwietering 1958).

$$N_{js} = Sv^{0.1} \left[\frac{g(\rho_s - \rho_l)}{\rho_l} \right]^{0.45} X^{0.13} d_p^{0.2} D^{-0.85} \qquad (6.1)$$

where:
  S is the Zwietering N_{js} constant (function of impeller and tank geometry)
  v is the kinematic viscosity (m^2/s)
  g is the gravitational constant (m/s^2)
  ρ_s is the volumetric mass density of the solid (kg/m^3)
  ρ_l is the volumetric mass density of the liquid (kg/m^3)
  X is the solids loading ((kg solids/kg liquid) $\times$ 100)
  d_p is the particle diameter (m)
  D is the impeller diameter (m)

To calibrate a new reactor geometry, N_{js} can be measured empirically and used to calculate the Zwietering constant, S. In the absence of detailed information regarding medium or microcarrier properties, Equation 6.1 can be reduced rationally to predict the fold change in N_{js} with respect to a known change of one or more factors. For example, Equation (6.2) details the predicted change in N_{js} corresponding to a change in the microcarrier concentration used.

$$\frac{N_{js2} = Sv^{0.1}\left[\dfrac{g(\rho_s - \rho_l)}{\rho_l}\right]^{0.45} X_2^{0.13} d_p^{0.2} D^{-0.85}}{N_{js1} = Sv^{0.1}\left[\dfrac{g(\rho_s - \rho_l)}{\rho_l}\right]^{0.45} X_1^{0.13} d_p^{0.2} D^{-0.85}} \rightarrow \begin{array}{l} N_{js2} = X_2^{0.13} \\ N_{js1} = X_1^{0.13} \end{array} \rightarrow N_{js2}$$

$$= N_{js1} \cdot \left[\frac{X_2}{X_1}\right]^{0.13} \tag{6.2}$$

6.2.2.2 Defining Maximum Agitation Limits

Operation above N_{js} promotes homogeneity of nutrients and overall cell culture conditions. Therefore, identifying the upper limit on agitation rate is also important to define the operational space of the bioreactor. In general, the upper limit is dictated by the shear sensitivity of the cells in culture that, in microcarrier-based systems, is more restrictive compared to systems in which cells are freely suspended (Wu et al. 1998). Within bioreactors, the length scale of turbulent eddies often ranges in size between that of the cells and of the microcarriers. Determining a detailed fluid mechanic model within bioreactor culture is computationally difficult (Aunins et al. 1993). So, the cellular damage associated with exposure to these eddies is described by the Kolmogorov turbulence model from 1941 (Equation 6.3). The high energy dissipation associated with turbulent eddies generates local shear-stress sufficient to damage cells and, in some cases, overcome cellular adhesion to the surface (Croughanet et al. 1989).

$$\lambda_k = \left(\frac{v^3}{\varepsilon_T}\right)^{0.25} \qquad\qquad \varepsilon_T = \frac{N_p N^3 D^5}{V} \tag{6.3}$$

where:
λ_k is the Kolmogorov eddy length (m)
v is the liquid kinematic viscosity (m²/s)
ε_T is the energy dissipation (W/kg)
N_p is the impeller power number (dimensionless)
N is the impeller rotational speed (rps)
D is the impeller diameter (m)
V is the liquid working volume (m³)

Cellular damage occurs maximally when the Kolmogorov eddy length is on the same order of magnitude of that of the suspended particle (Papoutsakis 1991, King and Miller 2007). This corresponds to the length scale of a cell or a microcarrier in free-suspension or microcarrier-adherent suspension, respectively. As shown in Equation (6.3), the Kolmogorov eddy length decreases with increasing impeller speed. Therefore, cells adhered to microcarriers experience shear-related damage at lower agitations than freely-suspended cells of the same type (King and Miller 2007). As a result, the Kolmogorov turbulence model helps determine the upper limit of agitation (Kolmogorov 1941). Growth of MSCs have been demonstrated to be unaffected when $\lambda_k \geq 2/3$ of the diameter of the microcarriers (Croughan et al. 1987, Hewitt et al. 2011).

However, cell damage is just one consideration when culturing human MSCs. The influence of shear on stem cell differentiation (Yeatts et al. 2013) as well as determination of stem cell fate (Kreke et al. 2005, 2008) are additional factors that complicate reactor operation in comparison to non-stem adherent cell lines, such as Vero and MDCK. Maintenance of key features of MSC, including differentiation potential, must be considered equally as important as supporting cell growth an expansion in these stirred-tank bioreactor systems for MSCs.

6.2.2.3   Controlling Dissolved Oxygen

As with other mammalian cells, maintaining an appropriate dissolved oxygen (DO) tension is critical when culturing human stem cells. It is standard practice to maintain culture at normoxic conditions (21% O_2). However, this concentration may be significantly higher than that required by MSCs since atmospheric conditions are not observed *in vivo* within the tissues that MSCs are derived from. For example, oxygen tension varies from 1% to 7% and 10% to 15% within the bone marrow and adipose tissue, respectively (Chow et al. 2001), (Bizzarri et al. 2006, Harrison et al. 2002). Hypoxic conditions may favor culture of MSCs. Indeed, by studying various hypoxic conditions of umbilical cord-derived MSCs, Lavrentieva et al. found increased self-renewal without sacrificing differentiation potential compared to normoxic conditions (Lavrentieva et al. 2010). The same group went on to link an over-expression of growth factors and growth factor receptors under hypoxic conditions while expression of differentiation factors was unaffected (Lonne et al. 2013). The increased rate of self-renewal was attributed to the over-expression of these cytokines. The interaction between oxygen tension and glucose levels and their cumulative effect on MSC culture have been reviewed by Haque et al. 2013 and by Sart et al. 2014. Their reviews indicated that various combinations of glucose and oxygen levels can induce or prevent differentiation into multiple phenotypes. Maintenance of oxygenation, at whichever desired level, is dependent on DO control through several methods, namely overlay, open pipe sparging, and microsparging. Oxygen dissolution via overlay within the reactor headspace produces the slowest mass transfer rates although adds no mechanical stress. It suits low-density cultures that consume oxygen less quickly than it can be replaced via interfacial diffusion alone. Of the sparging methods, microsparging is most effective due to the high surface area bubbles generated whereas open pipe sparging more effectively strips gas from the culture medium. In either method, the introduction of bubbles contributes to additional shear forces and foaming, both of which unfavorably affect cell culture. Extending the operation of human MSC culture into hypoxic conditions may contribute to solving certain challenges, such as slow cell growth rates, phenotypic instability, and low post-transplant engraftment rates, in the current field of MSC-based regenerative medicine.

6.2.3 Cell Harvest Approaches

6.2.3.1 Cell Detachment from Microcarriers

An important factor that distinguishes cell therapies from other biopharmaceuticals is that the cell itself is the final therapeutic product, rather than a vector by which to manufacture a drug substance. This creates a major point of divergence between downstream approaches for compounds such as monoclonal antibodies and cellular therapeutics. Although the use of microcarriers for anchorage-dependent cells, including Chinese Hamster Ovary for antibody manufacturing or VERO for viral vector production has been used routinely, these processes do not require detachment of cells at the point of final harvest. For microcarrier-based processes, the specific cell type, medium formulation and culture format all interact in how cell attachment, growth and detachment occur. A wide variety of microcarrier-based chemistries and functionalizations, including charge or proteinaceous coating, are available that will impact attachment and detachment of cells. For example, CellBind® surfaces (Corning) are modified with increased oxygen-containing functional groups to render the surface more hydrophilic and thus improve cellular attachment. Once attached, the cells themselves will secrete their own extracellular matrix (ECM) that further facilitates adherence during culture and will impact the detachment process.

In planar culture, typical cell detachment protocols include the use of enzymes such as trypsin and the chelating agent EDTA (Giard et al. 1986). Stirred-tanks enable the application of mild hydrodynamic forces to promote cell detachment, which may decrease the requirement for enzymes that can alter cell surface properties (Nienow et al. 2014). Recently, a thermoresponsive hydrogel surface material was described that sustained enzymatic-free passaging of adipose-derived MSCs with retention of key cell characteristics (Duffy et al. 2014). Although commercially-available options are still limited, the use of biodegradable, dissolvable, or stimulus-responsive materials as microcarriers shows promise for enzymatic-free cell detachment that could ease current challenges in cell-microcarrier separation (Yang et al. 2010). Because of the tight interplay between cell type, culture medium formulation, and attachment surface chemistry, downstream harvest conditions should be considered early when optimizing upstream parameters.

6.2.3.2 Separation of Cells from Microcarriers

Anchorage-dependent cells cultivated in suspension systems must be physically separated from microcarriers after the detachment step. Microcarriers are generally ten times larger than cells (~150–200 μm compared to 15–20 μm, respectively) and can be filtered by size exclusion. Simple woven materials such as nylon filters are available for small scale, open processes and are widely used in research and

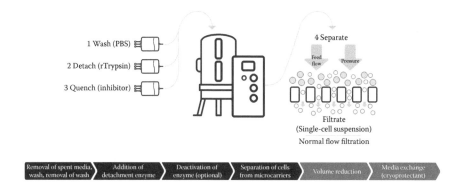

FIGURE 6.3 Detachment and separation of cells from microcarriers at the end of the expansion process.

early development (Goh et al. 2013, Frauenschuh et al. 2007, Weber et al. 2007). Larger scale processes for cell manufacturing require single-unit operation assemblies that can be integrated into closed processes. For example, after detachment of cells from microcarriers within a bioreactor, the separation device should be seamlessly attached to the bioreactor either through sterile welding or a sterile connecting device such as the Lynx® line of sterile connectors by EMD Millipore. This allows for the microcarriers to be held back from the cells as they are harvested from reactor (Figure 6.3).

Simple and cost-effective options at moderate scales include normal flow (NF) filtration with larger nylon mesh filters (Harvestainer™, ThermoFisher Scientific) or capsule filters (Opticap® Polygard®-CR 100, EMD Millipore). At larger scales, tangential flow filtration (Cunha et al. 2015), available in hollow fiber or flat sheet formats, is a possible way to keep microcarriers from fouling a mesh screen, though scales above 200 L are not currently relevant to cell therapy or model cell applications. Moreover, many of these technologies are still not available in pre-sterilized formats and need to be autoclaved prior to use, thus increasing the burden to the end-user for additional process and validation efforts. Other cell types relevant for allogeneic uses such as iPSC, can be grown on microcarriers or as aggregates. In the latter case, although cell-microcarrier separation is not required, the removal of large cell aggregates will be necessary during scale-up as well as harvest. A simple normal flow filtration device can address both applications.

6.2.3.3 Concentration of Cells to Meet Therapy Dosing Requirements

Regardless of culture surface or vessel, most therapies will require a means by which to exchange buffer for final formulation and to concentrate the cells to a small volume. Cell doses as high as 10 million cells/kg, equating to hundreds of millions of cells per dose, have been reported (Maziarz et al. 2015). Although larger dose volumes may be suitable for intravenous infusions, more concentrated doses at smaller volumes may be required for local administration of therapeutic cells. For example, to accommodate a 5 or 10 mL injection, cell concentrations of 20–50 million cells/mL

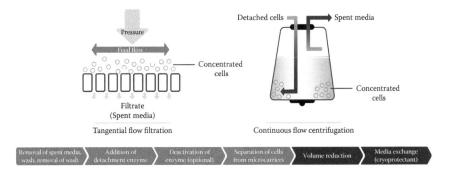

FIGURE 6.4   Cell concentration and medium exchange by tangential flow filtration of continuous flow centrifugation.

may be needed. In open systems, standard centrifuges may be operated under a biosafety cabinet. However, this approach becomes un-manageable when processing more than a few liters of material. Alternatively, a continuous flow centrifugation supports a closed, single-use process (Figure 6.4). Continuous flow centrifuges (kSep®, Sartorius Stedim) enable process volumes up to 6000L and can sediment up to 1.2×10^{12} cells using single-use contact surfaces. Although greatly dependent on the cell type and medium formulation, volume reduction (fold concentration) of 20–100 fold may be possible for cells expanded in a 200 L bioreactor utilizing this approach.

Tangential flow filtration is an alternative technology to centrifugation that can be used to reduce volume and concrete cell therapy products (Figure 6.4) (Hassan et al. 2015). The advantage to TFF lies in the orientation of the fluid flow to pressure across the membrane. Whereas in normal flow filtration with both flow and pressure in the same direction across the membrane, the tangential flow of liquid in TFF devices helps reduce membrane fouling by cells as they are concentrated. Flat sheet TFF products such as the Pellicon® or Prostack™ cassettes from EMD Millipore utilize large flow paths for the cells to flow through while permeate is removed, though pre-sterilization is a hurdle that needs to be overcome by vendors for these devices to best serve the cell therapy industry. TFF technology also is available as hollow fiber membrane and is currently available in pre-sterilized options than can concentrate the cells after they have been separated from the microcarriers. Regardless of the approach, the concentration technology must be gentle enough on the cells to maintain cell viability and function in a short processing window, while also minimizing yield losses.

6.3   CASE STUDY – 50 L EXPANSION OF MESENCHYMAL STEM/STROMAL CELLS

The adoption of both autologous and allogeneic cell-based therapies in the field of personalized medicine drives the need for efficient and reliable expansion of human MSCs at scale. This case study details the process development of determining the bioreactor parameters optimal for expanding MSCs in a 50 L single-use stirred-tank bioreactor (Lawson et al. 2017). The parameters investigated include the agitation strategy, DO and pH control settings, as well as culture medium formulation. A 3 L system was used to evaluate scalable parameters such as pH and DO whereas both theoretical modeling and empirical testing were used to determine the agitation strategy at scale. The results of this case study outline a useful parameter space for the expansion of human MSCs.

6.3.1 DETERMINATION OF PROCESS PARAMETERS

The acceptable agitation range was defined by using the Zwietering correlation to determine the minimum agitation (see Equation 6.1). To determine N_{js} within the Mobius® 50 L Bioreactor, collagen-coated microcarriers (ρ_s = 1.026 g/cm^3, d_p = 170 μm) were suspended at a concentration of 7.0 g/L in simulated cell culture medium (1 x PBS, 0.2% Pluronic® F-68) (ρ_l = 1.004 g/cm^3 and v = 1.24 × 10^{-6} m^2/s). The profile of microcarrier concentration as a function of height was determined. N_{js} corresponds to the agitation rate at which the relative difference in microcarrier suspension between the top and bottom is 20%. In this reactor system, N_{js} was determined to be 67 rpm. For a microcarrier concentration of 15 g/L, N_{js} was then calculated to be 75 rpm.

Damage to anchorage-dependent cells is not observed if the Kolmogorov eddy length is greater than 2/3 the diameter of the suspended particle (Hewitt et al. 2011, Croughan et al. 1987). Although N_{js} does not depend on reactor volume, maximum agitation does (see Equation 6). Solving Equations 6.1 and 6.3 over the operational volume of the reactor produced an operating space for utilizing 15 g/L of microcarriers within the 50 L Mobius® Bioreactor. However, the growth of cells modulates particle size and density over time, making the agitation requirements a function of cell growth.

Initial adhesion is also influenced by agitation and must be considered. Whereas some sources indicate that shear forces unfavorably impact the adhesion of human MSC to microcarriers (Ferrari et al. 2012), the overall interaction at the cell-microcarrier interface is also governed by the microcarrier surface chemistry and morphology. In this case study, cell adhesion was promoted by maintaining a sub-N_{js} agitation rate of 64 rpm for four hours in an initial volume of 20 L, after which agitation was increased to N_{js}. 64 rpm is sufficient to allow pH, DO, and temperature control while generating weaker hydrodynamic forces on the cells. At the first volume addition, in which 20 L of medium maintaining a 15 g/L microcarrier concentration are added, agitation was increased to 85 rpm to compensate for the change in size of the combined microcarrier-cell particle. A final volume addition of 10 L brings the working volume to the 50 L limit and the agitation rate can be increased if needed.

6.3.2 PILOT SCALE CELL EXPANSION

The theoretical operating window was tested with the operation of a Mobius® 50 L bioreactor to expand MSCs. To evaluate various pH and DO control strategies, and initial experiment was performed using a simple control strategy. Air supplemented with 5% CO_2 provided medium oxygenation via overlay in the reactor headspace at a constant flow rate. When glucose levels dropped below 0.5 g/L, feeds were initiated which occurred on days 7 and 9 by the addition of 20 and 10 L of fresh medium with microcarriers, respectively. After 11 days of culture, the reactor yielded 5 × 10^9 cells representing four cumulative population doublings (cPD) of the 3×10^8 cells that were initially seeded in the reactor. Over this time, pH decreased from 7.5 to below 7.2 due to lactic acid production which was not neutralized since the control strategy did not implement base addition. DO levels decreased from 100% to 50% due to oxygen consumption of the increased cell population.

To address the lack of information regarding pH and DO settings optimal for MSC cultivation, the 3-L bioreactor system was chosen as a scale-down model for microcarrier based culture. Bioreactors with pH set-points of 7.2, 7.4, 7.6, or 7.8 were operated over a 10-day culture. Glucose was monitored, and the reactors required feed on days 5 and 7 to prevent glucose from dropping to concentrations below 1 g/L. During pH optimization, the DO set-point was held constant at 80% (relative to operation under ambient air). Given these conditions, the highest cell yields were observed at pH 7.4 and 7.6, while the extremities of the range showed poorer growth.

A similar process was utilized in determining the optimal DO set-point. pH 7.5 was selected moving forward based on the previous results. Bioreactors were operated with DO set points at 20%, 50%, and 80%. Each of the three conditions was comparable in cell growth although microcarrier aggregation was visually observed to be lower at higher DO levels. A DO set-point of 80% was selected for future operation since it used less O_2 without sacrificing cell growth.

Lastly, the culture medium itself was assessed. The initial run utilized Dulbecco's Modified Eagle's Medium, low-glucose (DMEM)/10% fetal bovine serum (FBS). However, there is a growing body of data in favor of human platelet lysate (PL) supplemented Minimum Essential Medium Alpha Modification (αMEM), also with low-glucose (Chen et al. 2015, Doucet et al. 2005). Although still an undefined medium, replacing FBS with PL removes a xenogenic component and its associated concerns. Previous work indicated that αMEM supplemented with 5% PL performed more favorably than 10% FBS. In 3-L bioreactors, the four combinations of basal media and supplement (DMEM/FBS, DMEM/PL, αMEM/FBS, and αMEM/PL) were compared and it was determined that αMEM/PL was optimal.

6.3.3   RESULTS AND DISCUSSION

Two 50-L bioreactor cycles were performed using the parameters optimized in the 3-L system. In these cycles, pH was controlled at 7.45 with a 0.05 dead band. DO was maintained between 75% and 85%. Either αMEM supplemented with 5% PL, or DMEM supplemented with 10% FBS were used as culture media. Cell yield under the improved reactor conditions improved over two-fold to produce 1×10^{10} cells at day 11. Over this time frame, 5.5 cPD were observed, an increase from 4 in the initial run. Unlike the initial run, sodium bicarbonate was used to increase pH during pH control. Additionally, pure oxygen was mixed with air in the overlay gassing in a cascading control scheme to maintain DO levels.

At four hours, only 13% of cells had attached to microcarriers whereas nearly all cells were attached after 24 hours. Calcein AM staining confirmed the increase in the number of cells attached on microcarriers between 4 and 24 hours. Additionally, nuclear staining via DAPI revealed bare microcarriers at day 3 but almost every microcarrier had adherent cells by day 10. Because 60% of the total number of microcarriers within the bioreactor was added in volume additions post-inoculation, the ability of MSCs to transfer between microcarriers in this system was verified.

It is critical to verify the phenotype and functionality of the expanded cells if their intended use is therapeutic. MSC phenotype was analyzed by flow cytometry to detect the expression of positive surface markers (CD105, CD90, CD73) as well as negative surface markers (HLA-DR, CD19, CD34, CD11b, CD79a, CD45, CD14). The expression profile from cells expanded in the improved bioreactor run agreed with those outlined by the International Society for Cellular Therapy (Dominici et al. 2006). The multipotency of cells post-harvest was confirmed by inducing adipogenesis, chondrogenesis, and osteogenesis and staining for lipid vacuoles, glycoconjugates, and calcium deposition, respectively.

The approach outlined in this case study illustrates the scalability of parameters, such as microcarrier suspension, optimal DO and pH, as well as culture medium that can be investigated at small-scale. The process development outlined herein culminated in a cell yield more than double that of the initial, un-optimized run whilst preserving stem cell characteristics including differentiation potential, surface marker expression profiles, and immunomodulation capacity. Importantly, the study demonstrates scalability of performance between the smaller-scale bioreactors and the larger capacity 50-L system (Figure 6.5), relevant to the manufacturing of allogeneic cell therapy and regenerative medicine applications.

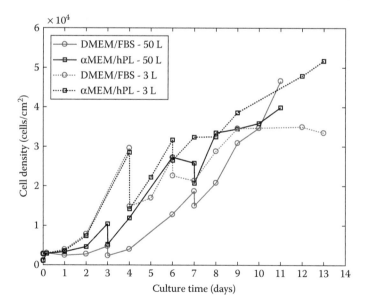

FIGURE 6.5 MSCs derived from bone marrow can be expanded in 3 and 50 L stirred-tank bioreactors with similar cell densities (as cells per cm^2). Extra surface area was added on feed days by increasing the volume of medium and microcarriers (days 4 and 6 for the 3 L process; days 3 and 7 for the 50 L process). The expansion of cells onto newly provided microcarriers was apparent by microscopy and the increasing cell growth. (From Lawson, T. et al., *Biochem. Eng. J.,* 120, 49–62, 2017.)

6.4 REGULATORY PERSPECTIVES

This section offers insight into current and expected industry approaches and regulatory requirements for single-use bioreactor-based manufacturing processes for cell therapy, such as more detailed extractables and material qualification.

6.4.1 CELL THERAPIES AND GOOD MANUFACTURING PRACTICES

The manufacture of cell therapy products should comply with the principles of good manufacturing practices (GMP). In the United States, these regulations are detailed in Title 21 of the Code of Federal Regulations (CFR) in several sections (21CFR210, 211, 610, and 820), including the use of human tissue and cell products (21CFR1271). In Europe, this is covered in Commission Directive 2003/94/EC of 8 October 2003. The FDA, European Medicines Agency (EMA), and *Medicines and Healthcare Products Regulatory Agency* (MHRA, UK) have all published similar guidelines. Both the EMA and MHRA consider cell therapy products to be advanced therapy medicinal products (ATMPs). Additional guidance for ATMPs are found in Regulation (EC) No. 1394/2007.

All regulations cover principles and guidelines of GMP with respect to medicinal products for human use and investigational medicinal products for human use. These regulations must be adapted based on the characteristics of specific cell therapy products and associated manufacturing processes.

It is important to understand these regulations early in the product development phase in order to ensure that compliance can be achieved, and potential challenges can be addressed prior to production.

Considerations for developing a GMP-compliant process should be to reduce the risk of contamination of the product, establish documentation to verify that the entire process is correctly performed, and minimize variability in the process while maintaining the salient characteristics and function of the cells of interest. Where practical, the use of single-use, disposable materials and closed systems for manipulating cells is preferred by regulators. Reagents, media and supplements, and cytokines should be manufactured under GMP, or US Pharmacopeia or European guidelines. If not available in these forms, then additional testing may be needed to ensure the appropriate level of purity and lot-to-lot consistency.

6.4.2 QUALIFICATION OF SINGLE-USE MATERIALS
FOR BIOREACTOR-BASED MANUFACTURING

Cell therapies pose many challenges with respect to keeping the product sterile and free of impurities. Risks for exposure to microorganisms and other process impurities exist during cell therapy upstream processing, yet unlike in protein-based drug manufacturing, there is limited or no opportunity for their clearance before final formulation. As the cells are the product, harsh conditions cannot be used to inactivate any adventitious agents. Extensive processing, which may aid in clearance of other impurities, cannot be implemented because it may affect the viability and quality of cells. Terminal sterilization of the final product or sterile filtration based on size is also not feasible for the same reason. Risk identification followed by mitigation using appropriate sourcing from safe regions, treatment and testing, physical containment and environmental controls are necessary tools for ensuring the safety of the product.

6.4.2.1 Raw or Ancillary Material Risk Mitigation

A wide variety of starting materials may be used in the manufacturing process, some of which are integral to the final product and, in some cases, contribute to its composition or are found in the final cell product as active ingredients or as excipients. Some materials used with the manufacturing process are ancillary materials, which, by definition, are components, reagents, or materials used during manufacture that exert an effect on the cell product but are not intended to be part of the final cell product. The term ancillary material (AM) is not globally recognized by regulators and is commonly referred to as raw material in some jurisdictions, such as in Europe. In a disposable bioreactor-based manufacturing process for cell therapies there are several materials whose quality needs to be carefully considered. The quality of a raw material can affect the stability, safety, potency, and purity of a cell product. For example, the mechanism by which a raw material exerts its effect may not be known, and the impact of normal variation on the quality and safety of the therapeutic product may not be understood. Alternatively, raw materials exposed to/consisting of components of human/animal origin may present an infectious disease transmission risk. Other materials, if administered to humans, may cause an immune reaction. Another important consideration in single-use systems is the likelihood of chemicals leaching from the plastic components into the culture and have an adverse effect on the growth or quality of cells. An impurity with toxic properties, which is introduced into a manufacturing process and is not adequately removed in subsequent processing steps, will expose the patient to a toxic substance and may impair the effectiveness of the therapeutic.

Evaluation of the impact of the raw material on the quality, safety, and efficacy of cell-based therapy products must be performed by the user of the raw material. A risk assessment must consider the biological origin and traceability of the raw material, the production steps applied to it and the ability of the drug product manufacturing process to control or remove the raw material from the final medicinal product (EP 5.2.12. Raw materials of biological origin for the production of cell-based and gene therapy medicinal products).

A tiered approach to risk categorization and appropriate measures for mitigation are recommended in USP General Chapter <1043>: Ancillary Materials for Cell, Gene and Tissue-Engineered Products. Materials that are considered lowest risk are classified as Tier 1. These are highly qualified materials that are well-suited for use in manufacturing. Recombinant insulin and human serum albumin intended for human use are Tier 1 materials, which can be used as cell culture medium additives. These require a low level of qualification. At the other end of the spectrum are Tier 4 materials that are considered highest risk level for ancillary materials. Extensive qualification is necessary prior to use in manufacturing. This risk level includes complex, animal-derived fluid materials not subjected to adventitious viral removal or inactivation procedures. These materials may require (a) an upgrade of manufacturing processes; (b) treatment to inactivate or remove adventitious agents, disease-causing substances, or specific contaminants (e.g., animal viruses, prions); (c) testing of each lot of material to ensure that it is free of adventitious agents, disease-causing substances, or specific contaminants; and (d) validation of the manufacturing process of the cell product to demonstrate reduction levels considered to be safe. Fetal bovine serum and porcine trypsin widely used in cell therapy processes are examples of Tier 4 materials. Alternative sources of reagents such as recombinant proteins instead of animal origin proteins for e.g. recombinant trypsin manufactured under GMP can be used instead of porcine trypsin for detachment of cells to minimize risk and ensure better traceability of materials.

In order to effectively mitigate risk around cell therapy development and patient safety, cell therapy manufacturers must work with their suppliers and regulators to qualify each raw/ancillary material to assess source, purity, identity, safety, and suitability in a given application. They should develop comprehensive and methodically sound qualification plans to ensure the traceability, consistency, suitability, purity, and safety of the material.

6.4.2.2 Plastics Component Qualification

Single-use components are mainly made of plastic materials. Bioreactors are lined with bags made of polymeric films. Polymeric films are made with additives, which can be antioxidants and processing additives such as slip aids. Sterilization of films in undertaken by gamma irradiation. The irradiation can result in the breakdown of some additives and release chemical species that may be detrimental to cellular growth or health, as shown in Chinese Hamster Ovary cultures (Hammond et al. 2013). Leachables and extractables from product-contact materials such as bioreactors and packaging components should be quantified, and limits should be established (USP General Chapter <1046>: Cell and Gene Therapy Products). Along with extractables studies with model solvents for a risk assessment, the user should develop a method to compare their cell culture in the bags to a control culture to rule out any adverse effects.

Although currently there is no specific guidance or standard for single-use bioprocess manufacturing systems, the industry has adopted standards for materials used for final drug product containers. The USP <661>: Container – Plastics describes tests for qualification of plastic materials including physicochemical tests and biological reactivity (USP <87>: Biological Reactivity Tests, *in vitro* and USP <88>: Biological Reactivity Tests, *in vivo*). A new chapter outlining the standards specific to plastic components and systems used in manufacturing systems is expected to be released in 2017, which will directly applicable and should be used for guidance.

USP <88> consists of *in vivo* testing of plastic materials in animals and a plastic class designation (Class I–VI) can be obtained whereas USP <87> is testing of an extract from the plastic on mammalian cell lines. Although there appears to be a trend in the industry to perform only USP <88>, as it is perceived to be a worst-case scenario, it might be advantageous to perform USP <87> when the bioreactors are used for cell therapies. In adherent cultures often microcarriers consisting of polymeric materials are used. The process of microcarrier selection generally includes an evaluation of cell growth and quality in comparison to control cultures. In addition to the evaluation of other parameters, it also ensures non-inclusion of microcarriers introducing leachables that adversely affect growth.

6.4.2.3 Particulates

As cells in suspension are not clear solutions, the final product particulate test is limited to foreign visible particles. A thorough evaluation of putative sources of particulates and a resulting mitigation plan should be implemented for each therapy due to the complex nature of the product and process (Clarke et al. 2016). This includes measures to limit the input of particles from materials and equipment used during manufacturing and verification of the ability of the manufacturing process to produce low particle products with simulated samples (without cells). In adherent cultures the tight control of the microcarrier size range should be maintained to ensure the absence of partial or significantly smaller particles that may escape any size–based removal steps. Bioreactor process conditions that may generate pieces of microcarriers should also be avoided.

6.4.3 RISK MITIGATION USING PHYSICAL CONTROLS

6.4.3.1 Closed Systems

The use of closed systems with aseptic connections between unit operations greatly minimizes the exposure of process starting materials, process intermediates and the product to the environment. The use of custom single-use assemblies connecting medium preparation to the bioreactor, bioreactor to harvest systems and other downstream operations can ensure closed operations. Use of closed systems can reduce the requirement of a higher classification clean room reducing the resources to create and maintain such a clean room as has been demonstrated for vaccine production (World Health Organization 2012).

6.4.3.2 Single-Use Systems

The qualification of single-use suppliers, materials, components, and completed assemblies require attention to several factors. It is important to use a scientific risk-based approach to quality both materials and completed assemblies for single-use systems in pharmaceutical development and manufacture.

The assessment of risk must be based on the complexity of the system and its intended use (e.g., product contact versus non-product contact, upstream versus downstream use, short-term contact versus long-term storage). In addition, scientific principles must be applied to (1) identify and monitor extractable materials from films and other components; (2) investigate potential interactions with product Critical Quality Attributes (CQA) or significant process parameters like cell culture media; and (3) assess the effects of assembly processes such as welding, fusing, mechanical stress, and sterilization (e.g., gamma irradiation) on the materials during manufacturing. In assessing the suitability of the single-use system with respect to potential leachables in the final product, extractables may be considered potential leachables. Finally, the design and functional integrity of the completed single-use system must be verified and maintained by appropriately trained end users via suitable processes for construction, packaging, shipping, and deployment.

6.4.3.3 Integrity Assurance

Ensuring structural integrity of is critical to implementing a closed single-use system. It is important to mitigate the risks including loss of valuable product from leaks in the system, contamination based on microbial ingress and threat to operator safety based on toxic or hazardous product.

Single-use systems used in the manufacturing of cell therapies may involve multiple components and may require significant amount of handling and other forms of stress. The integrity assurance level can be enhanced by employing comprehensive risk-based quality procedures.

Many factors impact single-use system integrity and include material selection, component design, assembly design, assembly manufacturing (process and facility), sub-component testing, final testing (supplier and end user), transportation/handling, and actual use. All of these must be qualified to ensure the integrity of the single-use assembly.

A phase appropriate sourcing, supply, testing, and validation of single-use systems should be planned and implemented for successful commercialization of cell therapies as outlined in (Clarke et al. 2012).

Specifically, for cell therapies, the manufacturer must work with the supplier of the single-use system to obtain extensive data with respect to purity, qualification, and validation to qualify the materials and components used to assemble the final single-use system. Clean rooms of Class C or above should be used to prepare the assemblies. Use of automation in component manufacture ensures consistent quality of products. In this regard, the manufacturer may use the documents from the supplier to meet most of the verification criteria required to make sure that the single-use system is fit for the intended application and purpose. This requires a thorough qualification of the supplier by the manufacturer to ascertain an acceptable supplier quality system, documentation, and an appropriate level of technical capability that extends all the way to the original sources of materials to the supplier.

6.4.4 IN-PROCESS CONTROLS AND TESTING

The manufacturing process needs to be controlled by several in-process controls at the level of critical steps or intermediate products. To demonstrate process control and monitor variability, assays should be developed to determine cell phenotype, genotype, and/or function. Intermediate cell products are products that can be isolated during the process; specifications of these products should be established in order to assure the reproducibility of the process and the consistency of the final product.

A critical manufacturing aspect is to establish that cell-based products are free from adventitious microbial agents (viruses, mycoplasma, bacteria, fungi). The contamination could originate from the starting or raw materials or adventitiously introduced during the manufacturing process. A thorough testing for the absence of bacteria, fungi, and mycoplasma should be performed at the level of finished product. These tests should be performed with the current methodologies described for cell-based products. In cases where the short shelf life of the cell therapy product is prohibitive for the testing of absence of microorganisms, alternative validated rapid testing methods may be employed, if justified.

6.5 CONCLUSIONS

The long-term view of regenerative medicine therapies predicts an increased need for expansion solutions that ease scalability, utilize animal origin-free materials, and are compatible with limited downstream processing steps. As more cell therapeutics progress through clinical testing, current *in vitro* culture methods are proving cumbersome to scale and lack robustness. The implementation of single-use bioreactors for a single cell expansion unit operation per batch will support the manufacturing of safe and effective cell therapies. Start-to-finish solutions for expansion and harvest are key enabling technologies for meeting regulatory criteria and success in commercializing these cutting edge therapeutic approaches.

REFERENCES

Augello, A., T. B. Kurth, and C. de Bari. 2010. Mesenchymal stem cells: A perspective from in vitro cultures to in vivo migration and niches. *European Cells and Materials* 20:121–133.

Aunins, J. G., K. Glazomitsky, and B. C. Buckland. 1993. Cell culture reactor design: Known and unknown. *Third International Conference on Bioreactor and Bioprocess Fluid Dynamics*, London, UK.

Bizzarri, A., H. Koehler, M. Cajlakovic, A. Pasic, L. Schaupp, I. Klimant, and V. Ribitsch. 2006. Continuous oxygen monitoring in subcutaneous adipose tissue using microdialysis. *Anal Chim Acta* 573–574:48–56. doi:10.1016/j.aca.2006.03.101.

Chamberlain, G., J. Fox, B. Ashton, and J. Middleton. 2007. Concise review: Mesenchymal stem cells: Their phenotype, differentiation capacity, immunological features, and potential for homing. *Stem Cells* 25 (11):2739–2749.

Chapman, C. M., A. W. Nienow, M. Cooke, and J. C. Middleton. 1983. Particle-gas-liquid mixing in stirred vessels. Part I: Particle-liquid mixing. *Chem Eng Res Des* 61 (2):71–81.

Chen, A. K., S. Reuveny, and S. K. Oh. 2013. Application of human mesenchymal and pluripotent stem cell microcarrier cultures in cellular therapy: Achievements and future direction. *Biotechnol Adv* 31 (7):1032–1046. doi:10.1016/j.biotechadv.2013.03.006.

Chen, A. K., Y. K. Chew, H. Y. Tan, S. Reuveny, and S. K. Weng Oh. 2015. Increasing efficiency of human mesenchymal stromal cell culture by optimization of microcarrier concentration and design of medium feed. *Cytotherapy* 17 (2):163–173. doi:10.1016/j.jcyt.2014.08.011.

Cherry, R. S., and E. T. Papoutsakis. 1988. Physical mechanisms of cell damage in microcarrier cell culture bioreactors. *Biotechnol Bioeng* 32 (8):1001–1014. doi:10.1002/bit.260320808.

Chisti, Y. 2001. Hydrodynamic damage to animal cells. *Crit Rev Biotechnol* 21 (2):67–110. doi:10.1080/20013891081692.

Chow, D. C., L. A. Wenning, W. M. Miller, and E. T. Papoutsakis. 2001. Modeling pO(2) distributions in the bone marrow hematopoietic compartment. I. Krogh's model. *Biophys J* 81 (2):675–684. doi:10.1016/S0006-3495(01)75732-3.

Clark, J. M., and M. D. Hirtenstein. 1981. Optimizing culture conditions for the production of animal cells in microcarrier culture. *Ann N Y Acad Sci* 369:33–46.

Clarke, D., D. Harati, J. Martin, J. Rowley, J. Keller, M. McCaman, M. Carrion et al. 2012. Managing particulates in cellular therapy. *Cytotherapy* 14 (9):1032–1040. doi:10.3109/14653249.2012.706709.

Clarke, D., J. Stanton, D. Powers, O. Karnieli, S. Nahum, E. Abraham, J. S. Parisse, and S. Oh. 2016. Managing particulates in cell therapy: Guidance for best practice. *Cytotherapy* 18 (9):1063–1076. doi:10.1016/j.jcyt.2016.05.011.

Croughan, M. S., E. S. Sayre, and D. I. Wang. 1989. Viscous reduction of turbulent damage in animal cell culture. *Biotechnol Bioeng* 33 (7):862–872. doi:10.1002/bit.260330710.

Croughan, M. S., J. F. Hamel, and D. I. Wang. 1987. Hydrodynamic effects on animal cells grown in microcarrier cultures. *Biotechnol Bioeng* 29 (1):130–141. doi:10.1002/bit.260290117.

Cunha, B., T. Aguiar, M. M. Silva, R. J. Silva, M. F. Sousa, E. Pineda, C. Peixoto, M. J. Carrondo, M. Serra, and P. M. Alves. 2015. Exploring continuous and integrated strategies for the up- and downstream processing of human mesenchymal stem cells. *J Biotechnol.* doi:10.1016/j.jbiotec.2015.02.023.

Dolley-Sonneville, P. J., L. E. Romeo, and Z. K. Melkoumian. 2013. Synthetic surface for expansion of human mesenchymal stem cells in xeno-free, chemically defined culture conditions. *PLoS One* 8 (8):e70263. doi:10.1371/journal.pone.0070263.

Dominici, M., K. Le Blanc, I. Mueller, I. Slaper-Cortenbach, F. Marini, D. Krause, R. Deans, A. Keating, Dj Prockop, and E. Horwitz. 2006. Minimal criteria for defining multipotent mesenchymal stromal cells. The international society for cellular therapy position statement. *Cytotherapy* 8 (4):315–317. doi:10.1080/14653240600855905.

Doucet, C., I. Ernou, Y. Zhang, J. R. Llense, L. Begot, X. Holy, and J. J. Lataillade. 2005. Platelet lysates promote mesenchymal stem cell expansion: A safety substitute for animal serum in cell-based therapy applications. *J Cell Physiol* 205 (2):228–236. doi:10.1002/jcp.20391.

Duffy, C. R., R. Zhang, S. E. How, A. Lilienkampf, P. A. De Sousa, and M. Bradley. 2014. Long term mesenchymal stem cell culture on a defined synthetic substrate with enzyme free passaging. *Biomaterials* 35 (23):5998–6005. doi:10.1016/j.biomaterials.2014.04.013.

Eibl, R., and D. Eibl. 2009. Application of disposable bag bioreactors in tissue engineering and for the production of therapeutic agents. *Adv Biochem Eng Biotechnol* 112:183–207. doi:10.1007/978-3-540-69357-4_8.

Engler, A. J., S. Sen, H. L. Sweeney, and D. E. Discher. 2006. Matrix elasticity directs stem cell lineage specification. *Cell* 126 (4):677–689. doi:10.1016/j.cell.2006.06.044.

Ferrari, C., F. Balandras, E. Guedon, E. Olmos, I. Chevalot, and A. Marc. 2012. Limiting cell aggregation during mesenchymal stem cell expansion on microcarriers. *Biotechnol Prog* 28 (3):780–787. doi:10.1002/btpr.1527.

Frauenschuh, S., E. Reichmann, Y. Ibold, P. M. Goetz, M. Sittinger, and J. Ringe. 2007. A microcarrier-based cultivation system for expansion of primary mesenchymal stem cells. *Biotechnol Prog* 23 (1):187–193. doi:10.1021/bp060155w.

Giard, D. J., W. S. Hu, and D. I. C. Wang. 1986. Detachment of anchorage-dependent cells from microcarriers. edited by European Patent Office (EPO). Cambridge, MA: Massachusetts Institute Technology.

Goh, T. K., Z. Y. Zhang, A. K. Chen, S. Reuveny, M. Choolani, J. K. Chan, and S. K. Oh. 2013. Microcarrier culture for efficient expansion and osteogenic differentiation of human fetal mesenchymal stem cells. *Biores Open Access* 2 (2):84–97. doi:10.1089/biores.2013.0001.

Hammond, M., H. Nunn, G. Rogers, H. Lee, A. L. Marghitoiu, L. Perez, Y. Nashed-Samuel, C. Anderson, M. Vandiver, and S. Kline. 2013. Identification of a leachable compound detrimental to cell growth in single-use bioprocess containers. *PDA J Pharm Sci Technol* 67 (2):123–134. doi:10.5731/pdajpst.2013.00905.

Haque, N., M. T. Rahman, N. H. Abu Kasim, and A. M. Alabsi. 2013. Hypoxic culture conditions as a solution for mesenchymal stem cell based regenerative therapy. *Sci World J* 2013:632972. doi:10.1155/2013/632972.

Harrison, J. S., P. Rameshwar, V. Chang, and P. Bandari. 2002. Oxygen saturation in the bone marrow of healthy volunteers. *Blood* 99 (1):394.

Hassan, S., A. S. Simaria, H. Varadaraju, S. Gupta, K. Warren, and S. S. Farid. 2015. Allogeneic cell therapy bioprocess economics and optimization: Downstream processing decisions. *Regen Med* 10 (5):591–609. doi:10.2217/rme.15.29.

Health, U.S. National Institutes of. ClinicalTrials.gov. National Library of Medicine (US) Accessed 27 March. https://clinicaltrials.gov/ct2/home. Accessed 27 March 2017.

Hewitt, C. J., K. Lee, A. W. Nienow, R. J. Thomas, M. Smith, and C. R. Thomas. 2011. Expansion of human mesenchymal stem cells on microcarriers. *Biotechnol Lett* 33 (11):2325–2335. doi:10.1007/s10529-011-0695-4.

Ibrahim, S., and A. W. Nienow. 2004. Suspension of microcarriers for cell culture with axial flow impellers. *Chem Eng Res Des* 82 (9 SPEC. ISS.):1082–1088.

Jing, D., N. Sunil, S. Punreddy, M. Aysola, D. Kehoe, J. Murrel, M. Rook, and K. Niss. 2013. Growth kinetics of human mesenchymal stem cells in a 3-L single-use, stirred-tank bioreactor. *BioPharm Int* 26 (4):28–38.

King, J. A., and W. M. Miller. 2007. Bioreactor development for stem cell expansion and controlled differentiation. *Curr Opin Chem Biol* 11 (4):394–398. doi:10.1016/j.cbpa.2007.05.034.

Kolmogorov, A.N. 1941. Dissipation of energy in locally isotropic turbulence. *Dokl. Akad. Nauk. SSSR* 32:19–21.

Kreke, M. R., L. A. Sharp, Y. W. Lee, and A. S. Goldstein. 2008. Effect of intermittent shear stress on mechanotransductive signaling and osteoblastic differentiation of bone marrow stromal cells. *Tissue Eng Part A* 14 (4):529–537. doi:10.1089/tea.2007.0068.

Kreke, M. R., W. R. Huckle, and A. S. Goldstein. 2005. Fluid flow stimulates expression of osteopontin and bone sialoprotein by bone marrow stromal cells in a temporally dependent manner. *Bone* 36 (6):1047–1055. doi:10.1016/j.bone.2005.03.008.

Lavrentieva, A., I. Majore, C. Kasper, and R. Hass. 2010. Effects of hypoxic culture conditions on umbilical cord-derived human mesenchymal stem cells. *Cell Commun Signal* 8:18. doi:10.1186/1478-811X-8-18.

Lawson, T., D. E. Kehoe, A. C. Schnitzler, P. J. Rapiejko, K. A. Der, K. Philbrick, S. Punreddy et al. 2017. Process development for expansion of human mesenchymal stromal cells in a 50L single-use stirred tank bioreactor. *Biochem Eng J* 120:49–62. doi:10.1016/j.bej.2016.11.020.

Lonne, M., A. Lavrentieva, J. G. Walter, and C. Kasper. 2013. Analysis of oxygen-dependent cytokine expression in human mesenchymal stem cells derived from umbilical cord. *Cell Tissue Res* 353 (1):117–122. doi:10.1007/s00441-013-1597-7.

Lopez, F., C. Di Bartolo, T. Piazza, A. Passannanti, J. C. Gerlach, B. Gridelli, and F. Triolo. 2010. A quality risk management model approach for cell therapy manufacturing. *Risk Anal* 30 (12):1857–1871. doi:10.1111/j.1539-6924.2010.01465.x.

Mauney, J. R., D. L. Kaplan, and V. Volloch. 2004. Matrix-mediated retention of osteogenic differentiation potential by human adult bone marrow stromal cells during ex vivo expansion. *Biomaterials* 25 (16):3233–3243. doi:10.1016/j.biomaterials.2003.10.005.

Mauney, J. R., V. Volloch, and D. L. Kaplan. 2005. Matrix-mediated retention of adipogenic differentiation potential by human adult bone marrow-derived mesenchymal stem cells during ex vivo expansion. *Biomaterials* 26 (31):6167–6175. doi:10.1016/j.biomaterials.2005.03.024.

Maziarz, R. T., T. Devos, C. R. Bachier, S. C. Goldstein, J. F. Leis, S. M. Devine, G. Meyers et al. 2015. Single and multiple dose MultiStem (multipotent adult progenitor cell) therapy prophylaxis of acute graft-versus-host disease in myeloablative allogeneic hematopoietic cell transplantation: A phase 1 trial. *Biol Blood Marrow Transplant* 21 (4):720–728. doi:10.1016/j.bbmt.2014.12.025.

McMurray, R. J., N. Gadegaard, P. M. Tsimbouri, K. V. Burgess, L. E. McNamara, R. Tare, K. Murawski, E. Kingham, R. O. Oreffo, and M. J. Dalby. 2011. Nanoscale surfaces for the long-term maintenance of mesenchymal stem cell phenotype and multipotency. *Nat Mater* 10 (8):637–644. doi:10.1038/nmat3058.

Morla, A., Z. Zhang, and E. Ruoslahti. 1994. Superfibronectin is a functionally distinct form of fibronectin. *Nature* 367 (6459):193–196. doi:10.1038/367193a0.

Mukhopadhyay, A., S. N. Mukhopadhyay, and G. P. Talwar. 1993. Influence of serum proteins on the kinetics of attachment of Vero cells to cytodex microcarriers. *J Chem Technol Biotechnol* 56 (4):369–374.

Nienow, A. W., Q. A. Rafiq, K. Coopman, and C. J. Hewitt. 2014. A potentially scalable method for the harvesting of hMSCs from microcarriers. *Biochem Eng J* 85 (0):79–88. doi:10.1016/j.bej.2014.02.005.

Papoutsakis, E. T. 1991. Fluid-mechanical damage of animal cells in bioreactors. *Trends Biotechnol* 9 (12):427–437.

Qian, L., and W. M. Saltzman. 2004. Improving the expansion and neuronal differentiation of mesenchymal stem cells through culture surface modification. *Biomaterials* 25 (7–8):1331–1337.

Rowley, J., E. Abraham, A. Campbell, H. Brandwein, and S. Oh. 2012. Meeting lot-size challenges of manufacturing adherent cells for therapy. *BioProcess Int* 10 (Suppl. 3):16–22.

Sart, S., S. N. Agathos, and Y. Li. 2014. Process engineering of stem cell metabolism for large scale expansion and differentiation in bioreactors. *Biochem Eng J* 84 (0):74–82. doi:10.1016/j.bej.2014.01.005.

Schätti, O., S. Grad, J. Goldhahn, G. Salzmann, Z. Li, M. Alini, and M. J. Stoddart. 2011. A combination of shear and dynamic compression leads to mechanically induced chondrogenesis of human mesenchymal stem cells. *Europ Cells Mat* 22:214–225. doi:10.22203/eCM.v022a17.

Schnitzler, A. C., A. Verma, D. E. Kehoe, D. Jing, J. R. Murrell, K. A. Der, M. Aysola, P. J. Rapiejko, S. Punreddy, and M. S. Rook. 2016. Bioprocessing of human mesenchymal stem/stromal cells for therapeutic use: Current technologies and challenges. *Biochem Eng J* 108:3–13. doi:10.1016/j.bej.2015.08.014.

Simaria, A. S., S. Hassan, H. Varadaraju, J. Rowley, K. Warren, P. Vanek, and S. S. Farid. 2014. Allogeneic cell therapy bioprocess economics and optimization: Single-use cell expansion technologies. *Biotechnol Bioeng* 111 (1):69–83. doi:10.1002/bit.25008.

Soininen, A., E. Kaivosoja, T. Sillat, S. Virtanen, Y. T. Konttinen, and V. M. Tiainen. 2014. Osteogenic differentiation on DLC-PDMS-h surface. *J Biomed Mater Res B Appl Biomater* 102 (7):1462–1472. doi:10.1002/jbm.b.33125.

Sun, L. Y., S. Z. Lin, Y. S. Li, H. J. Harn, and T. W. Chiou. 2011. Functional cells cultured on microcarriers for use in regenerative medicine research. *Cell Transplant* 20 (1):49–62. doi:10.3727/096368910 × 532792.

Timmins, N. E., M. Kiel, M. Gunther, C. Heazlewood, M. R. Doran, G. Brooke, and K. Atkinson. 2012. Closed system isolation and scalable expansion of human placental mesenchymal stem cells. *Biotechnol Bioeng* 109 (7):1817–1826. doi:10.1002/bit.24425.

Weber, C., S. Pohl, R. Portner, C. Wallrapp, M. Kassem, P. Geigle, and P. Czermak. 2007. Expansion and harvesting of hMSC-TERT. *Open Biomed Eng J* 1:38–46. doi:10.2174/1874120700701010038.

World Health Organization. 2012. *Environmental Monitoring of Clean Rooms in Vaccine Manufacturing Facilities.* Draft Document: Points to consider for manufacturers of human vaccines. Geneva, Switzerland: World Health Organization.

Wu, S. C., W. C. Hsieh, and M. Y. Liau. 1998. Comparisons of microcarrier cell culture processes in one hundred mini-liter spinner flask and fifteen-liter bioreactor cultures. *Bioprocess Engineering* 19 (6):431–434.

Yang, H. S., O. Jeon, S. H. Bhang, S. H. Lee, and B. S. Kim. 2010. Suspension culture of mammalian cells using thermosensitive microcarrier that allows cell detachment without proteolytic enzyme treatment. *Cell Transplant* 19 (9):1123–1132. doi:10.3727/096368910 × 516664.

Yeatts, A. B., D. T. Choquette, and J. P. Fisher. 2013. Bioreactors to influence stem cell fate: Augmentation of mesenchymal stem cell signaling pathways via dynamic culture systems. *Biochim Biophys Acta* 1830 (2):2470–2480. doi:10.1016/j.bbagen.2012.06.007.

Zhao, W., X. Li, X. Liu, N. Zhang, and X. Wen. 2014. Effects of substrate stiffness on adipogenic and osteogenic differentiation of human mesenchymal stem cells. *Mater Sci Eng C Mater Biol Appl* 40:316–323. doi:10.1016/j.msec.2014.03.048.

Zwietering, Th N. 1958. Suspending of solid particles in liquid by agitators. *Chemical Eng Sci* 8 (3–4):244–253.

7 Bioreactors for the Cultivation of Hematopoietic Stem and Progenitor Cells

*Marta H.G. Costa, Frederico Castelo Ferreira,
Joaquim M. S. Cabral, and Cláudia Lobato da Silva*

CONTENTS

7.1 INTRODUCTION

Over 100 billion mature blood cells are produced every day by the hematopoietic system of adults (Hoggatt et al. 2016). However, mature blood cells have a limited lifespan (platelets, for instance, present a circulating lifespan of only around

10 days Daly 2011). To ensure that the pool of hematopoietic stem cells (HSC) gives rise to the entire hematopoietic system without leading to the exhaustion of the more primitive cells, a fine balance between their self-renewal and differentiation is required.

The hematopoietic system is organized in a hierarchical way, with long-term repopulating HSC (LT-HSC), positioned at the top of the hierarchy, able to give rise to short-term HSC (ST-HSC), which are still capable to generate all blood cell lineages but can only self-renew for a limited period of life time (Weissman 2000). The capability of the more primitive HSC to recreate the entire hematopoietic system, namely to originate both lymphoid and myeloid lineages (Figure 7.1), while retaining

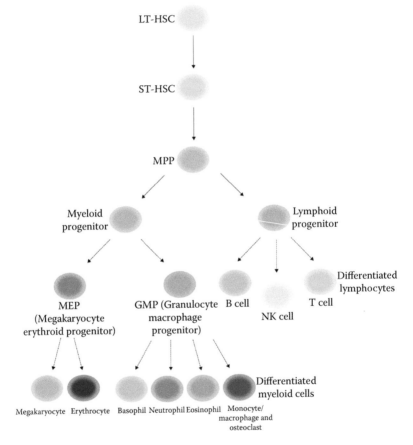

FIGURE 7.1 Hierarchical organization of the hematopoietic system. HSC are characterized by their long-term repopulating potential (LT-HSC), giving rise to short-term HSC (ST-HSC), whose self-renewal capability becomes limited in time, but are still capable to give rise to all lineages of the hematopoietic system, namely to the full spectrum of cells comprising the myeloid and lymphoid lineages. Myeloid progenitors can give rise to megakaryocyte erythroid progenitors (MEP) and granulocyte macrophage progenitors (GMP) while lymphoid progenitors originate NK cells, B and T cells. (From King, K.Y. and Goodell, M.A., *Nat. Rev. Immunol.*, 11, 685–692, 2011; Lampiasi, N. et al., *BioMed. Res. Int.*, 2016, 1–9, 2016.)

their self-renewal potential is key to maintaining a homeostatic hematopoiesis. HSC positioning within a tightly regulated bone marrow (BM) hematopoietic niche is thought to be essential to achieve this balance.

However, some hematopoietic cell-based clinical applications require the production of meaningful cell numbers in *ex-vivo* culture systems. BM transplantation, for instance, is usually performed with a minimum of 2×10^5 CD34$^+$ cells/kg of the patient (Weissman and Shizuru 2008) whereas neutrophil infusions are performed using 10^{10} cell doses (Timmins et al. 2009). Other clinical applications, such as generation of T cells and dendritic cells (dendritic cell vaccines, for instance, usually comprise around 10^7 cells per injection [Cui et al. 2013; Kamigaki et al. 2013] for adoptive immunotherapeutic purposes, as well as purging of contaminating tumor cells), would benefit from the development of successful strategies aimed at the expansion of hematopoietic stem and progenitor cells (HSPC).

The development of bioreactor systems and the identification of culture conditions, which can promote expansion and/or controlled differentiation of HSPC under conditions devoid of animal components will aid the generation of clinically relevant cell numbers.

7.2 UMBILICAL CORD BLOOD FOR TRANSPLANTATION—CLINICAL MOTIVATION

Hematopoietic cell transplantation (HCT) was firstly performed in 1957, when healthy BM from an identical twin was used to treat a leukemia patient exposed to radiation and chemotherapy (Thomas et al. 1957). However, it was not until 1968 that the first allogeneic transplant was performed (Gatti et al. 1968). Although, historically, HCT has been performed using cells from the BM, more recently, both mobilized peripheral blood (mPB) and umbilical cord blood (UCB) have been used as alternative cell sources, with mPB surpassing BM as a cell source for allogeneic HCT (Baldomero et al. 2011; Bensinger 2012). UCB offers some advantages over BM and mPB, particularly when autologous transplantation is not feasible. Besides avoiding invasive harvesting, the use of UCB as a source of HSPC requires less stringent human leukocyte antigen (HLA) matching, is associated with long-term immune recovery and reduced incidence of graft-*versus*-host disease (GvHD) (Rocha et al. 2000). On the other hand, cells retrieved from UCB are less likely to be contaminated with blood-borne viruses, which would represent a threat to patients subjected to HCT. Nevertheless, since the first UCB transplant performed in 1988 (Gluckman et al. 1990), much work has been performed to fully exploit the potential of this HSPC source; however, the low cell numbers available in UCB units has restricted the use of UCB transplantation mainly to pediatric patients. Indeed, the recommended clinical cell doses required to achieve successful HCT with UCB cells range between 1.5 and 14.0×10^6 CD34$^+$ cells/kg (Pedrazzoli et al. 2002; Rocha et al. 2000) and $2.0–2.5 \times 10^7$ mononuclear cells (MNC)/kg, a value below the limited cell dose available in an UCB unit (whose content ranges between 0.4 and 1.0×10^9 MNC) (Andrade-Zaldívar et al. 2008). To achieve clinically relevant cell numbers, double unit transplants have been explored to treat adult patients, although only one unit is thought to contribute to long-term engraftment of hematopoietic cells (Barker et al. 2005; Brunstein et al. 2010).

Nonetheless, simultaneous administration of two unmanipulated UCB units is still associated with suboptimal hematopoietic recovery as the median time to neutrophil and platelet recovery ranges between 12–32 and 41–105 days, respectively (Sideri et al. 2011). The reported clinical limitations have prompted researchers to exploit strategies envisioning the expansion of hematopoietic cells *ex-vivo*. Several studies point out the importance that *ex-vivo* expanded hematopoietic cells can assume in a dual transplant strategy (Delaney et al. 2010; Shpall et al. 2002). Whereas non-cultured UCB units might contribute to long-term engraftment, *ex-vivo* expanded hematopoietic cell populations could constitute a valuable therapeutic tool to abrogate UCB transplant-related neutropenia due to the presence of more committed progenitors. Recently, Casamayor-Genescà and colleagues achieved clinically relevant doses of hematopoietic cells committed to the granulocytic lineage as well as neutrophil precursors upon culture of CD34$^+$-enriched cells in non-stirred gas-permeable cell culture bags using a rate-controlled feeding strategy (Casamayor-Genescà et al. 2017). Indeed, a 14-day expansion in such a good manufacturing practice (GMP)-compliant cell culture scheme allowed clinical doses up to 2×10^6 CD34$^+$ cells/kg to be produced. This led the authors to suggest that the generated cell product could be used as part of a dual transplant strategy. While a non-cultured UCB unit would be involved on the definitive engraftment of more primitive cell populations, transient neutrophil engraftment achieved upon administration of the expanded cell product would work as a "bridge transplant," therefore being instrumental to reduce aplastic periods (Cabrera et al. 2005; Fernández et al. 2003).

Nevertheless, the fact that only unmanipulated UCB units seem to contribute to the long-term engraftment of the administered hematopoietic cells (de Lima et al. 2012) should be addressed in *ex-vivo* expansion strategies to enhance the success of HCT. Indeed, the methods typically applied to expand UCB cells involve the use of cytokines that promote proliferation/differentiation of hematopoietic progenitors at the expense of their multipotency (Figure 7.2). To circumvent the loss of long-term repopulating activity when HSPC are cultured *ex-vivo* (McNiece et al. 2002), mimicking the BM niche through *ex-vivo* cultivation is thought to be crucial and several strategies have been followed to tailor culture conditions towards that purpose.

7.3 HSPC CULTIVATION—STATIC *VERSUS* DYNAMIC

Currently, most hematopoietic cell culture studies and clinical trials are supported by static systems, such as T-flasks, multi-well plates, and gas-permeable blood bags with poorly controlled culture settings. In such systems, gas exchange takes place at the medium/gas interface and nutrient diffusion occurs in an unmixed environment. Different cytokine cocktails (with or without the presence of additional chemical factors) have been exploited to expand HSPC in such static systems. However, and despite the higher simplicity and ease of handling of static systems for research purposes, the use of cell-based products in clinical settings requires scale-up of highly controlled cell cultures that static systems could not offer.

Continuous culture monitoring and the use of automated control systems to stabilize culture conditions within predetermined ranges could play a relevant role

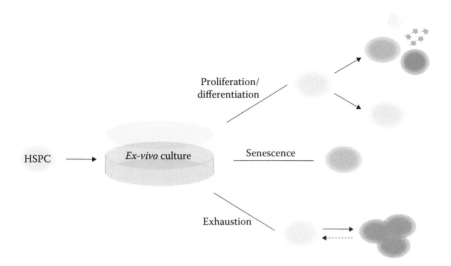

FIGURE 7.2 Culture of HSPC outside their natural hematopoietic niche typically results in cellular differentiation, senescence and exhaustion of the stem cell pool. (From Glimm, H. and Eaves, C. J., *Blood,* 94, 2161–2168, 1999; Hofmeister, C.C. et al., *Bone Marrow Transplant.,* 39, 11–23, 2007; Orford, K.W. and Scadden, D.T., *Nat. Rev. Gen.,* 9, 115–128, 2008; Wilson, A. et al., *Cell,* 135, 1118–1129, 2008.)

in achieving functional hematopoietic cell products. Dynamic cell cultures offer control over a set of parameters, including agitation rates and feeding regimens, which could limit the development of concentration gradients (metabolites, pH, dissolved oxygen, cytokines) while producing high cell numbers. Importantly, the reduced handling requirement in bioreactor cultures reduces the risk of contamination and increases the safety of the final product. Additionally, large-scale cell manufacture could lead to the development of more cost-effective processes relative to static cell cultures.

7.4   *EX-VIVO* CULTURE PARAMETERS

Several cell culture parameters must be balanced to potentiate the therapeutic potency of cultured hematopoietic cells in a cost-effective and GMP-compliant manner. Besides the use of different bioreactor configurations, which can impact the autocrine and paracrine cues provided by the cultured cells, culture medium formulation, combining a multitude of biological factors, is one of the primary parameters that researchers have long explored to maximize cell proliferation or differentiation into specific cell lineages. Adequate oxygen tension and maintenance of pH and shear stress values within cell amenable ranges should also be considered during the bioprocess optimization efforts. The presence of stromal cells and biomaterials, as biomimetics of the BM niche, can equally contribute to help tailor the cellular outcomes in terms of both cell number and function.

7.4.1 INITIAL CELL POPULATION ISOLATION AND CHARACTERIZATION

HSPC can be isolated from different adult (BM, mPB) and neonatal (UCB) cell sources, typically through cell sorting (either magnetic- or fluorescence-based methods) based on the expression of surface antigens (Wognum et al. 2003).

Several surface markers are currently used to determine the lineage commitment of hematopoietic cells. CD34 is the most common antigen used to isolate HSPC for transplantation. CD34 is thought to be associated with more primitive populations, and the number of CD34+ cells available for a HCT seems to correlate with the success of the transplant (Wagner et al. 2002). Nonetheless, the expression of this marker is not limited to HSPC populations, and it is also present on the vascular endothelium, for instance (Baumhueter et al. 1994). CD34+CD90+ cells represent a more primitive subset of hematopoietic cells, as studies evidence lower cellular outputs per transplanted stem cells and higher difficulty of engraftment for cells lacking the expression of CD90 (Majeti et al. 2007). Nevertheless, Notta and co-workers, for instance, reported that single CD49f+ cells were capable to efficiently generate long-term multilineage grafts (Notta et al. 2011). HSPC are also thought to be enriched in the CD34+CD133+ subset of hematopoietic cells (Pastore et al. 2008), which turns the CD133+ cells into a desirable population to favor engraftment and repopulation in transplant models (Bauer et al. 2008, 133; Yin et al. 1997, 133), with isolation of CD133+ cells being clinically used as an alternative to CD34+-isolated cells (Gordon et al. 2003).

Cell surface marker enrichment prior to expansion could also be important to promote tumor purging and T cell depletion, particularly in UCB transplant settings. However, each step in cell purification and enrichment in specific surface markers results in loss of primitive cells. The ability to expand cells without the need of a pre-enrichment step (e.g. culture of mononuclear cells (MNC)) would reduce the processing efforts, therefore decreasing the associated costs and limiting cell loss (Koller et al. 1995).

7.4.2 CULTURE MEDIUM FORMULATION

The development of clinical-grade HSPC products requires serum-free, well-defined medium formulations to decrease variability and prevent the immunological rejection of the transplanted cells. Several cytokines and growth factors have been incorporated in *ex-vivo* hematopoietic cell cultures, with stem cell factor (SCF), Flt-3 ligand (Flt3L), thrombopoietin (TPO), interleukins (IL)-1, IL-3, IL-6, IL-11, angiopoietin-1 osteopontin, erythropoietin (EPO), granulocyte- (G-), macrophage- (M-), and granulocyte/macrophage- (GM-) colony stimulating factors (CSF) being the most commonly used to promote expansion of HSPC (Audet et al. 2002; Conneally et al. 1997; Majumdar et al. 2000; Nilsson et al. 2005; Oostendorp et al. 2005; Qian et al. 2007; Sacchetti et al. 2007; Wagner et al. 2007; Zhang and Lodish 2004). Amongst these, SCF, Flt3L, and TPO are the most frequently used in most clinical and preclinical studies (Flores-Guzman et al. 2013).

Newly tested growth factors have also been reported to promote the expansion of HSPC. Some of these factors include insulin-like growth-factor binding protein-2 (IGFBP-2), which has been demonstrated to synergize with members of the angiopoietin-like protein family in combination with early-acting cytokines (Fan et al. 2014;

Zhang et al. 2008), promoting the expansion of cells with repopulating ability. Other cytokines include insulin-like growth factor-2 (IGF-2), fibroblast growth factor-1 (FGF-1), pleiotrophin, and neurotrophin (Çelebi et al. 2012; Himburg et al. 2010; Walenda et al. 2011). Similarly, medium formulations have been supplemented with ligands like the Notch ligand Delta-1 (Csaszar et al. 2014a; Delaney et al. 2010), the copper chelator tetraethylenepentamine (TEPA) (Peled et al. 2004), and several small molecules such as StemRegenin-1 (SR-1) or UM171/729 (Boitano et al. 2010; Fares et al. 2014; Pabst et al. 2014) to promote expansion of HSC and progenitors.

Whether biological factors are delivered as soluble cues or immobilized within a culture system, they can also impact the function and potency of the provided signals. For instance, SCF, GM-CSF, and Notch-1 ligand present different functional effects depending on their soluble or matrix bound state (Delaney et al. 2005; Ohishi et al. 2002, 1; Worrallo et al. 2017; Mahadik et al. 2015).

Nonetheless, and even though different combinations of cytokines and/or small molecules have been used to culture HSPC, they generally lead to the exhaustion of the stem cell pool. The optimal combination and concentration of factors able to preserve and even expand undifferentiated HSC while maintaining their engraftment capacity has not been achieved yet.

7.4.3 Oxygen Tension and pH

The oxygen tension in *ex-vivo* culture systems is usually set at 20%–21% O_2, the atmospheric oxygen level. Nonetheless, the mean oxygen tension in the BM microenvironment is reported to be around 5% (Jež et al. 2015). More primitive HSC and progenitors are thought to reside in the most hypoxic (i.e., relatively to atmospheric air) regions of the BM (0.1%–6% O_2 levels (Eliasson and Jönsson 2010; Guitart et al. 2010; Spencer et al. 2014)), while more proliferative and committed progenitors occupy the regions at higher oxygen tension (Mohyeldin et al. 2010; Parmar et al. 2007; Spencer et al. 2014). Formation of erythrocytes and megakaryocytes, for instance, has been reported to be favored at an oxygen tension of 20% O_2 (Mostafa et al. 2000; Laluppa et al. 1998).

More recent measurements in the BM of mice also uncovered heterogeneities in the different areas of the BM, with oxygen levels detected situated between 1.3%–4.2% (Spencer et al. 2014). Protection from oxidative stress and induction of quiescence are two possible effects of hypoxia over HSC (Eliasson and Jönsson 2010; Shima et al. 2009). On the other hand, it is known that low oxygen tensions have the capability to stabilize hypoxia inducible factor-1α (HIF-1α) (Ivan et al. 2001), a major regulator of several genes involved in the cellular trophic activity and maintenance of more primitive HSC.

The impact of oxygen tension on determining the proliferative state and fate of HSPC has been investigated in *ex-vivo* culture systems. Several studies found that low oxygen tension inhibits the differentiation of HSPC, delaying the differentiation and apoptosis of megakaryocytes (5% O_2) (Mostafa et al. 2000) and inhibiting the terminal phase of natural killer cells differentiation (1% O_2) (Yun et al. 2011). Whereas, higher oxygen tension is actually primordial to ensure megakaryocyte maturation and platelet production (Mostafa et al. 2000; Eliades et al. 2012). Low oxygen tensions (0.1%) were also shown to regulate the return of cycling UCB CD34$^+$ HSPC into a non-proliferative, more quiescent state (Hermitte et al. 2006). Moreover, once transplanted into immunocompromised mice, HSPC previously exposed to hypoxia evidenced higher numbers of BM-repopulating cells, suggesting the beneficial role of low oxygen tensions on promoting cellular survival and enhanced engraftment (Danet et al. 2003; Shima et al. 2009). Culture of UCB-derived CD34$^+$ cells under an oxygen concentration of 3% better maintained the self-renewal and reconstitution ability of the expanded HSPC relative to culture performed under 20% O_2 (Ivanovic et al. 2004). Also, in co-culture with BM mesenchymal stem/stromal cells (MSC), a 10% O_2 environment resulted in a higher expansion of UCB CD34$^+$CD90$^+$ cells when compared to 2%, 5%, or 21% O_2 (Andrade et al. 2015), which has been attributed to an enhanced proliferative potential of HSPC. When cultured at 10% O_2 rather than to cell death at the O_2 tensions tested more akin to fresh, uncultured, cells (Wang et al. 2016). Therefore, as the hematopoietic reconstitution ability and differentiation capacity of HSPC could be tailored according to exposure to different oxygen tensions, *ex-vivo* culture settings might benefit from bioreactor systems where oxygen is a controlled parameter.

The sensitivity of hematopoietic cells to pH changes in culture should also be considered, as accumulation of metabolic byproducts could lead to acidification of the culture medium, thereby inhibiting cellular proliferation. Myeloid differentiation, for instance, is thought to be favored in the pH range 7.2–7.4, whereas erythroid lineage commitment occurs optimally within the pH range 7.1–7.6 (Kowalczyk et al. 2011).

7.4.4 SHEAR STRESS

The need to provide cells with an adequate oxygen supply and a homogeneous distribution of metabolites and cytokines must be balanced with appropriate hydrodynamic profiles within a bioreactor culture to minimize the detrimental effects associated to shear stress. Indeed, in stirred systems, air bubbles created by headspace gas can cause shear-related cell damage (Michaels et al. 1996). On the other hand, shear stress has been shown to be essential to induce platelet production from megakaryocytes differentiated from UCB-derived CD34$^+$ cells (Avanzi et al. 2016). Importantly, although a stirring condition in spinner flasks was shown to enhance erythroid cell maturation from UCB-derived CD34$^+$ cells, cell shielding through culture inside macroporous microcarriers and porous scaffolds reduced the shear stress induced by medium flow, therefore increasing cell viability (Lee et al. 2015). Actually, the sensitivity of erythroid cells to shear stress has been shown in a previous study where the generated red blood cells lacked an intact morphology upon culture in a stirred bioreactor (Timmins et al. 2011).

Hosseinizand and colleagues investigated the effect of different agitation speeds on the expansion and differentiation of spinner flask-cultured UCB HSPC, showing that, whereas an agitation of 20 rpm is associated with insufficient mixing to create an uniform environment and a 60 rpm agitation speed is related to increased shear stress decreasing cell expansion, a 40 rpm agitation results in optimized cell expansion levels (Hosseinizand, Ebrahimi, and Abdekhodaie 2016). Similar observations were reported in other studies, where higher expansion folds of MNC cultured in stirred flasks were achieved at an agitation speed of 30 rpm while increased agitation speeds of 60 and 80 rpm were detrimental for cell growth and viability (Jing et al. 2013; Collins et al. 1998a). Additionally, impeller design has also been shown to impact cell expansion in stirred systems, with a flat blade agitator rendering higher expansion of total hematopoietic cells relative to a magnetic stir bar-based agitation in spinner flasks (Collins et al. 1998a).

7.4.5   STROMA-BASED CO-CULTURES *VERSUS* LIQUID CULTURES

In 1978, the concept of a hematopoietic niche was introduced by Schofield (Schofield 1978), who suggested that the behavior of a stem cell could be determined by its cross-talk with the surrounding environment, including neighboring stromal cells. *In vivo*, HSC are in the BM, where they localize nearby or in direct contact with several cell types. Initial attempts to recreate the hematopoietic niche exploited long-term cultures of feeder layer of BM stromal cells (Dexter-type cultures) to support the culture of hematopoietic cells (Dexter et al. 1977). The presence of stromal cells, capable of conditioning the culture environment with hematopoietic supportive cues and adding to their ability to promote cell-cell and cell-extracellular matrix (ECM) interactions, is thought to improve the survival and functionality of co-cultured HSPC (Kim 1998). Particularly, MSC are known to be a key player in the BM hematopoietic niche (Méndez-Ferrer et al. 2010) and the crosstalk between HSPC and MSC is thought to regulate cell fate (Baksh et al. 2005; Reikvam et al. 2015; da Silva et al. 2009).

Bioreactor systems, such as packed- and fluidized-bed bioreactors, provide suitable configurations to enhance the cell-cell contact between stromal supportive cells and HSPC (Jelinek et al. 2002; Mantalaris et al. 1998; Meissner et al. 1999). Miniaturized systems, such as a chip-based 3D co-culture of UCB hematopoietic cells and MSC added to a perfused microcavity array, were also shown to allow integration of HSPC within the supportive 3D network of stromal cells through β-catenin and N-cadherin intercellular junctions (Wuchter et al. 2016). Moreover, when compared to monolayer co-cultures, this microbioreactor setting evidenced enhanced capability to better preserve the primitiveness of hematopoietic progenitor cells. Nonetheless, whether or not direct contact between HSPC and stromal supportive cells is essential to promote HSPC expansion, differentiation or an increase in their homing capabilities still needs to be determined.

Alternatively, the use of strategies such as immobilization of growth factors in suspension cell culture platforms could constitute a valuable alternative to cell-cell contact in order to impact the hematopoietic cell activity. Depending on being immobilized or in a soluble state, factors like SCF and Notch-1 ligand have distinct effects on cell activity (Delaney et al. 2005; Fox et al. 2000; Varnum-Finney et al. 2003). Immobilized GM-CSF has also been shown to provide an increase of a two to three order of magnitude potency relative to their soluble counterpart on the growth response of hematopoietic cell lines (Worrallo et al. 2017). In addition, the use of immobilization techniques might help to decrease the quantity of growth factors required to regulate cellular responses, which would ultimately reduce manufacturing costs (Worrallo et al. 2017).

Liquid suspension cultures, generally performed in plastic culture flasks or disposable bags (Boiron et al. 2006; Shpall et al. 2002; de Lima et al. 2008), are easier to implement and are more prone to standardization. There is no risk of contamination with detrimental products derived from stromal feeder layers, while the complexity inherent in culture systems that, in addition to hematopoietic cells, must support adherent stromal cells, is avoided. Therefore, liquid cultures, through incorporation of well-defined chemical factors might offer some advantages when it comes to establish GMP-compliant platforms. Nonetheless, the biological cues provided by stromal co-cultured cells might be required to sustain hematopoietic cell expansion and functionality in *ex-vivo* culture systems.

7.4.6 CELL DENSITY

Contrary to static cultures, dynamic systems can support growth of high cell densities, particularly if coupled to controlled medium dilution strategies (Csaszar et al. 2012; Choi et al. 2010). Hematopoietic cell proliferation is limited by factors such as pH and lactate (Patel et al. 2000), which could reach inhibitory levels when high cell densities are attained. Simultaneously, inhibitory factors that accumulate as cells expand and tend to differentiate in *in vitro* cultures could also hinder stem cell self-renewal (Madlambayan et al. 2005).

On the other hand, and although initiation of cultures at low cell densities could maximize cell expansion yields, low density cultures of CD34+-enriched hematopoietic cells under agitation have been reported to lead to cell stress and damage (Timmins et al. 2009). Accordingly, Collins and collaborators have observed that minimum initial cell densities of 3×10^5 MNC/mL should be attained in serum-free medium when cultures are established in stirred systems (Collins et al. 1998a)—an observation further corroborated in serum-containing medium (Levee et al. 1994; Kim 1998). Zandstra and colleagues have also reported that the long-term expansion potential of BM MNC was enhanced when initial cell densities inoculated in spinner flasks increased from 1×10^5 cells/mL to a 10-fold higher level of 1×10^6 cells/mL (Zandstra et al. 1994).

7.5 REACTOR CONFIGURATIONS

Several of the studies mentioned in the previous sections have relied on static, small-scale systems to expand hematopoietic cells. Contrary to static systems, where the development of metabolite, oxygen tension and pH gradients increases the heterogeneity of cell culture platforms and hinders growth of high cell densities, bioreactor systems allow more controlled monitoring and feeding regimes while handling is limited (as in Csaszar et al. 2012). This enables cell culture to be performed without the periodic fluctuations in medium composition associated with standard medium change protocols.

Since the first report of BM HSPC expansion performed in suspension culture (Zandstra et al. 1994), attempts to expand hematopoietic cells, either derived from BM or from other sources such as UCB and mPB, have been performed. Although some studies have reported the cultivation of hematopoietic cells in stirred systems, several other bioreactor configurations could be used to culture HSPC. Some configurations, such as hollow-fibers and packed-bed bioreactors, are more prone to recreate the 3D hematopoietic niche environment, especially when incorporating scaffolds and/or stromal co-cultures (Cho et al. 2008; Jelinek et al. 2002; Meissner et al. 1999). Such configurations could favor cell-cell, cell-ECM interactions and provide spatio-temporal cues in contrast to the microenvironmental homogeneity associated to cell culture in stirred suspension bioreactors.

Direct comparison between different types of bioreactors is not straightforward due to the wide variety of culture conditions and parameters applied amongst studies (including changes in medium composition, hematopoietic cell source, cell isolation and purification processes, initial cell seeding density, feeding regimens, presence of supportive cells and/or biomaterials, oxygen, and pH levels). In the following subsections, we will depict different bioreactor configurations exploited for the cultivation of hematopoietic cells, highlighting their main features, as well as advantages and associated limitations. Furthermore, direct comparison between different bioreactor systems and culture methodologies are limited in their ability to unravel the impact of individual culture parameters on cellular outcomes. In Table 7.1, we summarize the culture conditions and proliferation levels obtained in numerous bioreactor platforms.

We will focus on several bioreactor configurations (Figure 7.3) currently used to expand hematopoietic cells, with a focus on UCB-derived cells.

7.5.1 STIRRED BIOREACTORS

HSPC cultivation in stirred bioreactors was first reported in the early 1990s (Zandstra et al. 1994; Sardonini and Wu 1993). Shortly after, in 1998, culture under serum-free conditions was achieved in a stirred bioreactor (Collins et al. 1998a).

Spinner flasks are often explored as a screening system prior to the use of stirred-tank reactors (Jelinek et al. 2002; Xiong et al. 2002; Sardonini and Wu 1993; Collins et al. 1998a; Levee et al. 1994; Kim 1998). Xiong and colleagues, for instance, co-cultured UCB MNC with microcarriers coated with BM stromal cells in spinner flasks under serum-free conditions (Xiong et al. 2002). Whereas Sardonini and Wu performed a comparative analysis of BM expansion in static *versus* dynamic systems, including spinner flasks, with suspension cell cultures rendering the highest levels of total cell expansion (Sardonini and Wu 1993), Collins and collaborators reported that proliferation of PB CD34+ cells, as well as PB- and UCB-derived MNC in spinner flasks was dependent on the agitator design (Collins et al. 1998a). Particularly, the use of a stainless-steel stirrer could present detrimental effects on hematopoietic cell proliferation. To address this issue, the same authors adapted a stirred reactor with a Bellco spinner flask model 1965–250mL agitator to promote expansion of UCB and mPB MNC using a cell-dilution feeding protocol (Collins et al. 1998b). In addition, the rate of stirring is also known to impact the expansion fold of hematopoietic cells and, importantly, their clonogenic potential (Collins et al. 1998a; Jing et al. 2013).

Of notice, stirred systems have been shown to lead to changes on the gene expression profile (Li et al. 2006) and to contribute to a better engraftment and multilineage reconstitution ability of UCB CD34+ cells, regardless of the lower expansion of total cell number in comparison to static systems (Yang et al. 2007). Supplementation of stirred suspension bioreactors with stroma-conditioned medium could further support long-term expansion of human hematopoietic progenitors (Kim 1998).

Encapsulation of human BM cells in alginate microcapsules has also been explored in a spinner flask culture, where multilineage hematopoietic growth was achieved

TABLE 7.1

Bioreactor Culture Systems Used to Expand Human HSPC from Different Sources. The Cytokine/Molecular Cocktail and Culture Time Applied to Achieve the Depicted Fold Increase in the Number of Hematopoietic Cells Are Indicated

Type of Bioreactor	Input Cells	Cytokines/Molecules	Culture Time	Cell Fold Increase		Progenitor Cell Expansion	References
				TNC	CD34+, CD133+		
Stirred	BM MNC	IL-3, SCF	28	—	—	LTC-IC, 7; CFU, 22	Zandstra et al. (1994)
	BM MNC	IL-3, GM-CSF, c-kitL	14	15–27	—	CFU-GM, 1.5	Sardonini and Wu (1993)
	UCB/PB MNC	IL-3, IL-6, SCF, Flt-3L, G-CSF, GM-CSF, erythropoietin	12	>2 relatively to static culture	—	CFU-GM, BFU-E higher in stirred culture	Collins et al. (1998a)
	PB CD34+	IL-3, IL-6, SCF, Flt-3L, G-CSF, GM-CSF, erythropoietin	12	Comparable to static culture	—	Comparable to static culture	
	UCB/PB MNC	FBS, HS, IL-3, IL-6, SCF, G-CSF, GM-CSF, erythropoietin	12	201/397	—	CFU-GM, 29/13.5	Collins et al. (1998b)
	UCB MNC supported by BM stromal cells seeded on gelatin microcarriers	IL-3, IL-6, SCF	12	7.7	CD34+, 9.6	CFU, 23.3	Xiong et al. (2002)
	UCB MNC	FBS, IL-3, SCF, GM-CSF, erythropoietin	14	14	—	CFU, 9.2; CFU-GM, 10	De León et al. (1998)

(Continued)

TABLE 7.1 (Continued)

Bioreactor Culture Systems Used to Expand Human HSPC from Different Sources. The Cytokine/Molecular Cocktail and Culture Time Applied to Achieve the Depicted Fold Increase in the Number of Hematopoietic Cells Are Indicated

Type of Bioreactor	Input Cells	Cytokines/Molecules	Culture Time	Cell Fold Increase			References
				TNC	CD34+, CD133+	Progenitor Cell Expansion	
	Microencapsulated BM MNC	FBS, HS, IL-3, SCF, GM-CSF, erythropoietin	16–19	12–24	—	BFU-E, 11	Levee et al. (1994)
	UCB CD34+	FBS, IL-3, IL-6, SCF, Flt-3L, TPO	7	0.8	CD34+, 0.9; higher engraftment and multilineage reconstitution ability	CFU, 1.2	Yang et al. (2007)
	UCB CD34+	FBS, IL-6, SCF, Flt-3L, TPO	7	—	CD34+, 15	—	Luni et al. (2010)
	BM MNC in the presence of stroma-conditioned medium	FBS, HS, IL-3, SCF, GM-CSF, erythropoietin, MIP-1α	14	3	—	CFU-GM, 17	Kim (1998)
	UCB CD34+	FBS, HS, IL-3, SCF, TPO, Flt-3L	7	18	—	CFU, 8	Jelinek et al. (2002)
	UCB MNC	IL-3, IL-6, SCF, Flt-3L, TPO	6–9	Higher expansion at 30 rpm	Similar % of CD34+ cells between stirred culture at 30 rpm and static systems	Higher CFU at 30 rpm	Jing et al. (2013)
	UCB MNC	FBS, IL-3, IL-6, SCF	7	1.27	CD34+, 5.43	CFU, 10.6	Li et al. (2006)
Perfusion	BM MNC	IL-3, GM-CSF, erythropoietin, c-kitL	14	10	—	CFU-GM, 21; BFU-E (day 8), 12; LTC-IC, 7.5	Koller et al. (1993b)

(Continued)

TABLE 7.1 (Continued)
Bioreactor Culture Systems Used to Expand Human HSPC from Different Sources. The Cytokine/Molecular Cocktail and Culture Time Applied to Achieve the Depicted Fold Increase in the Number of Hematopoietic Cells Are Indicated

Type of Bioreactor	Input Cells	Cytokines/Molecules	Culture Time	Cell Fold Increase			References
				CD34+, CD133+	TNC	Progenitor Cell Expansion	
	BM MNC	FBS, HS, hydrocortisone, IL-3, GM-CSF, erythropoietin, c-kitL	14	—	20–25	CFU-GM, 10-30	Palsson et al. (1993)
	UCB CD34+ (heparin-chitosan scaffolds)	FBS, BSA, Flt-3L, TPO, SCF	7	CD34+ expansion similar to static conditions	—	CFU expansion similar to static conditions	Cho et al. (2008)
	PB MNC (grooves perpendicular to direction of flow)	IL-3, IL-6, SCF, G-CSF	10	—	stroma and stroma-free cultures—5.4 and 9.9	CFU-GM, 17-19; LTC-IC maintenance	Sandstrom et al. (1996)
	UCB MNC	HS, FBS, IL-3, IL-6, SCF	7	—	Increase of 93% relatively to static control (day 15)	CFU-GM, 18; CFU-MIX, 5.3; LTC-IC, 3	Koller et al. (1993a)
	BM MNC and CD34+	HS, FBS, IL-3, GM-CSF, erythropoietin, c-kitL	14	—	11.8 and 1298	CFU-GM, 14.3 and 41.4	Koller et al. (1995)
	UCB MNC	HS, FBS, PIXY321, erythropoietin, Flt-3L	12	—	3.9	CFU-GM, 298; LTC-IC, 350% increase relatively to unfed cultures	Koller et al. (1998)
	UCB MNC (phase I trial)	FBS, HS, PIXY321, Flt-3L, erythropoietin	12	CD34+Lin; 0.09-2.45	1–8.5	CFU-GM, 4.6-266.4	Jaroscak et al. (2003)

(Continued)

TABLE 7.1 (Continued)

Bioreactor Culture Systems Used to Expand Human HSPC from Different Sources. The Cytokine/Molecular Cocktail and Culture Time Applied to Achieve the Depicted Fold Increase in the Number of Hematopoietic Cells Are Indicated

Type of Bioreactor	Input Cells	Cytokines/Molecules	Culture Time	Cell Fold Increase			References
				TNC	CD34+, CD133+	Progenitor Cell Expansion	
	BM MNC (hydroxyapatite scaffold)	Mesencult supplements, human serum	42	—	—	CFU-GM, frequency of 13.3 versus no detection in static controls	Schmelzer et al. (2015)
	UCB Lin⁻	SCF, TPO, Flt-3L, IL-3	8	Similar TNC in TCPS and hollow fiber	Similar CD34+ in TCPS and hollow fiber	Equivalent human CD45+ BM repopulation of NSG mice between TCPS and hollow fiber cultured cells.	Xue et al. (2014)
Packed- and fluidized-bed	UCB CD34+ supported by macroporous collagen carriers seeded with a murine stromal cell line	FBS, HS, IL-3, SCF, TPO, Flt-3L	7	100	—	CFU-GM, 114; CAFC, 6.4-15	Jelinek et al. (2002)
	Macroporous microspheres seeded with BM MNC	Erytropoietin, SCF, IL-3, GM-CSF, IGF-1	28	43% erythroid cells versus 5.8% Dexter type cultures	—	—	Mantalaris et al. (1998)
	UCB MNC and murine stromal cells seeded on gelatin modified glass carriers	FBS, HS	8 of co-culture	—	—	CFU-MIX, 4.2; CFU-GM, 7; BFU-E, 1.8	Meissner et al. (1999)

(Continued)

TABLE 7.1 (Continued)

Bioreactor Culture Systems Used to Expand Human HSPC from Different Sources. The Cytokine/Molecular Cocktail and Culture Time Applied to Achieve the Depicted Fold Increase in the Number of Hematopoietic Cells Are Indicated

Type of Bioreactor	Input Cells	Cytokines/Molecules	Culture Time	Cell Fold Increase			References
				TNC	CD34+, CD133+	Progenitor Cell Expansion	
Roller bottle	UCB MNC	FBS, IL-3, IL-6, SCF, G-CSF, GM-CSF, Flt-3L, erythropoietin	5	1.04	—	CFU, 17	Andrade-Zaldivar et al. (2011)
RWV	UCB MNC with glass-coated microcarriers	SCF, Flt-3L, TPO, IL-3, G-CSF, GM-CSF	12	3.7	CD34+CD45+, 5.2	CFU, 5.1	Kedong et al. (2010)
	UCB MNC	FBS, HS, IL-3, SCF, G-CSF, GM-CSF, Flt-3L, TPO	8	436	CD34+, 33	CFU-GM, 22	Liu et al. (2006)
	BM CD34+	FBS, SCF, IL-3, IL-6, MGDF	6	No expansion	No expansion	Higher total CFU and LTC-IC in RWV than in static systems	Plett et al. (2001)
	UCB MNC with microencapsulated rabbit BM MSC	SCF, Flt-3L, TPO, IL-3	7	107	CD34+, 26	CFU, 19	Song et al. (2011)
	UCB MNC encapsulated in alginate beads	FBS, SCF, Flt-3L, TPO	12	4.41	Higher % CD34+ (18.08%) in comparison to static systems (0.45%)	Higher CFU-G/GM than in static systems	Yuan et al. (2013)

BFU-E – erythroid burst-forming unit; c-kitL – ckit ligand; CFU-GM – colony forming-unit; CFU-MIX – multilineage colony forming-unit; BSA – bovine serum albumin; MGDF – megakaryocyte growth and development factor; TCPS – tissue culture polystyrene.

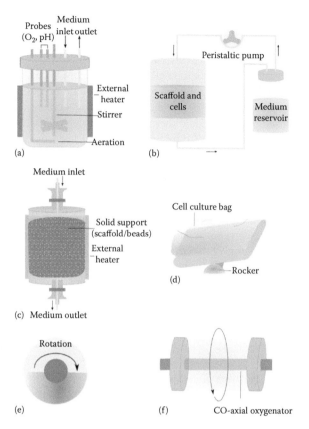

FIGURE 7.3   Schematic representation of bioreactors used to culture HSPC: (a) stirred; (b) perfusion; (c) packed-bed; (d) wave-mixed; (e) roller bottle; (f) RWV bioreactors.

(Levee et al. 1994). Additionally, scaled-down development platforms, such as the ambr™ system or the mini bioreactor developed by De León and colleagues (De León, Mayani, and Ramírez 1998), could represent valuable tools to optimize cell manufacturing processes prior to application of conventional stirred-tank technology (Ratcliffe et al. 2012). Worrallo and colleagues, for instance, showed that GM-CSF immobilized on magnetic particles could provide a two to three order of magnitude potency increase relative to the soluble factor, a functionality that was retained in the ambr™ microscale stirred-tank bioreactor (Worrallo et al. 2017). Interestingly, an array of bioreactors with controlled stirring has been developed at the microliter scale (Luni et al. 2010), thus widening the range of environmental conditions that could be tested at reduced costs. In contrast to the non-miniaturized bioreactors, in the Luni et al. study, a controlled stirring was achieved by buoyancy-driven thermos convection rather than by the mechanical stirring provided by impellers, which might lead to shear stress responsible for changes on cellular paracrine activity (Kurazumi et al. 2011) and on cell surface receptors (Lakhotia et al. 1993).

7.5.2 Perfusion Bioreactors

Perfusion systems and, particularly, hollow fiber membrane technology bioreactors, are suitable platforms to support cell culture at high density, consistent with the environment encountered by hematopoietic cells in the BM niche. In perfusion bioreactors, the continuous flow of medium at rates that could be optimized according to cell density levels allows nutrients to be replenished while waste products are removed. A perfused approach could therefore prove to be fundamental in high cell density cultures where removal of inhibitory factors secreted by cells overtime and nutrient supply are major requirements to enhance cell expansion.

In the 1990s, two flat-bed reactor systems, a multi-pass (Koller et al. 1993a) and a single-pass perfusion system (Koller et al. 1993b; Palsson et al. 1993) were used to expand UCB and BM MNC, respectively. A challenge posed to these systems relies on retention of cells while medium is perfused through the culture chamber. Addressing this issue, Sandstrom and colleagues modified a flat-bed perfusion chamber with grooves perpendicular to the direction of the bulk medium flow, therefore facilitating retention of non-adherent hematopoietic cells (Sandstrom et al. 1996). Additionally, this grooved bottom has been reported to protect cells from the primary flow environment, maintaining shear stress at negligible levels (Horner et al. 1998). Importantly, the single-pass system has been integrated into GMP closed and disposable reactor cassettes, allowing automated perfusion and clinical-grade cell recovery (Goltry et al. 2009; Mandalam et al. 1999).

Perfused systems might also benefit from the incorporation of strategies capable of retaining secreted growth factors in the cellular microenvironment. A perfused platform that allows accumulation of cell-secreted cytokines required for optimal growth is the hollow fiber bioreactor, which has been reported to enable cell densities 100-fold higher than traditional culture systems (Doran et al. 2009, 2012). The inclusion of membrane technology in hollow fiber culture of cells added to the intracapillary space helps retaining high molecular weight proteins within the fiber lumen (Doran et al. 2012). This prevents dilution of relevant cytokines involved on cellular autocrine and paracrine signaling, while oxygen and low molecular nutrients such as glucose are exchanged with the bulk medium. Additionally, the retention of proteins and other molecules, which are relevant for cell expansion in the intracapillary space of hollow fibers, in close contact with cultured cells, might contribute to significantly reduce the volume of culture medium required for the large-scale production of HSPC. Importantly, the ability of hollow fiber bioreactors to support the culture of stromal cells could help on establishing a biomimetic platform of the BM microenvironment. The resulting stroma-hematopoietic cellular interaction in a high cell density environment could therefore be explored to establish leukemia disease models (Usuludin et al. 2012) or to promote the *ex-vivo* expansion of HSPC (Xue et al. 2014). Nevertheless, a major limitation associated with hollow fiber cell culture is the difficulty associated with cell harvest and sampling along culture time. Sardonini and Wu, for instance, reported no observable expansion of HSPC in a hollow fiber reactor, in contrast to static control cultures (Sardonini and Wu 1993), which might be in part due to deficient cell recovery.

In a study where open-porous foamed hydroxyapatite scaffolds, mimicking the bone matrix, were coupled with a multi-compartment hollow fiber membrane, long-term culture (6 weeks) of whole human BM MNC was established (Schmelzer et al. 2015).

Although the presence of the scaffold did not seem to affect the BM cell population to a significant extent, bioreactor perfusion culture better supported the short- and long-term culture of hematopoietic progenitors, maintaining significantly higher colony numbers in comparison to static culture. Similarly, UCB CD34$^+$ cells cultured within a porous glycosaminoglycan (GAG)-chitosan scaffolds in a perfusion bioreactor showed higher levels of primitive progenitors relatively to static cultures (Cho et al. 2008).

The clinical-scale implementation of perfusion cultures has been performed in an automated Aastrom cell production system, rendering higher expansion of UCB samples comparatively to control static cultures (Koller et al. 1998). Importantly, Madlambayan and collaborators have designed a single-use, closed-system bioprocess that couples the ability of inline selection of UCB subpopulations to the capacity of promoting medium dilution (Madlambayan et al. 2006). This approach will likely inspire further studies where simultaneous removal of culture-generated mature blood cell populations (i.e. Lin(eage)$^+$ cells in the Madlambayan study) and dilution of endogenously produced negative regulators of HSC expansion could prove to be advantageous to expand HSC and progenitors with conserved engraftment potency and maintenance of multilineage potential.

7.5.3   PACKED- AND FLUIDIZED-BED BIOREACTORS

In the hematopoietic BM niche, a multitude of soluble factors, ECM components and supportive cells regulate the fate of HSC. Particularly, the interaction between stromal and hematopoietic cells is mediated by secreted factors, as well as by direct cell-cell contact. Several bioreactors, although promoting cytokine-mediated communication, lack the ability to favor physical interactions between cells. In this context, packed-bed (or fixed-bed) reactors can improve cell-cell contact between co-cultured HSPC and the stromal compartment, thought to be key regulators of hematopoiesis. Microcarriers (Jelinek et al. 2002; Mantalaris et al. 1998; Meissner et al. 1999) or nylon meshes (Naughton and Naughton 1989) provide 3D scaffolding for stromal cell adhesion while culture medium is perfused through the packed-bed culture system. The enhanced contact between stromal and UCB-isolated CD34$^+$ cells promoted by a fixed-bed reactor configuration resulted in a 6.4-fold higher expansion of the early progenitor cells relative to a suspension system (Jelinek et al. 2002). On the other hand, Mantalaris and co-workers exploited a similar packed-bed bioreactor to promote a high-density cell culture of MNC supported by porous microspheres. The intimate cell-cell interactions resembled the structure of the BM hematopoietic niche and generated a higher percentage of erythroid cells compared to cultures on conventional flasks (Mantalaris et al. 1998).

Besides packed-bed bioreactors, fluidized-bed technology can also be potentially applied to support the expansion of HSPC. Meissner and colleagues compared the two perfusion systems, packed-bed and fluidized-bed bioreactors, in a co-culture system where porous glass microcarriers were used to promote adhesion of stromal cells (Meissner et al. 1999). In the fluidized-bed system, the applied liquid flow rate caused an upward motion of the microcarriers, hindering the adhesion of HSPC to the microcarrier-supported stromal cells. In this system, no significant fold expansion of hematopoietic cells was achieved.

In an air-lift packed-bed bioreactor, sustained production of BM cells was supported by a stromal cell layer growing on a fiber-glass matrix packed in the bioreactor (Highfill et al. 1996).

7.5.4 WAVE-MIXED BIOREACTORS

The wave-mixed bioreactor (e.g. Wave®) employs a disposable culture bag on a rocking platform that can be partially filled with the culture medium and cells, and the headspace provides oxygenation. The rocking motion of the bag is responsible for the mass transfer and creation of waves at the air-liquid interface, therefore facilitating oxygen transfer. The easy scale-up and versatility associated to this bioreactor configuration, which can be operated in a batch, a fed-batch, or perfusion modes, turns it into a suitable alternative to cultivation of HSPC. Hami and colleagues, for instance, suggested the use of a 10-L volume wave bioreactor for the GMP manufacture of large numbers of highly pure T cells from a starting population of peripheral blood mononuclear cells (PBMC) for the treatment of chronic lymphocytic leukemia (CLL) patients (Hami et al. 2004). T lymphocytes have also been expanded in a wave-mixed bioreactor suspension culture to be used in the treatment of malignant melanoma (Sadeghi et al. 2011). A two-phase production of neutrophils from UCB- or mPB-derived CD34+ cells, initially cultured in a static platform (as the low density hematopoietic culture was reported to be particularly sensitive to agitation) and, after nine days, transferred to a 10-L scale wave-mixed bioreactor, resulted in significant production of clinical doses of neutrophils, important to address the period of neutropenia that arises from chemotherapy treatments (Timmins et al. 2009).

The ability of wave-mixed bioreactor to integrate continuous medium perfusion, allowing high cell densities to be attained, featuring disposability of cell culture bags, turns this system into a promising platform to support clinical-grade cell manufacturing.

7.5.5 ROLLER BOTTLES

Roller bottles have long been used for animal cell culture (Knight 1977; Kunitake et al. 1997), namely in the biotechnology industry to produce vaccines (Liu et al. 2003). However, it was not until 2011 that culture of human UCB HSPC was reported (Andrade-Zaldivar et al. 2011). In this study, Andrade-Zaldívar cultured UCB MNC in roller bottles and showed a 17-fold increase in colony-forming units (CFU) at day 5 of culture and a superior expansion relatively to static cultures (by a 10-fold).

The ability of roller bottles to prevent the formation of concentration gradients and provide superior gas exchange, while cells are maintained in suspension at low agitation rates, might be advantageous to culture shear-sensitive cells as hematopoietic cells. However, the incomplete filling of the vessel in roller bottles creates a headspace that results in turbulence and secondary bubble formation in the culture medium. Although a large-scale continuous perfusion roller bottle bioreactor has been reported that allows high cell densities to be obtained (Berson et al. 2002), roller bottles are most commonly operated in a batch or fed-batch mode, therefore limiting their production capacity. In addition, scale-up in roller bottles is frequently achieved by increasing the number of axially rotating vessels, which could pose space limitations.

Nonetheless, apart from the higher facility space requirements, the existence of several roller bottles operating in a single production run could simultaneously represent an advantage as contamination of one roller bottle would not necessarily lead to failure of the entire lot, a relevant parameter in risk contamination control.

7.5.6 ROTATING WALL VESSEL BIOREACTORS (RWV)

The Rotating Wall Vessel (RWV) bioreactor was originally conceived by the National Aeronautics and Space Administration (NASA) as a tool to study cellular responses to microgravity. The ability of the RWV technology to address some of the shortcomings of conventional cell culture bioreactors, namely the shear stress in stirred and perfusion systems, led researchers to exploit the RWV in hematopoietic cultures, either using cells from BM (Plett et al. 2001) or UCB (Kedong et al. 2010; Liu et al. 2006) sources. The RWV is a cylindrical bioreactor that operates on the principle of solid body rotation about a horizontal axis. The angular velocity of the system counteracts the sedimentation velocity of cells/aggregates allowing these to be kept in suspension. Unlike roller bottles, the culture chamber is completely filled with medium, avoiding the turbulence created by a headspace. As the culture medium rotates at the same angular velocity as the vessel wall, the large shear stresses associated with turbulent flow are avoided. Additionally, oxygen delivery is accomplished via a coaxial silicone membrane, thereby promoting a bubble-free aeration (Hammond and Hammond 2001). These two factors contribute to the minimization of the damaging effects associated with shear stress.

Liu and colleagues applied an increasing rotating speed from 0 to 6 rpm, allied with a dilution-type feeding protocol, to culture UCB MNC in RWV. Although the engraftment capacity of the expanded cells was not evaluated, a 33-fold expansion in the number of CD34+ cells was achieved upon eight days of culture (Liu et al. 2006).

Further exploiting the ability of RWV to sustain the culture of anchorage-dependent cell types, such as MSC, as well as cells in suspension, such as HSPC, Kedong and colleagues established a co-culture system where UCB-derived hematopoietic progenitors and MSC, seeded on microcarriers, were co-localized and able to proliferate, resulting in a 5.2-fold increase in the number of CD34+ cells (Kedong et al. 2010). Both MSC and HSPC presented a higher expansion capacity in the RWV than in a static system or in stirred spinner flask cultures. Interestingly, in this study, human adipose tissue (AT)-derived MSC encapsulated in alginate-chitosan beads were used as feeder cells due to their ability to secrete growth factors supporting UCB MNC. Also exploiting the support provided by microencapsulated rabbit BM MSC, Kedong Song and collaborators developed a co-culture system of UCB MNC and MSC where a cell-dilution feeding protocol was applied, resulting in a 26-fold expansion in the number of CD34+ cells (Song et al. 2011). The presence of co-cultured alginate microencapsulated MSC not only enhanced the expansion of total nucleated cells and CD34+ cells in RWV cultures, while the use of animal serum was avoided, but also favored the maintenance of a higher percentage of more undifferentiated hematopoietic cells. Similarly, a RWV culture system with human UCB MNC encapsulated in alginate beads has shown to provide higher amplification of human HSPC when compared to conventional 2-D culture systems (Yuan et al. 2013).

7.6 SELF-RENEWAL *VERSUS* LINEAGE COMMITMENT OF *EX-VIVO* CULTURE OF HEMATOPOIETIC STEM/PROGENITOR CELLS

During *ex-vivo* cell culture, together with maximizing total cell yields, it is essential to unravel the different subpopulations that comprise the final cell product to attain improved clinical outcomes once cellular transplantation is performed. High cell proliferation yields might be associated with cultures overpopulated with differentiated cells rather than expansion of more primitive HSC. Indeed, despite the importance of hematopoietic progenitors in reducing the period of pancytopenia in HCT patients, long-term hematopoiesis could only be sustained by administration of functional, self-renewing HSC.

The tight balance between self-renewal of HSC and their commitment to differentiated hematopoietic lineages, which exists *in vivo* in the hematopoietic niche, is challenged by the inadequacy of the biological signals provided in *ex-vivo* settings. The complexity of the factors involved in the regulation of the HSPC fate *in vivo* is hard to replicate in bioreactor settings, hindering the expansion levels of the primitive cells while more differentiated hematopoietic populations are originated. In part, this might be due to the inhibitory feedback signals generated from differentiating cells in culture, particularly by the monocyte-derived inhibitory factors (chemokines CCL3, CCL4, CXCL10, TGFB2, and TNFSF9) (Kirouac et al. 2010). Qiao and collaborators found that, while progenitor cells express signals involved on HSC self-renewal, mature cells, namely monocytes and granulocytes, produce ligands that promote differentiation (Qiao et al. 2014). To circumvent the inhibitory feedback mechanisms triggered by differentiated cells, Csaszar and co-workers proposed a controlled fed-batch culture of UCB HSPC where a syringe pumping system allowed medium dilution to be performed (Csaszar et al. 2012). After eight days of culture, the number of long-term severe combined immunodeficiency (SCID) repopulating cells (SRC) was 7.6- and 3.6-fold higher in the fed-batch system and under complete medium change control, respectively, relatively to fresh, uncultured cells. Of notice, Fares and colleagues showed that with the addition of a pyrimidoindole derivative, UM171, to this culture system featuring fed-batch medium-dilution, further stimulated the *in vitro* and *in vivo* expansion of UCB CD34$^+$ cells (Fares et al. 2014).

These studies corroborate the importance of providing adequate biological cues to *ex-vivo* hematopoietic cell culture systems while, simultaneously, endogenously produced signaling factors are adjusted in fed-batch feeding regimes. Real-time monitoring of cell culture could further contribute to enhance our control over dynamic systems. Indeed, a higher *ex-vivo* expansion of UCB-derived HSPC has been achieved when quantum dot barcoded microbeads were incorporated in a fed-batch operated bioreactor to maintain the concentration of transforming growth factor-β1 (TGF-β1), a potent inhibitor of HSPC expansion (Csaszar et al. 2012), below a critical inhibitory level (Csaszar et al. 2014b).

Besides the development of methods to expand HSC while blocking their differentiation, manufacture of mature lineage cells, such as platelets, erythrocytes, and lymphoid effector cells capable to generate a competent immune system, is also of utmost importance. Biomimetic cues of the BM native microenvironment have been incorporated in the design of bioreactors when the generation of differentiated hematopoietic cells is targeted. For instance, signaling cues provided by the ECM composition, the architecture of blood vessels and shear stress are essential for the production of functional platelets. Although it has been since 1995 that the first human platelets were generated from HSC *in vitro* under static conditions (Choi et al. 1995), bioreactors could possibly contribute to an increase in platelet yields and enhance their functionality. Additionally, biomaterial-based approaches might also be exploited, as reported by Sullenbarger and co-workers who used polyester and hydrogel scaffolds coated with fibronectin and inserted in multiple bioreactor modules to promote platelet formation from hematopoietic progenitor cells (Sullenbarger et al. 2009).

Avanzi and colleagues showed that expansion of UCB CD34$^+$ cells early in the culture process could promote platelet production to a higher extent than expansion of more differentiated cells, at a later stage of the cell culture protocol (Avanzi et al. 2016). On the other hand, medium perfusion in silk microtubes with ECM components and stromal cell-derived factor-1α (SDF-1α) entrapped in the microtube walls, mimicking a blood vessel, successfully supported platelet formation *ex-vivo* (Di Buduo et al. 2015).

By mimicking the cell-cell contacts occurring in the BM erythroblastic microenvironment, a 3-D aggregate culture system of late erythroblasts differentiated from UCB CD34$^+$ cells enhanced the production of mature red blood cells in comparison to 2-D cultures, by using spinner flasks to promote a more efficient medium supply to the cellular aggregates (Lee et al. 2015). Moreover, entrapping cells inside macroporous microcarriers and porous scaffolds protected cells from the shear stress induced by stirring in the spinner flask. Besides the impact on cellular viability, tailoring the hydrodynamics of bioreactor systems is thought to impact hematopoietic differentiation. The high shear stress and turbulent flows that could be generated in mechanically stirred systems contrast with the lower shear stress and laminar flow that is generally associated with culture in rotating vessels. These differences might be exploited, for instances, to achieve lineage commitment into specific hematopoietic cell populations.

In this context, manufacture of red blood cells from UCB-derived CD34$^+$ cells has been shown to benefit from culture in a stirred micro-bioreactor (ambr^{TM}), where a 24% and 42% medium volume and culture time reduction, respectively, were achieved in comparison to non-optimized flask culture (Bayley et al. 2016).

A relevant question when terminally differentiated blood cell products are manufactured resides on the extent to which proliferation of early-stage progenitors should be maximized prior to providing differentiating cues to the cultured cells. Non-expanded or Delta1 Notch ligand-expanded CD34$^+$ cells, for instance, have been studied as potential cell sources to promote production of erythrocytes in a micro-scale ambr^{TM} suspension bioreactor (Glen et al. 2013). The use of small-scale prototypes could help in achieving an efficient production of cells, not only fully functional but also produced in a cost-effective way.

7.7 FROM THE BENCH TO THE CLINIC—APPLICATIONS OF EXPANDED CELLS

Although most HCT performed with UCB cells have exploited unmanipulated and/ or double transplantation of UCB units, clinical-grade expansion of HSPC might help to improve the success of HCT. Importantly, clinical trials testing the infusion of *ex-vivo* expanded hematopoietic cells from UCB have attested to the safety of these cell products. The clinical-grade *ex-vivo* expansion of UCB cells has been attempted through the following methods: (1) using cytokines alone (Shpall et al. 2002); (2) blocking the differentiation of HSPC using (a) nicotinamide analogs (Horwitz et al. 2014), (b) copper chelators (de Lima et al. 2008; Stiff et al. 2013), (c) small molecules, such as an aryl hydrocarbon receptor antagonist (Wagner et al. 2016), (d) activating Notch signaling pathways (Delaney et al. 2010); or (3) exploiting co-culture systems with stromal cells (de Lima et al. 2012).

The use of cytokines alone in static systems to expand HSPC has shown limited success in UCB transplants (Shpall et al. 2002), which prompted researchers to develop *ex-vivo* HSPC culture strategies that would favor expansion of more primitive HSPC while blocking their differentiation. In a phase I trial enrolling 11 patients subjected to a double UCB transplantation, CD133$^+$-enriched cells were expanded in culture bags for 21 days in the presence of nicotinamide and a cytokine cocktail. The negative fraction containing non-cultured T cells was cryopreserved. The cultured CD133$^+$-enriched hematopoietic cells and the thawed CD133-negative fraction were infused (NiCord) together with a second unmanipulated UCB unit (Horwitz et al. 2014). The expanded unit contributed to partial or complete chimerism in seven patients and hematopoietic recovery was accelerated. In a phase I/II trial, de Lima and colleagues cultured UCB-CD133$^+$ cells with the copper chelator TEPA and "early acting cytokines," reporting engraftment in 9 out of 10 patients with hematological malignancies (de Lima et al. 2008). In a subsequent phase II/III trial involving 101 patients (StemEx®), a fraction of a single UCB unit was expanded for 21 days with cytokines and TEPA and infused in patients along with the unmanipulated UCB fraction (Stiff et al. 2013). The median time to neutrophil and platelet engraftment was significantly improved compared to a registry cohort undergoing double unit UCB transplantation. Exploring a double UCB transplantation, SR-1, an aryl hydrocarbon receptor antagonist, was applied in the presence of Flt-3L, SCF, TPO, and IL-6 to expand a CD34$^+$-enriched fraction of an UCB unit. The CD34-depleted fraction was cryopreserved. Together with an unmanipulated UCB unit, the expanded fraction and the unexpanded, cryopreserved T cell-containing fraction of the UCB unit were applied in a phase I/II trial. Improved neutrophil and platelet recovery was reported (Wagner et al. 2016). Delaney and co-workers exploited a Notch-mediated expansion of UCB-CD34$^+$ cells in a phase I trial enrolling 10 patients (Delaney et al. 2010). A decrease of the neutrophil recovery time was observed when an unmanipulated UCB unit was infused four hours prior to infusion of the expanded unit in comparison to patients who received two unmanipulated UCB units. Nonetheless, one-year post transplantation, only the unexpanded UCB unit contributed to engraftment.

Exploring the ability of co-cultured MSC to be part of the hematopoietic niche and regulate hematopoietic cell fate (Méndez-Ferrer et al. 2010), unselected UCB cells have been expanded in the presence of a cytokine cocktail and supported by stromal cells. In a phase I trial involving 31 patients, transplantation of HSPC co-cultured with MSC (in combination with a second, non-manipulated UCB unit) proved to be safe and led to a more rapid recovery of neutrophils and platelets when compared to historical control patients receiving two unmanipulated UCB units (de Lima et al. 2012).

To date, the majority of the clinical trials performed with *ex-vivo* expanded UCB units exploited static culture systems, namely cellbags. Nevertheless, a phase I trial at Duke University Medical Center evaluated the use of an automated continuous perfusion bioreactor, the AastromReplicell System (Vericel Corp, Cambridge, MA), to expand a small portion of an UCB unit while most of the cells from the same unit were unmanipulated and infused in 27 patients (Jaroscak et al. 2003). However, the administration of expanded UCB cells in addition to the conventional graft did not render improved neutrophil or platelet engraftment, at 12 days post-transplantation.

Overall, a successful HCT using *ex-vivo* expanded cell products would require not only cultivation protocols capable of producing a sufficient number of cells to be infused in the patient, but also able to preserve a pool of undifferentiated HSC while allowing expansion of committed myeloid progenitors.

7.8   FUTURE PERSPECTIVES

The *ex-vivo* expansion of HSPC, particularly of UCB-derived hematopoietic cells, whose biological potency and easy access relative to other cell sources are counterbalanced by their limited cell numbers, is frequently achieved at the expense of a selective expansion of differentiated progenies and short-term reconstituting HSC. Strategies that would promote maintenance of the more primitive HSC pool could contribute to the success of HCT. To accomplish this goal, many technical and regulatory challenges must be overcome. Amongst those, heterogeneous cell starting inoculum and the transient nature of cell subpopulations limit our ability to systematically obtain robust cell products.

Bioreactors able to incorporate the several dimensions present in the hematopoietic niche, whether these are related to cytokine concentrations, cell-cell or cell-ECM interactions, oxygen and nutrient gradients would certainly contribute to high quality cell products in a cost-effective, controlled, and reproducible manner.

Defining optimal culture conditions is frequently achieved in static systems, which would not necessarily represent the optimal conditions when cell culture is performed in larger-scale culture platforms. Smaller-scale bioreactors equipped with online monitoring would allow us to perceive how our acquired understanding of biological phenomena could translate into optimal cell production systems, accelerating bioprocess optimization. Additionally, high-throughput culture platforms might help overcome challenges associated with the scale-up of cell cultures, namely the need to minimize costs of production while quality and consistency of the cell product is maintained.

To date, the development of laboratory-scale bioreactors, although representing proof-of-concept basic research tools, lack standardized manufacture and operation processes, and they are not amenable for high throughput industrial scale operation of cell products. Computer-monitored culture systems, capable of maintaining cytokine concentrations within an optimal range, can help us establish an improved milieu of growth factors for HSPC. On the other hand, the frequency of medium exchange was identified to be the primary economical constraint to maximize cell growth (Bayley et al. 2016), highlighting the role that controlled bioreactor systems could play to improve cell manufacturing efficiency.

In the future, the identification of adequate bioreactor configurations and feeding regimens coupled to the use of optimized serum- and animal component-free and well-defined media would certainly contribute to the implementation of more efficient bioprocesses envisioning the application of *ex-vivo* expanded hematopoietic cells in the clinical arena. In addition, automated systems able to isolate and process biological samples in a closed circuit (e.g. BioSafe Sepax) would also contribute to achieve a GMP-compliant, clinical-grade cell manufacturing.

Coupled to upstream processing, downstream processing of cellular products would also require process parameters to be optimized and the design of bioreactors should not neglect this aspect of the cell product formulation.

The use of bioreactors for HSPC cultivation could help to reduce manufacturing variability, particularly by setting process limits and allowing feedback loops to be established in a controlled way. Moreover, automated approaches, associated with cell culture in bioreactors, can help to control variation and enhance the comparability of cell culture protocols.

REFERENCES

Andrade, P. Z., A. M. de Soure, F. dos Santos, A. Paiva, J. M. Cabral, and C. L. da Silva. 2015. *Ex Vivo* expansion of cord blood hematopoietic stem/progenitor cells under physiological oxygen tensions: Clear-cut effects on cell proliferation, differentiation and metabolism: Oxygen tension effects on HSC-MSC co-cultures. *Journal of Tissue Engineering and Regenerative Medicine* 9 (10): 1172–1181. doi:10.1002/term.1731.

Andrade-Zaldívar, H., L. Santos, and A. León Rodríguez. 2008. Expansion of Human Hematopoietic Stem cells for transplantation: Trends and perspectives. *Cytotechnology* 56 (3): 151–160. doi:10.1007/s10616-008-9144-1.

Andrade-Zaldivar, H., M. A Kalixto-Sánchez, A. P. B. de la Rosa, and A. De León-Rodríguez. 2011. Expansion of human hematopoietic cells from Umbilical Cord Blood using roller bottles in CO_2 and CO_2-free atmosphere. *Stem Cells and Development* 20 (4): 593–598.

Audet, J., C. L. Miller, C. J. Eaves, and J. M. Piret. 2002. Common and distinct features of cytokine effects on hematopoietic stem and progenitor cells revealed by dose-response surface Analysis. *Biotechnology and Bioengineering* 80 (4): 393–404. doi:10.1002/bit.10399.

Avanzi, M. P., O. E. Oluwadara, M. M. Cushing, M. L. Mitchell, S. Fischer, and W. B. Mitchell. 2016. A Novel Bioreactor and culture method drives high yields of platelets from stem cells. *Transfusion* 56 (1): 170–178. doi:10.1111/trf.13375.

Baksh, D., J. E. Davies, and P. W. Zandstra. 2005. Soluble factor cross-talk between human bone marrow-derived hematopoietic and mesenchymal cells enhances *in vitro* CFU-F and CFU-O growth and reveals heterogeneity in the mesenchymal progenitor cell compartment. *Blood* 106 (9): 3012–3019. doi:10.1182/blood-2005-01-0433.

Baldomero, H., M. Gratwohl, A. Gratwohl, A. Tichelli, D. Niederwieser, A. Madrigal, and K. Frauendorfer. 2011. The EBMT activity survey 2009: Trends over the past 5 years. *Bone Marrow Transplantation* 46 (4): 485–501.

Barker, J. N., D. J. Weisdorf, T. E. DeFor, B. R. Blazar, P. B. McGlave, J. S. Miller, C. M. Verfaillie, and J. E. Wagner. 2005. Transplantation of 2 partially HLA-matched Umbilical Cord blood units to enhance engraftment in adults with hematologic malignancy. *Blood* 105 (3): 1343–1347. doi:10.1182/blood-2004-07-2717.

Bauer, N., A. Fonseca, M. Florek, D. Freund, J. Jászai, M. Bornhauser, C. A. Fargeas, and D. Corbeil. 2008. New insights into the cell biology of hematopoietic progenitors by studying prominin-1 (CD133). *Cells Tissues Organs* 188 (1–2): 127–138. doi:10.1159/000112847.

Baumhueter, S., N. Dybdal, C. Kyle, and L. A. Lasky. 1994. Global vascular expression of murine CD34, a sialomucin-like endothelial ligand for L-selectin. *Blood* 84 (8): 2554–2565.

Bayley, R., F. Ahmed, K. Glen, M. McCall, A. Stacey, and R. Thomas. 2016. The productivity limit of manufacturing blood cell therapy in scalable stirred bioreactors. *Journal of Tissue Engineering and Regenerative Medicine*. doi:10.1002/term.2337.

Bensinger, W. I. 2012. Allogeneic transplantation: Peripheral blood versus bone marrow. *Current Opinion in Oncology* 24 (2): 191–196. doi:10.1097/CCO.0b013e32834f5c27.

Berson, R. E., W. J. Pieczynski, C. K. Svihla, and T. R. Hanley. 2002. Enhanced mixing and mass transfer in a recirculation loop results in high cell densities in a roller bottle reactor. *Biotechnology Progress* 18 (1): 72–77. doi:10.1021/bp0101482.

Boiron, J., B. Dazey, C. Cailliot, B. Launay, M. Attal, F. Mazurier, I. K. McNiece et al. 2006. Large-scale expansion and transplantation of CD34+ hematopoietic cells: *In vitro* and *in vivo* confirmation of neutropenia abrogation related to the expansion process without impairment of the long-term engraftment capacity. *Transfusion* 46 (11): 1934–1942. doi:10.1111/j.1537-2995.2006.01001.x.

Boitano, A. E., J. Wang, R. Romeo, L. C. Bouchez, A. E. Parker, S. E. Sutton, J. R. Walker et al. 2010. Aryl hydrocarbon receptor antagonists promote the expansion of human hematopoietic stem cells. *Science* 329 (5997): 1345–1348. doi:10.1126/science.1191536.

Brunstein, C. G., J. A. Gutman, D. J. Weisdorf, A. E. Woolfrey, T. E. DeFor, T. A. Gooley, M. R. Verneris, F. R. Appelbaum, J. E. Wagner, and C. Delaney. 2010. Allogeneic hematopoietic cell transplantation for hematologic malignancy: Relative risks and benefits of double Umbilical Cord blood. *Blood* 116 (22): 4693–4699. doi:10.1182/blood-2010-05-285304.

Cabrera, J. R., I. Krsnik, R. Fores, E. Ruiz, G. Bautista, B. Navarro, J. Gayoso et al. 2005. Post-engraftment infections in adult patients transplanted with single cord blood units supported by co-infusion of mobilized purified hematopoietic stem cells from a third party donor. *Blood* 106: 3240.

Casamayor-Genescà, A., A. Pla, I. Oliver-Vila, N. Pujals-Fonts, S. Marín-Gallén, M. Caminal, I. Pujol-Autonell et al. 2017. Clinical-scale expansion of CD34+ cord blood cells amplifies committed progenitors and rapid SCID repopulation cells. *New Biotechnology* 35 (March): 19–29. doi:10.1016/j.nbt.2016.10.011.

Çelebi, B., D. Mantovani, and N. Pineault. 2012. Insulin-like growth factor binding protein-2 and neurotrophin 3 synergize together to promote the expansion of hematopoietic cells *ex vivo*. *Cytokine* 58 (3): 327–331. doi:10.1016/j.cyto.2012.02.011.

Cho, C. H., J. F. Eliason, and H. W. T. Matthew. 2008. Application of porous glycosaminoglycan-based scaffolds for expansion of human cord blood stem cells in perfusion culture. *Journal of Biomedical Materials Research Part A* 86A (1): 98–107. doi:10.1002/jbm.a.31614.

Choi, E. S., J. L. Nichol, M. M. Hokom, A. C. Hornkohl, and P. Hunt. 1995. Platelets Generated *in vitro* from proplatelet-displaying human megakaryocytes are functional. *Blood* 85 (2): 402–413.

Choi, Y., S. Noh, S. Lim, and D. Kim. 2010. Optimization of *ex vivo* hematopoietic stem cell expansion in intermittent dynamic cultures. *Biotechnology Letters* 32 (12): 1969–1975. doi:10.1007/s10529-010-0355-0.

Collins, P. C., L. K. Nielsen, S. D. Patel, E. T. Papoutsakis, and W. M. Miller. 1998b. Characterization of hematopoietic cell expansion, oxygen uptake, and glycolysis in a controlled, stirred-tank bioreactor system. *Biotechnology Progress* 14 (3): 466–472.

Collins, P. C., W. M. Miller, and E. T. Papoutsakis. 1998a. Stirred culture of peripheral and cord blood hematopoietic cells offers advantages over traditional static systems for clinically relevant applications. *Biotechnology and Bioengineering* 59 (5): 534–543.

Conneally, E., J. Cashman, A. Petzer, and C. Eaves. 1997. Expansion *in vitro* of transplantable human cord blood stem cells demonstrated using a quantitative assay of their lympho-myeloid repopulating activity in nonobese diabetic–SCID/SCID mice. *Proceedings of the National Academy of Sciences* 94 (18): 9836–9841.

Csaszar, E., D. C. Kirouac, M. Yu, W. Wang, W. Qiao, M. P. Cooke, A. E. Boitano, C. Ito, and P. W. Zandstra. 2012. Rapid expansion of human hematopoietic stem cells by automated control of inhibitory feedback signaling. *Cell Stem Cell* 10 (2): 218–229. doi:10.1016/j.stem.2012.01.003.

Csaszar, E., K. Chen, J. Caldwell, W. Chan, and P. W. Zandstra. 2014b. Real-time monitoring and control of soluble signaling factors enables enhanced progenitor cell outputs from human cord blood stem cell cultures. *Biotechnology and Bioengineering* 111 (6): 1258–1264.

Csaszar, E., W. Wang, T. Usenko, W. Qiao, C. Delaney, I. D. Bernstein, and P. W. Zandstra. 2014a. Blood stem cell fate regulation by delta-1–mediated rewiring of IL-6 paracrine signaling. *Blood* 123 (5): 650–658.

Cui, Y., X. Yang, X. Zhu, J. Li, X. Wu, and Y. Pang. 2013. Immune response, clinical outcome and safety of dendritic cell vaccine in combination with cytokine-induced killer cell therapy in cancer patients. *Oncology Letters*, June. doi:10.3892/ol.2013.1376.

Daly, M. E. 2011. Determinants of platelet count in humans. *Haematologica* 96 (1): 10–13. doi:10.3324/haematol.2010.035287.

Danet, G. H., Y. Pan, J. L. Luongo, D. Bonnet, and M. C. Simon. 2003. Expansion of human SCID-repopulating cells under hypoxic conditions. *Journal of Clinical Investigation* 112 (1): 126–135. doi:10.1172/JCI200317669.

De León, A., H. Mayani, and O. T. Ramírez. 1998. Design, characterization and application of a mini bioreactor for the culture of human hematopoietic cells under controlled conditions. In *Cell Culture Engineering VI*, pp. 127–138. Springer. doi:10.1007/978-94-011-4786-6_14.

Delaney, C., B. Varnum-Finney, K. Aoyama, C. Brashem-Stein, and I. D. Bernstein. 2005. Dose-dependent effects of the notch ligand delta1 on *ex vivo* differentiation and *in vivo* marrow repopulating ability of cord blood cells. *Blood* 106 (8): 2693–2699. doi:10.1182/blood-2005-03-1131.

Delaney, C., S. Heimfeld, C. Brashem-Stein, H. Voorhies, R. L. Manger, and I. D. Bernstein. 2010. Notch-mediated expansion of human cord blood progenitor cells capable of rapid myeloid reconstitution. *Nature Medicine* 16 (2): 232–236. doi:10.1038/nm.2080.

Dexter, T. M., T. D. Allen, and L. G. Lajtha. 1977. Conditions controlling the proliferation of haemopoietic stem cells *in vitro*. *Journal of Cellular Physiology* 91 (3): 335–344.

Di Buduo, C. A., L. S. Wray, L. Tozzi, A. Malara, Y. Chen, C. E. Ghezzi, D. Smoot et al. 2015. Programmable 3D silk bone marrow niche for platelet generation *ex vivo* and modeling of megakaryopoiesis pathologies. *Blood* 125 (14): 2254–2264. doi:10.1182/blood-2014-08-595561.

Doran, M. R., B. D. Markway, A. Clark, S. Athanasas-Platsis, G. Brooke, K. Atkinson, L. K. Nielsen, and J. J. Cooper-White. 2009. Membrane bioreactors enhance microenvironmental conditioning and tissue development. *Tissue Engineering Part C: Methods* 16 (3): 407–415.

Doran, M. R., I. A. Aird, F. Marturana, N. Timmins, K. Atkinson, and L. K. Nielsen. 2012. Bioreactor for blood product production. *Cell Transplantation* 21 (6): 1235–1244. doi:10.3727/096368911×627363.

Eliades, A., S. Matsuura, and K. Ravid. 2012. Oxidases and reactive oxygen species during hematopoiesis: A focus on megakaryocytes. *Journal of Cellular Physiology* 227 (10): 3355–3362. doi:10.1002/jcp.24071.

Eliasson, P., and J. Jönsson. 2010. The hematopoietic stem cell niche: Low in oxygen but a nice place to be. *Journal of Cellular Physiology* 222 (1): 17–22. doi:10.1002/jcp.21908.

Fan, L., J. Li, Z. Yu, X. Dang, and K. Wang. 2014. The hypoxia-inducible factor pathway, pro-lyl hydroxylase domain protein inhibitors, and their roles in bone repair and regenera-tion. *BioMedical Research International* 2014: 1–11. doi:10.1155/2014/239356.

Fares, I., J. Chagraoui, Y. Gareau, S. Gingras, R. Ruel, N. Mayotte, E. Csaszar et al. 2014. Cord blood expansion. Pyrimidoindole derivatives are agonists of human hematopoietic stem cell self-renewal. *Science* 345 (6203): 1509–1512. doi:10.1126/science.1256337.

Fernández, M. N., C. Regidor, R. Cabrera, J. A. García-Marco, R. Forés, I. Sanjuán, J. Gayoso et al. 2003. Unrelated umbilical cord blood transplants in adults: Early recoveryof neu-trophils by supportive co-transplantation of a low number of highlypurified peripheral blood CD34+ cells from an HLA-haploidentical donor. *Experimental Hematology* 31: 535–544.

Flores-Guzman, P., V. Fernandez-Sanchez, I. Valencia-Plata, L. Arriaga-Pizano, G. Alarcon-Santos, and H. Mayani. 2013. Comparative *in vitro* analysis of dif-ferent hematopoietic cell populations from human cord blood: In search of the best option for clinically oriented *ex vivo* cell expansion. *Transfusion* 53 (3): 668–678. doi:10.1111/j.1537-2995.2012.03799.x.

Fox, R. A., M. Sigman, and K. Boekelheide. 2000. Transmembrane versus soluble stem cell factor expression in human testis. *Journal of Andrology* 21 (4): 579–585.

Gatti, R. A., H. J. Meuwissen, H. D. Allen, R. Hong, and R. A. Good. 1968. Immunological reconstitution of sex-linked lymphopenic immunological deficiency. *Lancet* 2: 1366–1369.

Glen, K. E., V. L. Workman, F. Ahmed, E. Ratcliffe, A. J. Stacey, and R. J. Thomas. 2013. Production of erythrocytes from directly isolated or delta1 notch ligand expanded CD34+ hematopoietic progenitor cells: Process characterization, monitoring and implications for manufacture. *Cytotherapy* 15 (9): 1106–1117. doi:10.1016/j.jcyt.2013.04.008.

Glimm, H., and C. J. Eaves. 1999. Direct evidence for multiple self-renewal divisions of human *in vivo* repopulating hematopoietic cells in short-term culture. *Blood* 94 (7): 2161–2168.

Gluckman, E., A. Devergié, H. Bourdeau-Esperou, D. Thierry, R. Traineau, A. Auerbach, and H. E. Broxmeyer. 1990. Transplantation of umbilical cord blood in Fanconi's anemia. *Nouvelle Revue Francaise D'Hematologie* 32: 423–425.

Goltry, K. L., B. S. Hampson, N. A. Venturi, and R. L. Bartel. 2009. Large-scale production of adult stem cells for clinical use. In: Lakshmipathy U, Chesnut JD, Thyagarajan B, (Eds.). *Emerging Technology Platforms for Stem Cells.* John Wiley and Sons, Hoboken, NJ, pp. 153–168.

Gordon, P. R., T. Leimig, A. Babarin-Dorner, J. Houston, M. Holladay, I. Mueller, T. Geiger, and R. Handgretinger. 2003. Large-scale isolation of CD133+ progenitor cells from G-CSF mobilized peripheral blood stem cells. *Bone Marrow Transplantation* 31 (1): 17–22. doi:10.1038/sj.bmt.1703792.

Guitart, A. V., C. Debeissat, F. Hermitte, A. Villacreces, Z. Ivanovic, H. Boeuf, and V. Praloran. 2010. Very low oxygen concentration (0.1%) reveals two FDCP-mix cell subpopulations that differ by their cell cycling, differentiation and P27KIP1 expression. *Cell Death & Differentiation* 18 (1): 174–182.

Hami, L. S., C. Green, N. Leshinsky, E. Markham, K. Miller, and S. Craig. 2004. GMP pro-duction and testing of xcellerated T cells™ for the treatment of patients with CLL. *Cytotherapy* 6 (6): 554–562.

Hammond, T. G., and J. M. Hammond. 2001. Optimized suspension culture: The rotating-wall vessel. *American Journal of Physiology-Renal Physiology* 281 (1): F12–F25.

Hermitte, F., P. Brunet de la Grange, F. Belloc, V. Praloran, and Z. Ivanovic. 2006. Very low O_2 concentration (0.1%) favors G_0 return of dividing CD34+ cells. *Stem Cells* 24 (1): 65–73. doi:10.1634/stemcells.2004-0351.

Highfill, J. G., S. D. Haley, and D. S. Kompala. 1996. Large-scale production of murine bone mar-row cells in an airlift packed bed bioreactor. *Biotechnology and Bioengineering* 50: 514–520.

Himburg, H. A., G. G. Muramoto, P. Daher, S. K. Meadows, J. L. Russell, P. Doan, J. Chi et al. 2010. Pleiotrophin regulates the expansion and regeneration of hematopoietic stem cells. *Nature Medicine* 16 (4): 475–482. doi:10.1038/nm.2119.

Hofmeister, C. C., J. Zhang, K. L. Knight, P. Le, and P. J. Stiff. 2007. *Ex vivo* expansion of umbilical cord blood stem cells for transplantation: Growing knowledge from the hematopoietic niche. *Bone Marrow Transplantation* 39 (1): 11–23. doi:10.1038/sj.bmt.1705538.

Hoggatt, J., Y. Kfoury, and D. T. Scadden. 2016. Hematopoietic stem cell niche in health and disease. *Annual Review of Pathology: Mechanisms of Disease* 11 (1): 555–581. doi:10.1146/annurev-pathol-012615-044414.

Horner, M., W. M. Miller, J. M. Ottino, and E. T. Papoutsakis. 1998. Transport in a grooved perfusion flat-bed bioreactor for cell therapy applications. *Biotechnology Progress* 14 (5): 689–698.

Horwitz, M. E., N. J. Chao, D. A. Rizzieri, G. D. Long, K. M. Sullivan, C. Gasparetto, J. P. Chute et al. 2014. Umbilical cord blood expansion with nicotinamide provides long-term multilineage engraftment. *Journal of Clinical Investigation* 124 (7): 3121–3128. doi:10.1172/JCI74556.

Hosseinizand, H., M. Ebrahimi, and M. J. Abdekhodaie. 2016. Agitation increases expansion of cord blood hematopoietic cells and promotes their differentiation into Myeloid Lineage. *Cytotechnology* 68 (4): 969–978. doi:10.1007/s10616-015-9851-3.

Ivan, M., K. Kondo, H. Yang, W. Kim, J. Valiando, M. Ohh, A. Salic, J. M. Asara, W. S. Lane, and W. G. Kaelin. 2001. HIFalpha targeted for VHL-mediated destruction by proline hydroxylation: Implications for O_2 sensing. *Science* 292: 464–468.

Ivanovic, Z., F. Hermitte, P. B. de la Grange, B. Dazey, F. Belloc, F. Lacombe, G. Vezon, and V. Praloran. 2004. Simultaneous maintenance of human cord blood SCID-repopulating cells and expansion of committed progenitors at low O_2 concentration (3%). *Stem Cells* 22 (5): 716–724.

Jaroscak, J., K. Goltry, A. Smith, B. Waters-Pick, P. L. Martin, T. A. Driscoll, R. Howrey et al. 2003. Augmentation of umbilical cord blood (UCB) transplantation with *ex vivo*-expanded UCB cells: Results of a phase 1 trial using the aastromReplicell system. *Blood* 101 (12): 5061–5067. doi:10.1182/blood-2001-12-0290.

Jelinek, N., S. Schmidt, U. Hilbert, S. Thoma, M. Biselli, and C. Wandrey. 2002. Novel bioreactors for the *ex vivo* cultivation of hematopoietic cells. *Engineering in Life Sciences* 2: 15–18.

Jež, M., P. Rožman, Z. Ivanović, and T. Bas. 2015. Concise review: The role of oxygen in hematopoietic stem cell physiology. *Journal of Cellular Physiology* 230 (9): 1999–2005. doi:10.1002/jcp.24953.

Jing, Q., H. Cai, Z. Du, Z. Ye, and W. Tan. 2013. Effects of agitation speed on the *Ex Vivo* expansion of cord blood hematopoietic stem/progenitor cells in stirred suspension culture. *Artificial Cells, Nanomedicine, and Biotechnology* 41 (2): 98–102. doi:10.3109/10731199.2012.712043.

Kamigaki, T., T. Kaneko, K. Naitoh, M. Takahara, T. Kondo, H. Ibe, E. Matsuda, R. Maekawa, and S. Goto. 2013. Immunotherapy of autologous tumor lysate-loaded dendritic cell vaccines by a closed-flow electroporation system for solid tumors. *Anticancer Research* 33 (7): 2971–2976.

Kedong, S., F. Xiubo, L. Tianqing, H. M. Macedo, J. LiLi, F. Meiyun, S. Fangxin, M. Xuehu, and C. Zhanfeng. 2010. Simultaneous expansion and harvest of hematopoietic stem cells and Mesenchymal stem cells derived from Umbilical Cord blood. *Journal of Materials Science: Materials in Medicine* 21 (12): 3183–3193. doi:10.1007/s10856-010-4167-5.

Kim, B. 1998. Production of human hematopoietic progenitors in a clinical-scale stirred suspension bioreactor. *Biotechnology Letters* 20 (6): 595–601.

King, K. Y., and M. A. Goodell. 2011. Inflammatory modulation of HSCs: Viewing the HSC as a foundation for the immune response. *Nature Reviews Immunology* 11 (10): 685–692. doi:10.1038/nri3062.

Kirouac, D. C., C. Ito, E. Csaszar, A. Roch, M. Yu, E. A. Sykes, G. D. Bader, and P. W. Zandstra. 2010. Dynamic interaction networks in a hierarchically organized tissue. *Molecular Systems Biology* 6 (417). doi:10.1038/msb.2010.71.

Knight, E. 1977. Multisurface glass roller bottle for growth of animal cells in culture. *Applied and Environmental Microbiology* 33 (3): 666–669.

Koller, M. R., I. Manchel, B. S. Newsom, M. A. Palsson, and B. Ø. Palsson. 1995. Bioreactor expansion of human bone marrow: Comparison of unprocessed, density-separated, and CD34-enriched cells. *Journal of Hematotherapy* 4 (3): 159–169.

Koller, M. R., I. Manchel, R. J. Maher, K. L. Goltry, R. D. Armstrong, and A. K. Smith. 1998. Clinical-scale human umbilical cord blood cell expansion in a novel automated perfusion culture system. *Bone Marrow Transplantation* 21 (7): 653–663.

Koller, M. R., J. G. Bender, W. M. Miller, and E. T. Papoutsakis. 1993a. Expansion of primitive human hematopoietic progenitors in a perfusion bioreactor system with IL-3, IL-6, and stem cell factor. *Nature Biotechnology* 11: 358–363.

Koller, M. R., S. G. Emerson, and B. O. Palsson. 1993b. Large-scale expansion of human stem and progenitor cells from bone marrow mononuclear cells in continuous perfusion cultures. *Blood* 82 (2): 378–384.

Kowalczyk, M., K. Waldron, P. Kresnowati, and M. K. Danquah. 2011. Process challenges relating to hematopoietic stem cell cultivation in bioreactors. *Journal of Industrial Microbiology & Biotechnology* 38 (7): 761–767. doi:10.1007/s10295-011-0951-6.

Kunitake, R., A. Suzuki, H. Ichihashi, S. Matsuda, O. Hirai, and K. Morimoto. 1997. Fully-automated roller bottle handling system for large scale culture of mammalian cells. *Journal of Biotechnology* 52 (3): 289–294.

Kurazumi, H., M. Kubo, M. Ohshima, Y. Yamamoto, Y. Takemoto, R. Suzuki, S. Ikenaga et al. 2011. The effects of mechanical stress on the growth, differentiation, and paracrine factor production of cardiac stem cells. *PLoS One* 6 (12): e28890. doi:10.1371/journal.pone.0028890.

Lakhotia, S., K. D. Bauer, and E. T. Papoutsakis. 1993. Fluid-mechanical forces in agitated bioreactors reduce the CD13 and CD33 surface protein content of HL60 cells. *Biotechnology and Bioengineering* 41 (9): 868–877.

Laluppa, J. A., E. T. Papoutsakis, and W. M. Miller. 1998. Oxygen tension alters the effects of cytokines on the Megakaryocyte, Erythrocyte, and Granulocyte lineages. *Experimental Hematology* 26: 835–843.

Lampiasi, N., R. Russo, and F. Zito. 2016. The alternative faces of macrophage generate osteoclasts. *BioMed Research International* 2016: 1–9. doi:10.1155/2016/9089610.

Lee, E., S. Y. Han, H. S. Choi, B. Chun, B. Hwang, and E. J. Baek. 2015. Red blood cell generation by three-dimensional aggregate cultivation of late erythroblasts. *Tissue Engineering Part A* 21 (3–4): 817–828. doi:10.1089/ten.tea.2014.0325.

Levee, M. G., G. Lee, S. Paek, and B. O. Palsson. 1994. Microencapsulated human bone marrow cultures: A potential culture system for the clonal outgrowth of hematopoietic progenitor cells. *Biotechnology and Bioengineering* 43 (8): 734–739.

Li, Q., Q. Liu, H. Cai, and W. Tan. 2006. A comparative gene-expression analysis of CD34+ hematopoietic stem and progenitor cells grown in static and stirred culture systems. *Cellular and Molecular Biology Letters* 11 (4). doi:10.2478/s11658-006-0039-x.

Lima, M. de, I. McNiece, S. N. Robinson, M. Munsell, M. Eapen, M. Horowitz, A. Alousi et al. 2012. Cord-blood engraftment with *ex vivo* mesenchymal-cell coculture. *New England Journal of Medicine* 367 (24): 2305–2315. doi:10.1056/NEJMoa1207285.

Lima, M. de, J. McMannis, A. Gee, K. Komanduri, D. Couriel, B. S. Andersson, C. Hosing et al. 2008. Transplantation of *ex vivo* expanded cord blood cells using the copper chelator tetraethylenepentamine: A phase I/II clinical trial. *Bone Marrow Transplantation* 41 (9): 771–778. doi:10.1038/sj.bmt.1705979.

Liu, Y. L., K. Wagner, N. Robinson, D. Sabatino, P. Margaritis, W. Xiao, and R. W. Herzog. 2003. Research report optimized production of high-titer recombinant adeno-associated virus in roller bottles. *Biotechniques* 34 (1): 184–189.

Liu, Y., T. Liu, X. Fan, X. Ma, and Z. Cui. 2006. *Ex Vivo* expansion of hematopoietic stem cells derived from umbilical cord blood in rotating wall vessel. *Journal of Biotechnology* 124: 592–601.

Luni, C., H. C. Feldman, M. Pozzobon, P. De Coppi, C. D. Meinhart, and N. Elvassore. 2010. Microliter-bioreactor array with buoyancy-driven stirring for human hematopoietic stem cell culture. *Biomicrofluidics* 4 (3): 034105. doi:10.1063/1.3380627.

Madlambayan, G. J., I. Rogers, D. C. Kirouac, N. Yamanaka, F. Mazurier, M. Doedens, R. F. Casper, J. E. Dick, and P. W. Zandstra. 2005. Dynamic changes in cellular and microenvironmental composition can be controlled to elicit *in vitro* human hematopoietic stem cell expansion. *Experimental Hematology* 33: 1229–1239.

Madlambayan, G. J., I. Rogers, K. A. Purpura, C. Ito, M. Yu, D. Kirouac, R. F. Casper, and P. W. Zandstra. 2006. Clinically relevant expansion of hematopoietic stem cells with conserved function in a single-use, closed-system bioprocess. *Biology of Blood and Marrow Transplantation* 12 (10): 1020–1030. doi:10.1016/j.bbmt.2006.07.005.

Mahadik, B. P., S. Pedron Haba, L. J. Skertich, and B. A. C. Harley. 2015. The use of covalently immobilized stem cell factor to selectively affect hematopoietic stem cell activity within a gelatin hydrogel. *Biomaterials* 67 (October): 297–307. doi:10.1016/j.biomaterials.2015.07.042.

Majeti, R., C. Y. Park, and I. L. Weissman. 2007. Identification of a hierarchy of multipotent hematopoietic progenitors in human cord blood. *Cell Stem Cell* 1 (6): 635–645. doi:10.1016/j.stem.2007.10.001.

Majumdar, M. K., M. A. Thiede, S. E. Haynesworth, S. P. Bruder, and S. L. Gerson. 2000. Human marrow-derived mesenchymal stem cells (MSCs) express hematopoietic cytokines and support long-term hematopoiesis when differentiated toward stromal and osteogenic lineages. *Journal of Hematotherapy & Stem Cell Research* 9 (6): 841–848.

Mandalam, R., M. Koller, and A. Smith. 1999. *Ex Vivo* hematopoietic cell expansion for bone marrow transplantation. In: Nordon KSaR (Ed.). *Ex Vivocell Therapy*. New York: Academic Press, pp. 273–291.

Mantalaris, A., P. Keng, P. Bourne, A. Y. C. Chang, and J. H. D. Wu. 1998. Engineering a human bone marrow model: A case study on *ex vivo* erythropoiesis. *Biotechnology Progress* 14 (1): 126–133.

McNiece, I. K., G. Almeida-Porada, E. J. Shpall, and E. Zanjani. 2002. *Ex vivo* expanded cord blood cells provide rapid engraftment in fetal sheep but lack long-term engrafting potential. *Experimental Hematology* 30: 612–616.

Meissner, P., B. Schröder, C. Herfurth, and M. Biselli. 1999. Development of a fixed bed bioreactor for the expansion of human hematopoietic progenitor cells. *Cytotechnology* 30 (1–3): 227–234.

Méndez-Ferrer, S., T. V. Michurina, F. Ferraro, A. R. Mazloom, B. D. MacArthur, S. A. Lira, D. T. Scadden, A. Ma'ayan, G. N. Enikolopov, and P. S. Frenette. 2010. Mesenchymal and haematopoietic stem cells form a unique bone marrow niche. *Nature* 466 (7308): 829–834. doi:10.1038/nature09262.

Michaels, J. D., A. K. Mallik, and E. T. Papoutsakis. 1996. Sparging and agitation-induced injury of cultured animal cells: Do cell-to-bubble interactions in the bulk liquid injure cells? *Biotechnology and Bioengineering* 51: 399–409.

Mohyeldin, A., T. Garzón-Muvdi, and A. Quiñones-Hinojosa. 2010. Oxygen in stem cell biology: A critical component of the stem cell niche. *Cell Stem Cell* 7 (2): 150–161. doi:10.1016/j.stem.2010.07.007.

Mostafa, S. S., W. M. Miller, and T. Papoutsakis. 2000. Oxygen tension influences the differentiation, maturation and apoptosis of human megakaryocytes. *British Journal of Haematology* 111: 879–889.

Naughton, B. A., and G. K. Naughton. 1989. Hematopoiesis on nylon mesh templates comparative long-term bone marrow culture and the influence of stromal support cells A. *Annals of the New York Academy of Sciences* 554 (1): 125–140.

Nilsson, S. K., H. M. Johnston, G. A. Whitty, B. Williams, R. J. Webb, D. T. Dernhardt, I. Bertoncello, L. J. Bendall, P. J. Simmons, and D. N. Haylock. 2005. Osteopontin, a key component of the hematopoietic stem cell niche and regulator of primitive hematopoietic progenitor cells. *Blood* 106 (4): 1232–1239. doi:10.1182/blood-2004-11-4422.

Notta, F., S. Doulatov, E. Laurenti, A. Poeppl, I. Jurisica, and J. E. Dick. 2011. Isolation of single human hematopoietic stem cells capable of long-term multilineage engraftment. *Science* 333 (6039): 218–221. doi:10.1126/science.1201219.

Ohishi, K., B. Varnum-Finney, and I. D. Bernstein. 2002. Delta-1 enhances marrow and thymus repopulating ability of human CD34$^+$CD38$^-$ cord blood cells. *Journal of Clinical Investigation* 110 (8): 1165–1174. doi:10.1172/JCI200216167.

Oostendorp, R. A. J., C. Robin, C. Steinhoff, S. Marz, R. Bräuer, U. A. Nuber, E. A. Dzierzak, and C. Peschel. 2005. Long-term maintenance of hematopoietic stem cells does not require contact with embryo-derived stromal cells in cocultures. *Stem Cells* 23 (6): 842–851. doi:10.1634/stemcells.2004-0120.

Orford, K. W., and D. T. Scadden. 2008. Deconstructing stem cell self-renewal: Genetic insights into cell-cycle regulation. *Nature Reviews Genetics* 9 (2): 115–128. doi:10.1038/nrg2269.

Pabst, C., J. Krosl, I. Fares, G. Boucher, R. Ruel, A. Marinier, S. Lemieux, J. Hébert, and G. Sauvageau. 2014. Identification of small molecules that support human Leukemia stem cell activity *ex vivo*. *Nature Methods* 11 (4): 436–442. doi:10.1038/nmeth.2847.

Palsson, B. O., S. Paek, R. M. Schwartz, M. Palsson, G. Lee, S. Silver, and S. G. Emerson. 1993. Expansion of human bone marrow progenitor cells in a high cell density continuous perfusion system. *Nature Biotechnology* 11 (3): 368–372.

Parmar, K., P. Mauch, J. A. Vergilio, R. Sackstein, and J. D. Down. 2007. Distribution of hematopoietic stem cells in the bone marrow according to regional hypoxia. *Proceedings of the National Academy of Sciences* 104 (13): 5431.

Pastore, D., A. Mestice, T. Perrone, F. Gaudio, M. Delia, F. Albano, A. R. Rossi et al. 2008. Subsets of CD34+ and early engraftment kinetics in allogeneic peripheral SCT for AML. *Bone Marrow Transplantation* 41: 977–981.

Patel, S. D., E. T. Papoutsakis, J. N. Winter, and W. M. Miller. 2000. The lactate issue revisited: Novel feeding protocols to examine inhibition of cell proliferation and glucose metabolism in hematopoietic cell cultures. *Biotechnology Progress* 16 (5): 885–892. doi:10.1021/bp000080a.

Pedrazzoli, P., G. A. Da Prada, G. Giorgiani, R. Schiavo, A. Zambelli, E. Giraldi, G. Landonio, F. Locatelli, S. Siena, and G. R. D. Cuna. 2002. Allogeneic blood stem cell transplantation after a reduced-intensity, preparative regimen: A pilot study in patients with refractory malignancies. *Cancer* 94 (9): 2409–2415. doi:10.1002/cncr.10491.

Peled, T., E. Landau, J. Mandel, E. Glukhman, N. R. Goudsmid, A. Nagler, and E. Fibach. 2004. Linear polyamine copper chelator tetraethylenepentamine augments long-term *ex vivo* expansion of cord blood-derived CD34+ cells and increases their engraftment potential in NOD/SCID mice. *Experimental Hematology* 32 (6): 547–555.

Plett, P. A., S. M. Frankovitz, R. Abonour, and C. M. Orschell-Traycoff. 2001. Proliferation of human hematopoietic bone marrow cells in simulated microgravity. *In vitro Cellular & Developmental Biology-Animal* 37 (2): 73–78.

Qian, H., N. Buza-Vidas, C. D. Hyland, C. T. Jensen, J. Antonchuk, R. Månsson, L. A. Thoren, M. Ekblom, W. S. Alexander, and S. E. W. Jacobsen. 2007. Critical role of thrombopoietin in maintaining adult quiescent hematopoietic stem cells. *Cell Stem Cell* 1 (6): 671–684. doi:10.1016/j.stem.2007.10.008.

Qiao, W., W. Wang, E. Laurenti, A. L. Turinsky, S. J. Wodak, G. D. Bader, J. E. Dick, and P. W. Zandstra. 2014. Intercellular network structure and regulatory motifs in the human hematopoietic system. *Molecular Systems Biology* 10 (7): 741–741. doi:10.15252/msb.20145141.

Ratcliffe, E., K. E. Glen, V. L. Workman, A. J. Stacey, and R. J. Thomas. 2012. A novel automated bioreactor for scalable process optimisation of haematopoietic stem cell culture. *Journal of Biotechnology* 161 (3): 387–390. doi:10.1016/j.jbiotec.2012.06.025.

Reikvam, H., A. K. Brenner, K. M. Hagen, K. Liseth, S. Skrede, K. J. Hatfield, and Ø. Bruserud. 2015. The cytokine-mediated crosstalk between primary human acute myeloid cells and mesenchymal stem cells alters the local cytokine network and the global gene expression profile of the mesenchymal cells. *Stem Cell Research* 15 (3): 530–541. doi:10.1016/j.scr.2015.09.008.

Rocha, V., J. E. Wagner Jr, K. A. Sobocinski, J. P. Klein, M. Zhang, M. M. Horowitz, and E. Gluckman. 2000. Graft-versus-host disease in children who have received a cordblood or bone marrow transplant from an HLA-identical sibling. *New England Journal of Medicine* 342 (25): 1846–1854.

Sacchetti, B., A. Funari, S. Michienzi, S. Di Cesare, S. Piersanti, I. Saggio, E. Tagliafico et al. 2007. Self-renewing osteoprogenitors in bone marrow sinusoids can organize a hematopoietic microenvironment. *Cell* 131 (2): 324–336. doi:10.1016/j.cell.2007.08.025.

Sadeghi, A., L. Pauler, C. Annerén, A. Friberg, D. Brandhorst, O. Korsgren, and T. H. Tötterman. 2011. Large-scale bioreactor expansion of tumor-infiltrating lymphocytes. *Journal of Immunological Methods* 364 (1–2): 94–100. doi:10.1016/j.jim.2010.11.007.

Sandstrom, C. E., J. G. Bender, W. M. Miller, and E. T. Papoutsakis. 1996. Development of novel perfusion chamber to retain nonadherent cells and its use for comparison of human mobilized peripheral blood mononuclear cell cultures with and without irradiated bone marrow stroma. *Biotechnology and Bioengineering* 50: 493–504.

Sardonini, C. A., and Y. J. Wu. 1993. Expansion and differentiation of human hematopoietic cells from static cultures through small-scale bioreactors. *Biotechnology Progress* 9: 131–137.

Schmelzer, E., A. Finoli, I. Nettleship, and J. C. Gerlach. 2015. Long-term three-dimensional perfusion culture of human adult bone marrow mononuclear cells in bioreactors. *Biotechnology and Bioengineering* 112 (4): 801–810.

Schofield, R. 1978. The relationship between the spleen colony-forming cell and the haemopoietic stem cell. *Blood Cells* 4: 7–25.

Shima, H., K. Takubo, H. Iwasaki, H. Yoshihara, Y. Gomei, K. Hosokawa, F. Arai, T. Takahashi, and T. Suda. 2009. Reconstitution activity of hypoxic cultured human cord blood CD34-positive cells in NOG mice. *Biochemical and Biophysical Research Communications* 378 (3): 467–472. doi:10.1016/j.bbrc.2008.11.056.

Shpall, E. J., R. Quinones, R. Giller, C. Zeng, A. E. Barón, R. B. Jones, S. I. Bearman et al. 2002. Transplantation of *ex vivo* expanded cord blood. *Biology of Blood and Marrow Transplantation* 8: 368–376.

Sideri, A., N. Neokleous, P. B. De La Grange, B. Guerton, M. Le Bousse Kerdilles, G. Uzan, C. Peste-Tsilimidos, and E. Gluckman. 2011. An overview of the progress on double umbilical cord blood transplantation. *Haematologica* 96 (8): 1213–1220. doi:10.3324/haematol.2010.038836.

Silva, C. L. da, R. Gonçalves, F. dos Santos, P. Z. Andrade, G. Almeida-Porada, and J. M. S. Cabral. 2009. Dynamic cell-cell interactions between cord blood haematopoietic progenitors and the cellular niche are essential for the expansion of CD34+, CD34+CD38- and early lymphoid CD7+ cells. *Journal of Tissue Engineering and Regenerative Medicine* 4 (2): 149–158. doi:10.1002/term.226.

Song, K., Y. Liu, H. Wang, T. Liu, M. Fang, F. Shi, H. M. Macedo, X. Ma, and Z. Cui. 2011. *Ex vivo* expansion of human umbilical cord blood hematopoietic stem/progenitor cells with support of microencapsulated rabbit mesenchymal stem cells in a rotating bioreactor. *Tissue Engineering and Regenerative Medicine* 8 (3): 334–345.

Spencer, J. A., F. Ferraro, E. Roussakis, A. Klein, J. Wu, J. M. Runnels, W. Zaher et al. 2014. Direct measurement of local oxygen concentration in the bone marrow of live animals. *Nature* 508 (7495): 269–273. doi:10.1038/nature13034.

Stiff, P. J., P. Montesinos, T. Peled, E. Landau, N. Rosenheimer, J. Mandel, N. Hasson et al. 2013. StemEx®(Copper Chelation Based) *ex vivo* expanded umbilical cord blood stem cell transplantation (UCBT) accelerates engraftment and improves 100 Day survival in myeloablated patients compared to a registry cohort undergoing double unit UCBT: Results of a multicenter study of 101 patients with hematologic malignancies. *Blood* 122: 295.

Sullenbarger, B., J. H. Bahng, R. Gruner, N. Kotov, and L. C. Lasky. 2009. Prolonged continuous *in vitro* human platelet production using three-dimensional scaffolds. *Experimental Hematology* 37 (1): 101–110. doi:10.1016/j.exphem.2008.09.009.

Thomas, E. D., H. L. Lochte, W. C. Lu, and J. W. Ferrebee. 1957. Intravenous infusion of bone marrow in patients receiving radiation and chemotherapy. *The New England Journal of Medicine* 257: 491–496.

Timmins, N. E., E. Palfreyman, F. Marturana, S. Dietmair, S. Luikenga, G. Lopez, Y. L. Fung, R. Minchinton, and L. K. Nielsen. 2009. Clinical scale *ex vivo* manufacture of neutrophils from hematopoietic progenitor cells. *Biotechnology and Bioengineering* 104: 832–840. doi:10.1002/bit.22433.

Timmins, N. E., S. Athanasas, M. Günther, P. Buntine, and L. K. Nielsen. 2011. Ultra-high-yield manufacture of red blood cells from hematopoietic stem cells. *Tissue Engineering Part C: Methods* 17 (11): 1131–1137. doi:10.1089/ten.tec.2011.0207.

Usuludin, S. B. M., X. Cao, and M. Lim. 2012. Co-culture of stromal and erythroleukemia cells in a perfused hollow fiber bioreactor system as an *in vitro* bone marrow model for myeloid leukemia. *Biotechnology and Bioengineering* 109 (5): 1248–1258. doi:10.1002/bit.24400.

Varnum-Finney, B., C. Brashem-Stein, and I. D. Bernstein. 2003. Combined effects of notch signaling and cytokines induce a multiple log increase in precursors with lymphoid and myeloid reconstituting ability. *Blood* 101 (5): 1784–1789. doi:10.1182/blood-2002-06-1862.

Wagner, J. E., C. G. Brunstein, A. E. Boitano, T. E. DeFor, D. McKenna, D. Sumstad, B. R. Blazar et al. 2016. Phase I/II trial of stemRegenin-1 expanded umbilical cord blood hematopoietic stem cells supports testing as a stand-alone graft. *Cell Stem Cell* 18 (1): 144–155. doi:10.1016/j.stem.2015.10.004.

Wagner, J. E., J. N. Barker, T. E. DeFor, K. S. Baker, B. R. Blazar, C. Eide, A. Goldman et al. 2002. Transplantation of unrelated donor umbilical cord blood in 102 patients with malignant and nonmalignant diseases: Influence of CD34 cell dose and HLA disparity on treatment-related mortality and survival. *Blood* 100 (5): 1611–1618.

Wagner, W., C. Roderburg, F. Wein, A. Diehlmann, M. Frankhauser, R. Schubert, V. Eckstein, and A. D. Ho. 2007. Molecular and secretory profiles of human mesenchymal stromal cells and their abilities to maintain primitive hematopoietic progenitors. *Stem Cells* 25 (10): 2638–2647. doi:10.1634/stemcells.2007-0280.

Walenda, T., G. Bokermann, M. S. V. Ferreira, D. M. Piroth, T. Hieronymus, S. Neuss, M. Zenke, A. D. Ho, A. M. Muller, and W. Wagner. 2011. Synergistic effects of growth factors and mesenchymal stromal cells for expansion of hematopoietic stem and progenitor cells. *Experimental Hematology* 39: 617–628.

Wang, Z., Z. Du, H. Cai, Z. Ye, J. Fan, and W. Tan. 2016. Low oxygen tension favored expansion and hematopoietic reconstitution of CD34$^+$ CD38$^-$ cells expanded from human cord blood-derived CD34$^+$ cells. *Biotechnology Journal* 11 (7): 945–953. doi:10.1002/biot.201500497.

Weissman, I. L. 2000. Stem cells: Units of development, review units of regeneration, and units in evolution. *Cell* 100: 157–168.

Weissman, I. L., and J. A. Shizuru. 2008. The origins of the identification and isolation of hematopoietic stem cells, and their capability to induce donor-specific transplantation tolerance and treat autoimmune diseases. *Blood* 112 (9): 3543.

Wilson, A., E. Laurenti, G. Oser, R. C. van der Wath, W. Blanco-Bose, M. Jaworski, S. Offner, et al. 2008. Hematopoietic stem cells reversibly switch from dormancy to self-renewal during homeostasis and repair. *Cell* 135 (6): 1118–1129. doi:10.1016/j.cell.2008.10.048.

Wognum, A. W., A. C. Eaves, and T. E. Thomas. 2003. Identification and isolation of hematopoietic stem cells. *Archives of Medical Research* 34: 461–475.

Worrallo, M. J., R. L. L. Moore, K. E. Glen, and R. J. Thomas. 2017. Immobilized hematopoietic growth factors onto magnetic particles offer a scalable strategy for cell therapy manufacturing in suspension cultures. *Biotechnology Journal* 12 (2): 1600493. doi:10.1002/biot.201600493.

Wuchter, P., R. Saffrich, S. Giselbrecht, C. Nies, H. Lorig, S. Kolb, A. D. Ho, and E. Gottwald. 2016. Microcavity arrays as an *in vitro* model system of the bone marrow niche for hematopoietic stem cells. *Cell and Tissue Research*, January. doi:10.1007/s00441-015-2348-8.

Xiong, F., Z. Chen, H. Liu, Z. Xu, and X. Liu. 2002. *Ex vivo* expansion of human umbilical cord blood hematopoietic progenitor cells in a novel three-dimensional culture system. *Biotechnology Letters* 24 (17): 1421–1426.

Xue, C., K. Y. C. Kwek, J. K. Y. Chan, Q. Chen, and M. Lim. 2014. The hollow fiber bioreactor as a stroma-supported, serum-free *ex vivo* expansion platform for human umbilical cord blood cells. *Biotechnology Journal* 9 (7): 980–989. doi:10.1002/biot.201300320.

Yang, S., H. Cai, H. Jin, and W. Tan. 2007. Hematopoietic reconstitution of CD34+ cells grown in static and stirred culture systems in NOD/SCID mice. *Biotechnology Letters* 30 (1): 61–65. doi:10.1007/s10529-007-9517-0.

Yin, A. H., S. Miraglia, E. D. Zanjani, G. Almeida-Porada, M. Ogawa, A. G. Leary, J. Olweus, J. Kearney, and D. W. Buck. 1997. AC133, a novel marker for human hematopoietic stem and progenitor cells. *Blood* 90 (12): 5002–5012.

Yuan, Y., W. Sin, B. Xue, Y. Ke, K. Tse, Z. Chen, Y. Xie, and Y. Xie. 2013. Novel alginate three-dimensional static and rotating culture systems for effective *ex vivo* amplification of human cord blood hematopoietic stem cells and *in vivo* functional analysis of amplified cells in NOD/SCID mice: Novel 3D systems for human HSC amplification. *Transfusion* 53 (9): 2001–2011. doi:10.1111/trf.12103.

Yun, S., S. H. Lee, S. Yoon, P. Myung, and I. Choi. 2011. Oxygen tension regulates NK cells differentiation from hematopoietic stem cells *in vitro*. *Immunology Letters* 137 (1–2): 70–77. doi:10.1016/j.imlet.2011.02.020.

Zandstra, P. W., C. J. Eaves, and J. M. Piret. 1994. Expansion of hematopoietic progenitor cell populations in stirred suspension bioreactors of normal human bone marrow cells. *Nature Biotechnology* 12: 909–914.

Zhang, C. C., and H. F. Lodish. 2004. Insulin-like growth factor 2 expressed in a novel fetal liver cell population is a growth factor for hematopoietic stem cells. *Blood* 103 (7): 2513–2521. doi:10.1182/blood-2003-08-2955.

Zhang, C. C., M. Kaba, S. Iizuka, H. Huynh, and H. F. Lodish. 2008. Angiopoietin-like 5 and IGFBP2 stimulate *ex vivo* expansion of human cord blood hematopoietic stem cells as assayed by NOD/SCID transplantation. *Blood* 111: 3415–3423.

8 Quality Manufacturing of Mesenchymal Stem/Stromal Cells Using Scalable and Controllable Bioreactor Platforms

Sunghoon Jung, Krishna M. Panchalingam, Brian Lee, and Leo A. Behie

CONTENTS

8.1 INTRODUCTION

8.1.1 HUMAN MESENCHYMAL STEM CELL THERAPEUTIC POTENTIAL AND MANUFACTURING

Human Mesenchymal Stem Cells (hMSCs) are present in bone marrow, adipose tissue, umbilical cord, and other tissue/organs of the body and can be readily isolated and expanded through cell culture technologies. Culture-expanded hMSCs exhibit therapeutic properties such as immune modulation and tissue repair and regeneration, as well as multilineage differentiation potential. In addition, it has been demonstrated that hMSCs are immune evasive and thus therapeutically viable across allogeneic barriers; therefore, culture-expanded allogeneic MSCs could be used as an "off-the-shelf" therapeutic product. These characteristics have made hMSCs an attractive candidate for various therapeutic applications in both autologous and allogeneic settings. Clinical studies employing hMSCs are underway for the treatment of several diseases and injuries, such as myocardial infarction, osteogenesis imperfecta, graft-versus-host disease (GvHD), Crohn's disease, spinal cord injury, multiple sclerosis, and diabetes (www.clinicaltrials. gov). hMSC therapies are moving toward commercialization and, consequently, developing a commercial-scale manufacturing platform for hMSC products has become a major requirement in the field to make their therapeutic use a reality. Figure 8.1 illustrates an overview of the entire manufacturing process for hMSCs to summarize specific unit operations in tissue acquisition (e.g., bone marrow), upstream and downstream processes, and testing and release. Each of these operations should be optimized and streamlined to achieve a successful commercial cell-manufacturing platform. In this chapter, we focus specifically on upstream cell culture-related challenges, which is the primary hurdle to clear in order to develop scalable and speedy bioreactor-based cell manufacturing platforms. The main challenges for upstream process development for large-scale manufacturing of high quality hMSCs include the following:

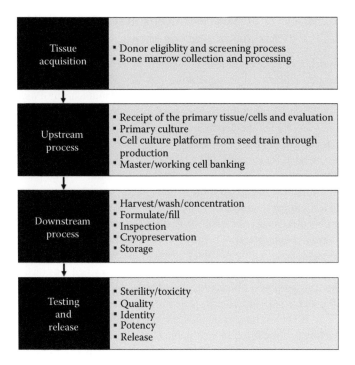

FIGURE 8.1 Overview of the hMSC manufacturing process.

1. Choosing the right platform to meet future commercial demand while understanding the performance of the following key platform components:
   a. Medium that exhibits high performance thereby enabling high volumetric cell concentration in culture. It is also desirable that the medium be serum-free, xeno (geneic)-free, or ideally chemically-defined conforming to both scientific and regulatory perspectives.
   b. Culture substrate (i.e., microcarrier) that not only supports the attachment and growth of hMSCs with high efficiency but also facilitates cell harvest.
   c. Bioreactor that provides an efficient fluid mixing mechanism for uniform suspension of microcarriers on which hMSCs grow under low shear and low power dissipation throughout the bioreactor vessel.
2. Conducting well-designed process development, optimization, and characterization studies to identify critical process parameters that affect cell attachment, growth, and harvest, determining their optimal values and ranges in order to maximize cell growth rate and yield while maintaining the critical quality attributes of the process.
3. Streamlining all the unit operations of the manufacturing process with a target of significantly reducing process time and documenting all the steps to make the entire manufacturing process efficient, reproducible, rapid, safe, and compliant.

8.1.2 REQUIREMENT OF SCALABLE AND CONTROLLABLE BIOREACTOR PLATFORMS FOR QUALITY MANUFACTURING FOR HUMAN MESENCHYMAL STEM CELLS

Moving forward, for therapeutic use on a commercial scale, the extensive expansion of hMSCs in culture is critical for generating 10^9–10^{12} of cells per batch. Also, the production of therapeutic cells in accordance with Good Manufacturing Practices (GMP) requires a scalable and controllable bioprocess that are desired to be operated in a closed system. hMSCs are anchorage-dependent cells; that is, they should attach to a plastic surface and spread out to proliferate. Based on this growth characteristic, hMSCs have mainly been generated for research and clinical applications through conventional static adherent culture methods using tissue culture flasks or multi-layered plates. Although being easy to implement, the generation of many cells using these approaches requires manipulating a significant number of vessels. This process is time consuming and labor intensive, and handling many vessels in an open system increases the possibility of contact with external contaminants. There are several allogeneic hMSC clinical trials to treat medical indications with unresolved medical needs, and many of these therapies would require a lot size of 10^{12} cells at a commercial scale when we consider the number of patients and proposed efficacious dose levels. As an example, for an indication with 200,000 patients per year, if each patient receives one treatment with hMSCs at a dose size of 2×10^8 cells (Jung et al., 2012b), the estimated number of cells that need to be manufactured to meet the annual demand would be 4×10^{13}. With this regard, contract manufacturing organizations for cell therapy, such as Lonza Walkersville Inc. (MD, US), have proposed the feasibility of manufacturing 1×10^{12} hMSCs as a doable lot size (i.e., 5,000 doses per lot with a dose level of 2×10^8 cells) and operating 40 batch-runs per year to meet the demand (Abraham et al., 2017).

The use of planar systems would be even unfeasible to realize cell manufacturing at the aforementioned lot size (i.e., 1×10^{12} cells per lot). For instance, thousands of 10-layer cell factories (CF10) would be needed, even with the use of high performance media, at the production stage only to produce a lot of 10^{12} hMSCs (Figure 8.2). Additional hundreds or thousands of CF10 are required for seed train stages, depending on medium performance. Figure 8.2 compares three different hMSC manufacturing upstream process models, which were hypothetically designed based on reported cell growth kinetics in two different media (a classical serum-based medium versus an optimized serum-free medium for hMSCs) in static versus microcarrier-based suspension cultures. In all cases, the starting material is 100 mL of bone marrow (BM) extract, which would contain ~1×10^9 mononuclear cells (MNCs). Assuming the frequency of primary hMSCs in BM-MNCs is 0.01%–0.001%, this corresponds with 1×10^4–10^5 hMSCs. With this starting material and an assumption that 90% of harvested cells are passaged at each passage, hMSCs should have undergone a total of 23.3–26.6 population doublings (PD) regardless of platforms (i.e., type of medium, microcarriers and culture vessels), including the PD in the primary culture, when the cell number reaches 1×10^{12} at the end of the production stage. The number of vessels, the amount of medium and the process time required for the entire upstream process, and the final passage level at the production stage are determined depending on which platform is employed for the manufacturing. These three hypothetical models are briefly described in the following to exemplify the number of vessels, passage numbers, and process time required for each platform.

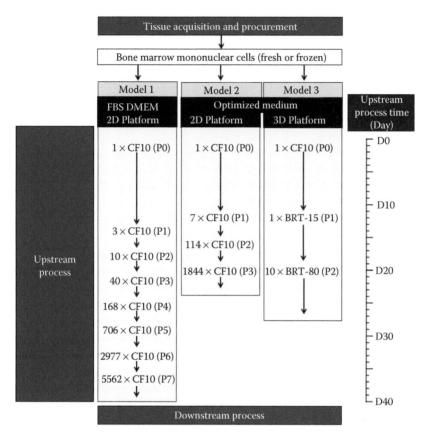

FIGURE 8.2 Hypothetical hMSC manufacturing platform models for the generation of a batch size of >1 trillion cells. The models are designed, based on the reported growth kinetics of hMSCs in two different media, 10%FBS DMEM vs. high performance PPRF-msc6 medium (Jung et al., 2010), and two different culture vessel types, planar systems vs. microcarrier-mediated suspension bioreactors. It is assumed that 100 mL of bone marrow extract is acquired and procured, which would generate ~1 billion bone marrow mononuclear cells (BM-hMNCs). After the primary culture in each model, if 90% of harvested cells are assumed to be passaged at each passage level, a batch size of >1 trillion cells can be manufactured after cells have undergone a total of 23.3 ~ 26.6 population doublings (PD), including PD in the primary culture (assuming that the frequency of primary hMSCs in BM-MNCs is 0.01%–0.001%), through a serial passage. Then, the harvested cells from the production stage of each model are processed through a down-stream process. Based on the selection of culture media and vessel types whose performance has been reported in literature, three different models of upstream process are compared, highlighting the number of culture vessels required at each passage and the upstream timelines. Cell growth rate and size are variable depending on the media used; that is, high performance medium leads to the production of hMSCs in a smaller size and at a higher growth rate, which results in the requirement of a lower number of passage and culture vessels and a reduced process time for a target lot size, compared to 10% FBS DMEM.

- *Model 1 (10%FBS DMEM + CF10)*: It would take about 40 days of upstream process, including ~14 days of primary culture (having undergone ~12.9 PD when the frequency of hMSCs in BM-MNCs is assumed to be 0.001%) and every four days per passage (with 2.23 PD/passage), to produce 1×10^{12} cells at passage 7 (Jung et al., 2012c). Due to the relatively large size of cells grown in this medium, the maximum cell density reached at each passage is low (Jung et al., 2012c). For this reason, a large number of passages and vessels are needed, requiring over $5,000 \times CF10$ at the final production stage.

- *Model 2 (high performance medium + CF10)*: This process model involves the use of a high-performance medium (for instance, an optimized serum-free medium, PPRF-msc6 (Jung et al., 2012c) that will be discussed in the following sections). This medium leads to a high cell growth rate and smaller cell sizes, resulting in a reduction of passage number and the required number of vessels significantly. Due to the high cell growth rate in high performance medium, the upstream process would require ~24 days, including ~12 days of primary culture (14.6 PD during the primary culture, and thereafter 4.17 PD per passage), and less than $2,000 \times CF10$ to produce 1×10^{12} cells at passage level 3. Although the use of high performance medium significantly reduces the number of vessels and process time compared to the case with the classical serum-based medium in Model 1, this platform involving a large number of planar vessels would be unable to support a lot size of 1×10^{12} hMSCs.

- *Model 3 (high performance medium + microcarrier-based suspension bioreactors enabling high-density culture)*: This model employs a microcarrier-based suspension bioreactor together with a high-performance medium. After the primary culture in a planar system for ~12 days (14.6 PD), this platform requires an additional 16 days (6.23 PD per passage in high-density culture) for a seed train stage (at 12 L working volume in a 15 L scale bioreactor) and a production stage (at a total of 750 L working volume in a 1,000 L scale bioreactor if feasible, or 10×80 L bioreactors) for the production of a lot size of 10^{12} cells at passage 3. This estimation is based on reported data that an optimized microcarrier-based suspension culture enables over 75-fold expansion of hMSCs with a final density of over 1.5 million cells/mL at harvest when inoculated at 20,000 cells/mL (Abraham et al., 2017).

These data-based cell manufacturing models clearly indicate that planar cell culture platforms, which suffer from the difficulty of scale-up, would not be an option for large commercial scale cell manufacturing (e.g., 1×10^{12} cells per lot) that is needed for most allogeneic hMSC therapies. Furthermore, these culture systems are not appropriate for controlling critical process parameters due to the heterogeneous nature of culture environment and again the need of employing many vessels. Therefore, despite their usefulness in basic research and preclinical studies, the utility of the planar vessels would not be desirable even for autologous and small-scale allogeneic settings.

8.1.3 IMPORTANCE OF EMPLOYING SCALABLE AND CONTROLLABLE BIOREACTORS DURING THE EARLY STAGE OF CELL THERAPY DEVELOPMENT

For successful and widespread implementation of hMSC therapy, developing an optimized, well-designed cell manufacturing process is essential to realize the consistent and reproducible production of high quality cells in a desired scale, while, at the same time, maintaining their key biological and therapeutic properties and controlling the cost of goods. A cell manufacturing bioprocess should be developed through a staged process development strategy, allowing for varying levels of production throughout the different stages of clinical testing. The staged development supports a gradual and efficient scale-up and improvement of the manufacturing process while establishing increasing definition of the product characteristics (i.e., identity, purity, safety, and potency) as it progresses to later-stage clinical trials. Hence, the use of "scalable" and "controllable" platforms throughout cell manufacturing process development is critical to minimize drug-comparability risks (i.e., the risks of altering the safety and/or efficacy of the product as the process is scaled-up), process development timelines, and costs (Rowley, 2010). It is highly recommended, therefore, that such scalable and controllable cell culture systems be employed at an early stage of cell therapy development.

8.2 CHOICE OF APPROPRIATE TECHNOLOGIES FOR LARGE-SCALE MANUFACTURING OF QUALITY-ASSURED CELLS

As demonstrated earlier, choosing the right platform is the primary requirement to realize large-scale, cost-effective and reproducible manufacturing of quality-assured cells. To this end, growth-promoting performance of key platform components, such as media and culture substrates (i.e., microcarriers), as well as bioreactors and their effect on critical quality attributes, and safety must be understood to select the best platform option. The importance of using a scalable and controllable bioreactor platform during the early stage of cell therapy development was pointed out earlier. It is also crucial for cell therapy product developers to select a defined high performance medium and a culture substrate, which is favorable in manufacturing perspective as well as regulatory viewpoint, as early as possible during their development stages. This is particularly true for hMSC therapies because their mechanism of action has not yet been clearly elucidated in many cases, and thus a process change at a late development stage should require a comprehensive comparability study. Changing platform components and associated raw materials may change various aspects of the biology of hMSCs, e.g., their secretory and immunomodulatory properties or differentiation potential as well as growth kinetics. Information found in literature should be useful to choose a high-performance platform. However, the experimental outcomes reported by different investigators have often demonstrated conflicting data. Therefore, it is important to understand the principles behind the current and future technologies of each platform component to choose the right platform. If possible, it is highly recommended to select competing options and evaluate them in parallel.

8.2.1 CULTURE MEDIA

Cell culture media are generally classified into serum-based, serum-free (SF), xeno-free (XF), animal component-free (ACF), chemically-defined (CD), or protein-free media by the source and characterization of growth supplements that are added to defined basal media. A traditional classification of cell culture media and their definitions have been described in the literature (Karnieli et al., 2017). The majority of hMSC clinical trials involves cell culture media supplemented with fetal bovine serum (FBS) at a concentration of 10%–20% for cell manufacturing (Mendicino et al., 2014). The level of FBS could be reduced when basal media are enriched with certain defined growth factors (Jung et al., 2010). The use of ill-defined non-human serum such as FBS is a regulatory concern, and subsequently much effort has been made to identify substitutes. Human-sourced materials, particularly human platelet lysate, have been the most common alternative for FBS to fulfill cell manufacturing in a xeno-free environment (Mendicino et al., 2014). SF, ACF, and CD media for hMSCs have also been reported in the literature (Jung et al., 2012a) or introduced into the marketplace.

8.2.1.1 Conventional Serum-Based Media

Conventional media for hMSC culture are made up of a defined basal medium (e.g., Dulbecco's modified Eagle's medium (DMEM) or alpha-minimum essential medium [α-MEM]) supplemented with FBS at 10%–20% (v/v). FBS contains nutritional and physiochemical compounds required for cell maintenance and growth and a high content of attachment and growth factors (Jung et al., 2012a). Although FBS-based media have been widely used to generate hMSCs in basic research and clinical studies, the use of FBS for cell therapy manufacturing raises the following serious concerns (Jung et al., 2012a; Karnieli et al., 2017).

1. *Safety*: FBS is produced under a strict manufacturing process with a set of quality controls to minimize the risk of contamination with harmful pathogens, such as viruses, mycoplasma, and prions. Nonetheless, the risk of contamination with unidentified zoonotic agents and transmission of these contaminants to cell therapy products still exist. In addition, FBS contains a high content of xenogeneic proteins that could be associated with the final cell product, which could cause an immune reaction in recipients. These safety issues with FBS are the primary drive to pursue serum-free media, or ideally chemically-defined media.
2. *Variability*: FBS suffers from a high degree of batch-to-batch variations and may contain growth inhibitors, cytotoxic substances, and/or differentiation agents, which could cause inconsistency in the generation of hMSCs. Figure 8.3 shows cell yields obtained from the primary culture of human bone marrow mononuclear cells using a classical FBS-containing medium, i.e., DMEM supplemented with 10% FBS, and an optimized hMSC medium (Jung et al., 2010). Two out of eight trials did not support cell growth over 21 days of culture when 10% FBS DMEM was used, while the optimized medium showed a consistent growth-promoting performance. The inherent variability of FBS and its lack of consistency

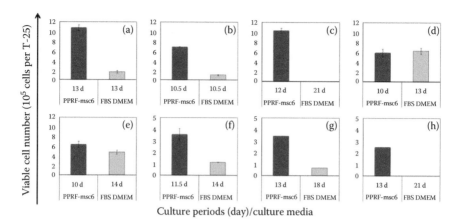

FIGURE 8.3 Primary culture of hMSCs in an optimized medium (PPRF-msc6) versus DMEM + 10% FBS (FBS DMEM). Cryopreserved human bone marrow monocular cells (MNCs from three donors, multiple aliquots per donor) were thawed and plated into either PPFF-msc6 or FBS DMEM at a seeding density of 150,000 cells/cm^2 (a–e) or 37,500 cells/cm^2 (f–h).

in generating hMSCs could make standardization of the manufacturing process difficult.

3. *Availability*: As hMSCs are on a path toward commercialization, the expected global demand for serum for mass cell production is very high but the production levels of qualified FBS are generally fixed (Brindley et al., 2012). Therefore, developing serum-free media, or at least serum-reduced media by enriching basal medium formulations, should be helpful for a secure supply chain of serum.

4. *Low performance for cell growth*: As described earlier (Figure 8.2), the poor growth-promoting performance of conventional FBS-containing media is a serious limitation to realize large-scale and reproducible manufacturing of hMSCs. In addition to the regulatory issues with serum, it is very important to consider the performance of media in manufacturing viewpoint as it affects the scale and design of cell manufacturing significantly.

8.2.1.2 Xeno-Free Media

To mitigate the safety and regulatory concerns raised by non-human serum, human blood-derived materials, such as human serum and platelet lysates, have been used for hMSC research and clinical trials as an alternative medium supplement (Tekkatte et al., 2011). Human autologous serum has been reported to support hMSC expansion (Stute et al., 2004; Shahdadfar et al., 2005; Tekkatte et al., 2011). It would be problematic, however, to acquire amounts sufficient to generate clinically-relevant numbers of hMSCs. The performance of human allogeneic serum is rather controversial because contradictory results have been reported (Kuznetsov et al., 2000; Shahdadfar et al., 2005; Le Blanc and Ringden, 2007; Tateishi et al., 2008; Poloni et al., 2009). By contrast, allogeneic human platelet lysate has demonstrated promising performance. It supported hMSC growth at a significantly higher rate, compared to conventional FBS-based media and the maintenance of the immunomodulatory properties, phenotype, and differentiation potential of the expanded hMSCs (Doucet et al., 2005; Muller et al., 2006; Capelli et al., 2007; Kocaoemer et al., 2007; Lange et al., 2007; Reinisch et al., 2007; Schallmoser et al., 2007; Bieback et al., 2009; Flemming et al., 2011). Indeed, human platelet lysate has been the most common alternative for FBS in regulatory submissions (Mendicino et al., 2014).

Although considered relatively safer than FBS for human therapeutic applications, the use of human-sourced supplements is still a matter of substantial debate, prompting some concerns (Reinhardt et al., 2011; Sensebé et al., 2011). There is a risk that allogeneic human growth supplements may be contaminated with human pathogenic viruses that might not be detected by routine screening of blood donors. Moreover, these crude blood derivatives are poorly defined and suffer from batch-to-batch variation, and thus their ability to maintain hMSC growth and therapeutic potential could be widely variable. Although the promising performance of human platelet lysate has been demonstrated in many studies, some reported conflicting data. For instance, a reduced osteogenic or adipogenic differentiation capacity (Gruber et al., 2004; Lange et al., 2007), an altered expression profile of surface molecules and a decreased immunosuppressive capacity (Abdelrazik et al., 2011) were reported when hMSCs were cultured in media supplemented with human platelet lysate. Considering the significance of qualification and comparability of the ill-defined products, the potential variability of human platelet lysate could be a major limitation for clinical-scale or commercial manufacturing of hMSCs.

8.2.1.3 Defined Serum-Free Media

A consensus in the MSC field is that, although many regulatory agencies tolerate the use of FBS to produce hMSCs for early phase studies, late phase trials and commercial manufacturing tend to require serum-free media (Tolar et al., 2010). A recent report by the FDA (Mendicino et al., 2014) indicated that many of the regulatory submissions for hMSC-based products described process developments to include replacing the use of animal-derived serum during manufacturing. Efforts to develop defined serum-free media for hMSCs have been reported (Marshak and Holecek, 1999; Liu et al., 2007). In addition, several companies introduced serum-free, xeno-free or chemically-defined media for hMSC expansion. Considering the ill-defined nature and batch-to-batch variation of FBS and human-sourced alternatives, chemically-defined media should be ideal to generate cells retaining desired qualities in a consistent and predictable manner, which is important for minimizing treatment failures.

Wu and co-workers (Wu et al., 2014) reported that a commercially-defined serum-free medium supported the growth of hMSCs from umbilical cord, and the culture-expanded cells exhibited strong immunosuppressive activities and secreted immunomodulatory cytokines at similar levels with those grown in an FBS-based control medium [DMEM/F12 supplemented with 10% FBS and 10 ng/mL Epidermal Growth Factor (EGF)]. The authors further administered cells expanded in the commercial medium *versus* the control medium into a rat model of pulmonary arterial hypertension. They observed a substantial reduction in the thickness of the pulmonary arterial wall and an increased survival rate of rats at comparable level in both cases. This finding suggests that producing "therapeutic" hMSC products using a defined medium formulation without the use of the classical supplements, such as FBS or human platelet lysate, is feasible. Therefore, it should be recommended for cell therapy developers to evaluate defined serum-free media commercially available or reported in literature and to further optimize the medium formulation, if necessary, at an early stage of development to ultimately realize a reproducible and compliant cell manufacturing platform.

Regarding the use of commercially defined media, the potential drawbacks would be the relatively low performance in promoting hMSC growth and their formulations being undisclosed. Although such media should contain defined growth and attachment factors for hMSC culture, it has been shown that primary culture of bone marrow cells using some commercial defined media failed to generate hMSCs or resulted in a very limited cell growth through a prolonged culture period (Jung et al., 2012a). This could be why most of studies involving commercial defined media use hMSCs that have previously derived from the primary culture in the presence of FBS. It was also observed that such commercial media would not support rapid expansion of hMSCs, with cells demonstrating a gradual decrease in growth when serially passaged without the use of additional supplements. For this reason, a low concentration (1% ~ 2%) of FBS or human platelet lysate is often added into the commercial media to enhance their growth-promoting performance. Generally, the proprietary formulations of commercial media are not disclosed, which may raise a concern with securing medium supply. Further, when a selected commercial medium is required to be further optimized by the end user to enhance its performance, the optimization study could be challenging without understanding their constituents.

We reported a systematic strategy to develop a defined serum-free medium for hMSCs from bone marrow (Jung et al., 2010). This medium, PPRF-msc6, supports consistent cell growth in the primary culture of bone marrow cells. hMSCs are expanded in both primary and subsequent cultures at a much higher cell growth rate in this medium, compared to a classical hMSC medium (DMEM + 10%FBS), over 10 passages without any noticeable decrease of cell growth rates. Specifically, average population doubling times of hMSCs derived from three separate donors and grown in PPRF-msc6 versus 10% FBS DMEM were 21–26 h and 35–38 h, respectively, at passage levels 1 through 10. Moreover, colony-forming unit-fibroblasts (CFU-F) assays demonstrated that culturing primary bone marrow cells in PPRF-msc6 resulted in a significantly higher number of colonies (Jung et al., 2012c). hMSC populations, typically isolated based on plastic adherence, are not homogeneous but contain a subpopulation of cells capable of forming colonies when plated at a low density (e.g., 1 cell per cm^2). A CFU-F assay does not specifically determine multilineage or self-renewal capacity of hMSCs; however, cells arising from a CFU-F have been shown to be highly proliferative and to regenerate specific mesenchymal lineages (Digirolamo et al., 1999). For this reason, the CFU-F assay has been widely used to assess the proliferative and differentiation potential of hMSCs. Therefore, finding that the use of PPRF-msc6 generates a high number of colonies may indicate that it rescues valuable hMSCs from the primary culture of bone marrow cells with high efficiency while a large portion of important cells could be washed out during the primary culture with the classical hMSC medium. Taken together with the high performance on hMSC growth, this capability of PPRF-msc6 to rescue a high number of colony-forming cells (i.e., corresponding to hMSCs) in the primary culture resulted in a significantly higher cell expansion (as demonstrated earlier in Figure 8.3).

Figure 8.4a and b shows growth profiles of BM hMSCs cultured in 10% FBS DMEM versus PPRF-msc6 in a planar system (i.e., T-flask) and a microcarrier-based bioreactor, respectively. In both cases, PPRF-msc6 enabled a significantly higher cell growth rate, demonstrating evidently reduced lag phases and greater cell yields. The high performance of PPRF-msc6 led to over 64 PDs of hMSCs from three donor BM cells for a two-month period, while 10% FBS DMEM resulted in 43 PDs during the same period (Figure 8.4c). Similar trends were observed for hMSCs from other sources; that is, hMSCs from adipose tissue (AT), umbilical cord (UC) and pancreatic tissue (Pan) were successfully isolated and serially expanded in PPRF-msc6 more rapidly and consistently compared to conventional FBS-based media (Figure 8.4d–f). In addition, a more homogeneous cell population in phenotype and size was obtained with the use of PPRF-msc6, and the average size of cells was significantly smaller than those generated in 10% FBS DMEM (Figure 8.5). Cells grown in PPRF-msc6 include such characteristics, such as high cell growth, reduced lag phase, and smaller size, and represent a benefit for mass production of cell products enabling a significantly reduced process time and number of culture vessels (as illustrated in Figure 8.2). The smaller size of cells could also be a benefit from a therapeutic viewpoint. It has been reported in animal model studies that the majority of hMSCs (when grown in conventional FBS-based media) tended to be trapped in the lungs (Schrepfer et al., 2007; Fischer et al., 2009; Lee et al., 2009). The generation of hMSCs with a smaller size may enhance their therapeutic effect since the small cells may travel through the lungs and home to the site of injury or disease at high efficiencies (Bartosh et al., 2010).

8.2.2 Culture Substrate (with Coating Materials)

hMSCs are anchorage-dependent, and therefore they need a culture substrate, normally a plastic surface, to attach to and spread out on in order to be maintained and proliferate. Another approach to culture hMSCs is a three-dimensional (3D) aggregate culture in suspension without the use of substrates such as microcarriers. When anchorage-dependent cells are inoculated as a single-cell suspension, neighboring cells tend to assemble to form aggregates or spheroids either spontaneously (i.e., cell aggregation in bulk culture or in hanging drops) or by

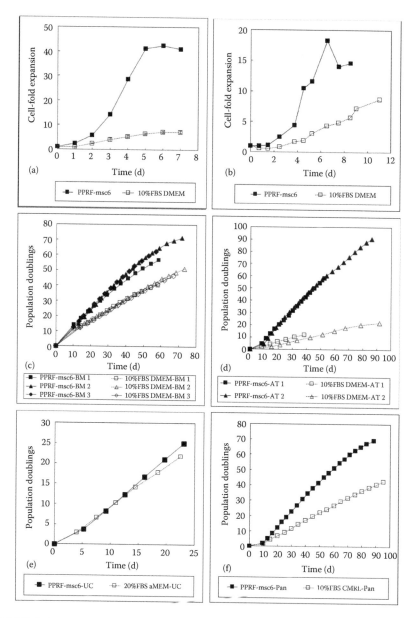

FIGURE 8.4 Growth of hMSCs in PPRF-msc6 in comparison with conventional media containing FBS. BM-hMSCs at passage 4 were expanded in a batch mode in planar system (a) and microcarrier-based bioreactors (b). hMSCs from different sources, including bone marrow from 3 different donors (*BM 1, BM2 and BM 3*), adipose tissue from 2 separate donors (*AT 1 and AT 2*), umbilical cord (*UC*) and pancreatic tissue (*Pan*), were serially passaged (c-f). The conventional FBS containing media used were 10%FBS DMEM (BM-hMSCs and AT-hMSCs), 20%FBS DMEM (UC-hMSCs) and 10%FBS CMRL-1066 (Pan-hMSCs). (Adapted from Jung, S. et al., *Biotechnol. Appl. Biochem.*, 59, 106–120, 2012b; Jung, S. et al., *J. Tissue Eng. Regen. Med.*, 6, 391–403, 2012c. With permission.)

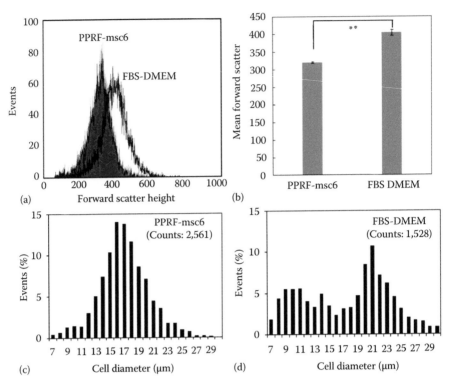

FIGURE 8.5 Comparison of size and size distribution of hMSCs cultured in PPRF-msc6 versus 10% FBS DMEM by flow cytometry (a,b) and ViCell Cell Analyzer (c,d). (Adapted from Jung, S., et al., *Stem Cells Int.*, 2012, 123030, 2012a. With permission.)

forced aggregation methods (Subramanian et al., 2011). Following the aggregation, population expansion may occur through subsequent cell division, depending on cell type and culture conditions. As cells divide, the daughter cells remain attached to each other because of the production of sticky extracellular matrix molecules. This eventually results in the formation of large spherical cell aggregates. By exploiting these characteristics, numerous cell types have been cultured as aggregates in suspension, including embryonic and induced pluripotent stem cells, adult stem/progenitor cells, such as neural stem cells and mammary epithelial stem cells, and cancer stem cells.

Attempts have also been made by several groups to culture hMSCs from various sources as aggregates (Potapova et al., 2008; Bartosh et al., 2010; Frith et al., 2010). These studies demonstrated the formation of 3D aggregates of hMSCs in bulk media or in hanging drops upon inoculation as a single cell population. However, the proliferation potential of hMSCs in aggregates is controversial. Baksh and co-workers described hMSC aggregates that formed in a suspension culture system were not capable of proliferation (Baksh et al., 2003). In contrast, others reported that hMSC aggregates remained viable in culture, while exhibiting proliferation potential (Frith et al., 2010; Dromard et al., 2011). Although these studies demonstrated proliferative characteristic of hMSCs embedded in aggregates by Bromodeoxyuridine (BrdU) staining, no notable expansion was reported. Therefore, identifying a culture protocol that allows for significant growth of hMSCs as aggregates in suspension culture while maintaining their key biological and therapeutic functionalities should be a subject of interest. Attempting to culture human pluripotent stem cells (hPSCs) in suspension culture, Olmer and colleagues demonstrated tightly controlled aggregate formation from single cells under well-defined parameters and strategies (Olmer et al., 2012). This study revealed that the tendency of cell aggregation was diminished and unpredictable when the single hPSCs were inoculated at a low density (i.e., 50,000 cells/mL), whereas a high inoculation density (i.e., 500,000 cells/mL) led to reproducible aggregate formation. As a different approach, Subramanian and colleagues reported that a population of primitive cells, known as multipotent adult progenitor cells (MAPCs) with extensive proliferation capacity and pluripotent differentiation potential, isolated from rat bone marrow were cultured as aggregates in suspension culture. In their study, the authors induced individual MAPCs to form aggregates using a forced aggregation method and maintained them in a static condition for two days, and then cultured the MAPC aggregates in an agitated suspension condition under a hypoxic environment, achieving a significant cell expansion (Subramanian et al., 2011). These results may indicate that the initial phase (i.e., self-assembly of cells) is critically important for successful suspension culture of hMSCs in aggregates as it is presumed that anchorage-dependent hMSCs can proliferate after their attachment to a substrate and spread out—for example, neighboring cells could play a role as a substrate in aggregate cultures. Therefore, cell inoculation protocols, such as the method of cell inoculation (individual cells versus aggregates) and cell inoculation density, can be optimized for a rapid and efficient self-assembly of hMSCs to generate well-organized hMSC aggregates in relatively consistent sizes during the initial culture periods. It has also been observed that self-assembly of other stem cell types (i.e., human neural stem cells) to 3D cell aggregates and their growth are affected significantly by medium components, oxygen tension, and pH (Baghbaderani et al., 2010; Baghbaderani et al., 2011). Once successfully developed, substrate-free aggregate-based suspension culture for cell expansion might be technically simple and easy to scale-up, even though realizing this seems to be a great challenge.

Since the growth of hMSCs as aggregates in substrate-free suspension culture is still very limited, the selection of an appropriate culture substrate should be seriously considered for successful cell manufacturing. Considering the scalability of suspension culture (such as stirred-tank, vertical-wheel, or wave bioreactors) to enable mass cell manufacturing at large scale, the use of microcarriers in hMSC culture are briefly reviewed here. Other systems including hollow-fiber system and scaffold-mediated fixed-/packed-bed systems are discussed in the following "bioreactor" section. Microcarriers are microscopic beads (with a diameter of 100–400 μm) that can be placed in medium to provide a surface upon which anchorage-dependent cells, such as hMSCs, can attach to and subsequently grow under pseudo-suspension conditions. Depending on the loading density of microcarriers in culture, this method can provide a significantly high ratio of the growth surface to medium volume. Therefore, microcarrier cultures in scalable bioreactors could offer a superior means of producing large quantities of adherent cells.

Various types of microcarriers are available in the market. It is important to select candidates based on the understanding of their characteristics, including surface topography, base materials, charged versus uncharged, and coated materials (Abraham et al., 2017), and testing the selected microcarriers for a particular hMSC population. Microcarriers are commonly classified into three groups, solid, microporous, and macroporous microcarriers according to their surface topographies:

Solid microcarriers do not have pores and are made up of plastic or glass. The plastic microcarriers are often charged or coated with proteins to enhance cell attachment and growth. It was generally considered that solid microcarriers are best used for shear-tolerant cells, while expanded cells can be readily dissociated from the microcarriers to facilitate cell harvest and passage. However, such microcarriers have also been successfully used for shear-sensitive hMSCs in small-scale suspension cultures even in xeno-free environment (dos Santos et al., 2011; Hervy et al., 2014). dos Santos and colleagues demonstrated comparable growth of hMSCs from bone marrow and adipose tissue in non-porous plastic microcarriers versus macroporous CultiSpher microcarriers (discussed later in this section). Moreover, exploiting the advantage of efficient cell harvest, Hervy and colleagues serially-expanded bone marrow hMSCs over seven passages using synthetic microcarriers in xeno-free conditions (Hervy et al., 2014).

Microporous microcarriers, such as Cytodex (GE Healthcare Inc.), have small microscopic pores to allow diffusion of nutrients and metabolites while offering their external surface for cell growth. They provide slightly increased surface area by having surface indentations, compared to solid microcarriers, which often favor better cell anchorage (Abraham et al., 2017). Cytodex 1, 2, and 3 have been used for the culture of human and animal MSCs (Frauenschuh et al., 2007; Jung et al., 2009; Schop et al., 2010).

Macroporous microcarriers such as CultiSpher (Percell Biolytica AB) or Cytopore (GE Healthcare Inc.) were originally designed to enable high density cell culture. They have pores that are large enough for cells to migrate into and grow within the internal space. This offers a very large surface for cell growth to achieve high density culture, and provide an environment in which fragile cells can be protected against detrimental shear stress caused by a mixing motion in the vessel. Macroporous microcarriers have been used to support hMSC growth (Yang et al., 2007; Sart et al., 2009; Eibes et al., 2010; Sart et al., 2010; dos Santos et al., 2011). High density culture of hMSCs using CultiSpher microcarriers has also been reported (Abraham et al., 2017). It should be pointed out, however, that cell harvest at high density may require a prolonged enzymatic dissociation of cells to obtain single cell suspension after harvest. This is particularly true for the culture of hMSCs in macroporous microcarriers as cells form multi-layered 3D structures while they grow. Of note, cells growing on solid or microporous microcarriers also tend to form aggregates, but the size of aggregates in culture with macroporous microcarriers is typically larger. To facilitate cell harvest from the macroporous microcarriers, digestible beads (i.e., CultiSpher) were introduced. The digestion time of gelatin-based CultiSpher microcarriers using collagenase is quick, supporting the dissociation of cells. However, hMSCs secrete a significant amount of sticky extracellular matrix molecules during growth on microcarriers, which induce cells to stick to each other generating very large aggregates over time. As a result, when harvesting cells at high density, the time needed to fully dissociate cells could significantly be increased. Optimizing the harvest protocol to enable obtaining completely dissociated cells should be a critical requirement for high density cell culture.

8.2.3 BIOREACTORS

For both allogeneic and autologous applications, hMSCs are currently generated for research and clinical applications through conventional static adherent culture methods using tissue culture flasks or multi-layered vessels in most cases. As discussed earlier (Figure 8.2), however, the use of such static platforms should not be an option and it is required to have scalable and controllable bioreactors, particularly for a large-scale allogeneic setting. In addition, it would be desirable to use pre-sterilized disposable bioreactors and associated instruments in a closed system to eliminate the painstaking steps of cleaning and sterilization. The use of disposable plastic linings that can be sterilized by gamma rays can cut down time between batches of cells with a quick turnaround, and thereby the number of cell manufacturing runs can be increased.

For mass production of hMSCs, several bioreactor types involving the use of microcarriers have been suggested, including the stirred-tank bioreactor (e.g., vendors: Eppendorf and GE Healthcare), the wave bioreactor (i.e., vendors: GE Healthcare and Sartorius) and the vertical-wheel bioreactor (i.e., vendor: PBS Biotech) (Jung et al., 2012b). In addition, other types of bioreactors, such as hollow-fiber and fixed-/packed-bed bioreactors, have been used as controllable high-density culture platforms. Each bioreactor type has its own specific features; therefore, in order to select the best option for a specific application, it is important to understand the characteristics of different bioreactors and evaluate their performance, preferably in side-by-side comparison, considering the followings:

- Growth-promoting performance to achieve both high cell growth rate and high cell yield
- Scalability from a scale-down model through the manufacturing scale
- Ability to incorporate probes to enable online monitoring and control of key process parameters (e.g., temperature, pH, dissolved oxygen (DO), agitation speed)
- Ability to accurately monitor cell growth—preferably using an online monitoring system
- Ease of sampling to monitor cell growth by direct cell counting methods to validate the online monitoring technology
- Ease of cell harvest
- Disposability
- Automation
- Simplicity of operation
- Effectiveness in terms of cost and time

First and foremost, the selected bioreactor should support the rapid growth of hMSCs toward high density culture. As discussed early, the bioreactor system should be scalable throughout the staged process development from a scale-down model at small-scale through the manufacturing platform at large-scale. Online real-time monitoring and control of key process variables is critical to maintain environmental input variables (e.g., pH, DO, agitation speed) within ranges established for optimal cell growth. Cell growth should be monitored throughout the culture period as accurately as possible. Ideally, this needs to be done using online monitoring technology (Abraham et al., 2017). Regardless of such online monitoring technology being available to estimate cell growth in bioreactors, it is important to be able to take representative samples at desired time points to accurately measure cell density and other important culture parameters (and to validate the online estimation). With this regard, it should be important to make the culture uniformly distributed in the vessel when a sample is taken. And, as the anchorage-dependent hMSCs must be dissociated and separated from the culture substrates (i.e., microcarriers, hollow-fibers, immobilizing scaffolds, etc.) at the end of culture in both seed train and manufacturing stage, the selected bioreactor process should support cell harvest at high efficiency to maximize cell yields and quality.

8.2.3.1   Hollow-Fiber Bioreactors

Hollow-fiber bioreactors have been used in a perfusion mode to achieve high density culture of anchorage-dependent cells as well as suspension cells (reviewed in Godara et al. [2008]). This system consists of bundles of synthetic, semi-permeable hollow-fiber micro-capillaries. Liquid medium flow through the intra-capillary space and adherent cells grow on the outer surface. The perfusion of medium through the fibers creates a relatively high pressure within the inner space, which permits the efficient exchange of nutrients and metabolites across the fiber wall through pores. The molecular weight cut-off of the pores is normally 6,000–10,000 Da, and thus cells grown in the outer space of the fiber are continuously supplied with nutrients and exposed to low concentration of waste products (Butler, 2004). This mechanism enables the growth of cells at very high densities (up to 10^9 cells/mL). Based on these features, this system seems to be attractive for high cell density tissue engineering and particularly high-yield protein production as the macromolecular proteins secreted by cells are concentrated in the outer space of the fibers. However, this process is complex and difficult to scale-up. Moreover, the pressure difference that may be generated along the long fibers could produce concentration gradients of nutrients and metabolites, which may cause uneven cell growth (Butler, 2004). Rojewski and colleagues reported a commercial hollow-fiber bioreactor system (Quantum Cell Expansion system) enabling the isolation and expansion of bone marrow hMSCs (Rojewski et al., 2013). Using a two-step approach, this closed process together with platelet lysate as culture supplement led to the generation of >100 × 10^6 hMSCs from bone marrow aspirate (~18.8–28.6 mL) during ~18-day culture period, providing a proof-of-principle to produce adherent hMSCs using a hollow-fiber bioreactor. Hanley and co-workers also used the hollow-fiber bioreactor to expand bone marrow hMSCs using DMEM supplemented with 5% human platelet lysate under a perfusion mode. As cells grow in culture, increasing the feed rate to meet the nutritional requirement of the increasing number of cells was crucial to maintain cell growth and quality, particularly for realizing the production of quality-assured cells at high density. Cell growth cannot be monitored by direct cell counts in the hollow-fiber bioreactor. Instead, the authors monitored lactate concentrations by off-line measurements as an indicator to change medium feed perfusion rate during culture (Hanley et al., 2014). Due to the capability to easily implement a perfusion mode (discussed later), the use of hollow-fiber bioreactors could be considered for small-scale cell manufacturing, particularly for autologous applications.

8.2.3.2 Fixed-Bed and Packed-Bed Bioreactors

Weber and colleagues used a fixed-bed bioreactor to assess their ability to expand hMSCs using an immortalized cell line, hMSC-TERT (Weber et al., 2010). Fixed-bed bioreactors contain packing materials, such as solid or porous glass beads, settled in a dense bed resting on penetrated base at the bottom of vessel (Freshney, 2000). The beads with a diameter of 2–5 mm provide a large surface for cell attachment and growth. In fixed-bed systems, the beads are not moved, and medium is perfused upward or downward through the packed beds in recirculation to supply the attached cells with nutrients. This process can support high cell densities providing a relatively favorable environment against mechanical shear forces, and thus could represent an effective production system for shear-sensitive cells. Using non-porous glass spheres in a disposable bioreactor, Weber and colleagues cultured a hMSC line up to a bed volume of 300 cm^2 (Weber et al., 2010).

Ma and colleagues used fibrous bed bioreactors to support long-term expansion of hMSCs under various flow conditions and demonstrated that hMSCs secreted extracellular matrix molecules extensively under low shear stress in the 3D fibrous scaffolds-based bioreactors. This led to better preservation of their stemness and proliferation potential compared to planar culture (Zhao and Ma, 2005; Kim and Ma, 2012). Osiecki and co-workers used polystyrene-based beds that were treated with air plasma to mimic traditional tissue culture plastic surface. Using this packed-bed bioreactor, a 10-fold expansion of placental-derived hMSCs was achieved after a one-week culture period (Osiecki et al., 2015). Despite its beneficial features, the scalability of the fixed-/packed-bed bioreactor system is limited and achieving efficient cell harvest should be a great challenge.

8.2.3.3 Stirred Bioreactors (with Microcarriers)

Stirred systems, such as stirred-tank bioreactors and their scale-down model spinner flasks, are cylindrical vessels with an impeller. The impeller is rotated by a magnetic stirrer or by a top-driven motor. The stirred bioreactors provide relatively uniform conditions throughout the vessel due to mixing induced by the rotated impeller. Moreover, the stirred bioreactor system can readily be integrated with online monitoring instruments, and thereby facilitate tight control of process variables such as pH, temperature, DO and carbon dioxide concentrations in the well-mixed environment. The stirred bioreactors have been widely used to culture mammalian cells due to the advantages of being scalable and controllable for a single cell suspension culture. These systems have also been used for the culture of anchorage-dependent cells growing on microcarriers or as aggregates under suspension conditions. Using a high loading density of microcarriers in culture, the microcarrier-based culture in stirred bioreactors (and other types of suspension bioreactors such as vertical-wheel bioreactors to be discussed later) provides a significantly high ratio of the growth surface to medium volume, and thereby offers a means of high density culture at large scale.

Microcarrier-mediated stirred culture has been used for the expansion of hMSCs from different sources including bone marrow, adipose tissue, umbilical cord, placenta and ear (Jung et al., 2012b; de Soure et al., 2017). Although the feasibility of expanding hMSCs in microcarrier-based stirred bioreactors has been demonstrated, the performance of this approach was relatively low (i.e., demonstrating lower cell growth rate) and highly variable, compared to conventional static cultures. It should be noted that, although stirred bioreactors provide a well-distributed culture of single cell suspension, keeping large particles (such as cells growing on microcarriers or as aggregates) uniformly distributed in a large vessel require a high rotating speed of impeller, which causes high shear stress to the cells in the culture medium and could harmfully affect the growth and quality of cells (Croughan et al., 2016). It is well known that anchorage-dependent cells growing on microcarriers or as aggregates are more sensitive to hydrodynamic shear stress than single cell suspension cultures (Croughan et al., 2016). Therefore, it is very challenging to achieve high density culture of quality-assured hMSCs in stirred bioreactors particularly at large scale.

8.2.3.4   Vertical-Wheel Bioreactors (with Microcarriers)

Recently, a novel single-use bioreactor system using a vertical-wheel technology has been introduced, which offers efficient fluid mixing, homogenous energy dissipation distribution, and uniform suspension of cell aggregates and microcarriers with low power input. The vertical-wheel single-use bioreactor system consists of a vessel with a U-shaped bottom, a wheel-shaped vertical impeller, and four flat, baffle-less walls that are completely enclosed within the vessel (Figure 8.6a). The entire single-use vessel is held within a stainless-steel housing that has a fully integrated controller system to monitor and provide the key bioreactor functions (Croughan et al., 2016). The vertical-wheel impeller occupies a large proportion of space inside the vessel, accounting for ~85% of the width of the U-shaped bottom, and the large surface area of the impeller helps dissipate its rotational energy to the liquid over a large contact area, which results in a minimum range of gradient for turbulent energy dissipation rate and gentle mixing. This is contrary to a typical stirred-type bioreactor, which has a relatively small impeller rotating at high speed. That leads to a wide distribution of local turbulent energy dissipation rates, spanning more than four orders of magnitude (i.e., >10,000-fold), with the highest rates in the impeller region. Paddles located on the rim of the impeller of vertical-wheel bioreactor direct fluid around the circumference. These paddles create tangential fluid flow around the wheel's circumference and along the sides of the vessel. The large vertical-wheel impeller sweeps fluid along the U-shaped bottom of the vessel, which prevents cell aggregates or microcarriers from settling at relatively low speed. The vertical-wheel is designed to be a combination of radial and axial flPaddles located on the rim of the impeller of vertical-wheel bioreactor direct fluid around the circumference. These paddletates, two oppositely oriented vanes within the wheel (Figure 8.6b) promote bi-directional fluid flow to generate cutting-and-folding action, in contrast to the unidirectional flow of impellers in stirred bioreactors. Together with the strong sweeping flow due to the large impeller

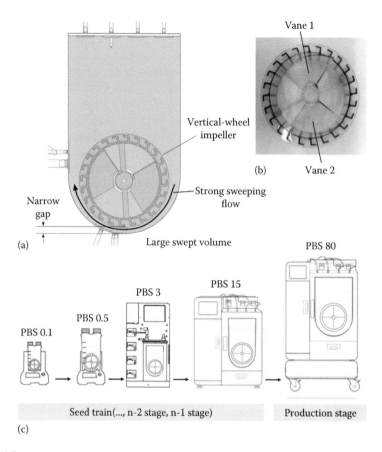

FIGURE 8.6 Single-use vertical-wheel bioreactors. Diagram and key features of PBS bioreactor vessel with enclosed vertical-wheel, U-shape round bottom and flat sides in the front and back (a). Vertical-wheel impeller showing the two oppositely pumping axial-flow vanes and radial-flow paddles (b). Family of single-use vertical-wheel bioreactors at a range of vessel sizes from 0.1 to 80 liters that can be used at train through production stages (c). (Adapted from Sousa, M.F. et al. *Biotechnol. Prog.*, 31, 1600–1612, 2015; Croughan, M.S. et al., Novel single-use bioreactors for scale-up of anchorage-dependent cell manufacturing of cell therapies, In Cabral, J.M.S., De Silva, C.L., Chase, L.G., Diogo, M.M. (Eds.), *Stem Cell Manufacturing*, Elsevier, Atlanta, GA, pp. 105–139, 2016.)

rotating to the U-shaped bottom, this unique mixing mechanism in the vertical-wheel bioreactors allows efficient liquid mixing with minimum dissipation energy gradients and low shear effects. The vertical-wheel bioreactors' various design features, such as size, position, and impellers, offer gentle and uniform fluid mixing, efficient particle suspension with low power input, and agitation speeds. Therefore, a favorable culture environment is created for shear-sensitive, anchorage-dependent cells such as hMSCs. Sousa and colleagues has demonstrated that bone marrow-derived hMSCs attach rapidly and uniformly onto microcarriers in a vertical-wheel bioreactor resulting in a higher percentage of microcarriers being attached with cells, compared to a stirred-tank bioreactor (Sousa et al., 2015). Moreover, when empty microcarriers were added to the both vertical-wheel and stirred cultures to provide additional surface area for cell growth, a higher percentage of microcarriers were occupied with cells in the vertical-wheel bioreactor. The higher homogeneous cell attachment and migration efficiency led to higher percentages of proliferative cells in the vertical-wheel bioreactor. Furthermore, the hMSCs produced in the vertical-wheel bioreactor demonstrated a significantly lower percentage of apoptotic cells.

Two types of vertical-wheel bioreactors, AirDrive and MagDrive, have been introduced. For the AirDrive bioreactor, the vertical wheel motion is driven by the buoyant force of gas bubbles that are introduced from the sparger at the bottom of the vessel and captured in the air cups on the wheel periphery. The gas bubbles are then released as the wheel turns, which drive further wheel rotation. The wheel rotational speed is determined by the main sparger gas flow rate. The AirDrive bioreactor is intended for cultures grown in medium that can be supplemented with anti-foaming agents, such as Pluronic F-68, as needed to eliminate problems with cell entrainment and damage from bubble bursts. The AirDrive units have been characterized and used for years for the cultivation of various mammalian cell types up to 250 L (Croughan et al., 2016). For the MagDrive bioreactor, the vertical wheel rotation is driven and controlled by external magnetic coupling between the vertical-wheel and the housing base unit. The advantage of the MagDrive system over the AirDrive is to eliminate the need for anti-foaming agents or shear protectants, and thus, is optimal for culturing shear sensitive, anchorage-dependent cells growing on microcarriers or as aggregates for cellular products. The MagDrive family are available at various scales including 0.1 and 0.5 L, named PBS-mini, as scale-down models without process control capability, and fully controlled bioreactors at 3 L, 15 L and 80 L scales. The innovative vertical mixing mechanism not only requires very low power input to suspend microcarriers or cell aggregates homogeneously, but also allows the low shear environment to remain constant across a full range of vessel sizes from 0.1 to 80 L. This is particularly beneficial since the small-scale vessels can effectively be used as a seed train as well as a scale-down model for large-scale bioreactors (Figure 8.6c), which will be discussed in the following section. In addition to supporting the growth of hMSCs at high efficiency, the benefit of such mixing mechanism in vertical-wheel bioreactors has been demonstrated as a superior system for the culture of various mammalian cell types, such as pluripotent stem cells, chondrocytes and other primary cells, which grow on microcarriers or as aggregates.

8.3 PROCESS DEVELOPMENT AND OPTIMIZATION

The use of scalable bioreactors together with microcarriers should represent a superior means of producing large quantities of hMSCs over current planar cell culture systems, providing significantly high ratios of growth surface per unit culture volume and process control capabilities. Indeed, studies have been demonstrating the feasibility of utilizing microcarrier-based suspension culture systems using either small-scale spinner flasks or instrumented bioreactors for the expansion of hMSCs from various sources. Despite much effort made by many groups, however, the performance of microcarrier-based suspension culture systems for expanding hMSCs has been often shown to be far from being optimal. Specifically, data in the literature on the microcarrier culture of hMSCs frequently showed prolonged lag growth phase and low growth rate. With this regard, efforts have been made to enhance hMSC yields from microcarrier suspension culture, emphasizing the importance of selecting appropriate platform components with high performance and conducting well-designed process optimization studies (Jung et al., 2012b; Abraham et al., 2017). In this section, various aspects of process development and optimization are briefly discussed.

8.3.1 Starting Materials

As discussed earlier, the choice of starting materials is an important consideration in cell therapy manufacturing process development. There are several important factors to consider: derivation of the material used (e.g., recombinant, human-derived, or xeno-derived sources), stability of the material, potential off-target effects, availability and ability to source material for long-term production. The first point concerns what is currently allowed by the regulatory agencies in different countries for cell therapy products and the allowable risk for potential adventitious and unknown pathogens. As described earlier (Section 2.1.1), FBS is used for the majority of currently ongoing hMSC trials, however, there is a risk of (1) lot-to-lot variability, (2) lack of detection of unknown zoonotic viruses, (3) presence of xenogeneic proteins, and (4) potential supply issues as processes are scaled-up for late-stage clinical trials and commercial markets and demand rises. Additionally, if you have a material that is used in your manufacturing process, which needs to be prepared on day of use, this can impact the cost and risk of using that material. This also applies to consumables used for the culture of cells. As we look to minimize cross-contamination and increase equipment reliability between different batches the move towards single-use and cell culture plastics is important. Therefore, when developing a manufacturing process that is moving towards the clinic it is important to firstly address the risks and potential high-risk materials that can be replaced with lower-risk materials.

8.3.2 Understanding Critical Quality Attributes and Critical Process Parameters

Quality by Design (QbD) has been an approach that has been implemented by the pharmaceutical industry for a number of years. It is an approach that emphasizes understanding of the process and performing continual monitoring and improvement to enhance the safety and efficacy of the final product (FDA report "Pharmaceutical Quality for the twenty-first century" [2007]). The approach starts with first defining what the Target Product Profile (TPP) is. Figure 8.7 provides an example of an hMSC process. As defined by the International Council for Harmonisation (ICH) of Technical Requirements for Pharmaceuticals for Human Use, the TPPs are a prospective summary of the quality characteristics of a drug product that ideally be achieved to ensure the desired quality, safety and efficacy. In the case of an

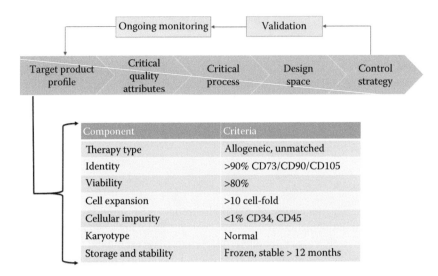

FIGURE 8.7   Quality by Design (QbD) approach for a robust, quality-assured cell therapy manufacturing process.

hMSC-based product, this would include the number of cells that would be expected to be obtained out of a given process scale (e.g., 5.0×10^{10} cells from 50 Cellstacks). Other measurable properties would also be included such as the identity of the cells (e.g., CD73+/CD90+/CD105+), viability, etc. Once these are set, then it would be possible to identify the Critical Quality Attributes (CQAs) of the process. These are parts of the cell therapy product that can cause harm (i.e., both safety and efficacy) if they are out of allowable limits. In the case of the mock hMSC-based product, this would include parts of TPPs as well as others outside of it. Cell expansion would be a CQA because if the cells did not maintain a similar fold expansion or rate of expansion between batches, this could indicate a serious issue with the efficacy/safety of the cell therapy product. Also, the specific expression of key markers, such as STRO-1, SSEA-4 (specific for BM MSCs), CD271, or CD146, could correlate with the therapeutic efficacy of cells *in vivo*. Conversely, at the end of the process, it is important to define the acceptable clearance of enzymatic agents (i.e., animal-derived enzymes). It is important to identify and define attributes that are measurable and can be tied to specifically contributing to the efficacy of the cell product. These CQAs will be tied to Critical Process Parameters (CPPs) that affect the CQAs of the process. CPPs can be culture parameters such as temperature, pH or DO level, or less obvious, potentially restrictive enzymatic exposure or processing time. These CPPs are parameters that – (1) can affect CQAs of the product and (2) can be controlled in the processes within a set range. This set range is defined by the Design Space, in which it is possible to understand what range of CPPs the process can withstand while achieving TPPs. And as the process is controlled within these ranges, it is possible to validate that the process can indeed maintain the established TPPs and perform ongoing improvement to increase the robustness and reliability of the process.

8.3.3 Process Parameters and Scale-Down Model Vessels

Designing an optimal culture system for a specific cell type first requires understanding the characteristics of the cells and the requirements for attachment to culture substrate and subsequent growth in culture. Investigating the effects of nutritional and physicochemical parameters (for example, medium formulation, medium pH, and oxygen tension) and basic process input variables (such as cell seeding density, medium volume, and medium feeding) on cell growth and other characteristics are often sufficient for static planar culture systems. In dynamic suspension culture environment, however, fluid-mechanics induced by the rotating impeller of bioreactors could also significantly affect cells. Moreover, the incorporation of microcarriers into the suspension bioreactors will make the culture system very complex. The microcarrier-based dynamic culture environment induces shear forces through the interaction of microcarrier beads with small turbulent eddies and bead-to-bead collisions, which can significantly influence cell growth and quality and could even cause cellular damage. For these reasons, even though the microcarrier technology has several advantages over static cultures for large-scale biological production, many industrial processes to culture mammalian cells still employ simpler systems such as roller bottles or multi-layered vessels (Croughan and Hu, 2006; Jung et al., 2012b). Such cellular damage caused by fluid-mechanics in microcarrier-based bioreactor culture is particularly concerning for hMSCs as these primitive cells are fragile and thus can be highly sensitive to the shear stress and mechanical impacts. Therefore, in order to develop a robust cell manufacturing platform for hMSC therapies, it is crucial to recognize CPPs critical for hMSC growth in microcarrier-based suspension culture and carry out systematic process development and optimization studies, performing a detailed investigation for the effect of such process input variables on cell growth and other characteristics. Table 8.1 shows a list of the variables for the growth of anchorage-dependent cells in dynamic microcarrier-based suspension bioreactors. The variables are grouped under seven different categories including medium type, medium composition, medium feeding, microcarrier, bioreactor, physicochemical parameters, and others. Depending on cell types, some of these parameters may have major impact on cell attachment/growth and other properties, while others have minor effects. It has been reported that the attachment and growth of hMSCs could be significantly affected by medium type and formulation, microcarrier type, bioreactor and impeller type, agitation speed, cell seeding density, and microcarrier loading density (Abraham et al., 2017). Therefore, it would be rational to investigate the effect of these variables at an early stage of hMSC process development to identify CPPs. Once optimized, it is important to standardize all the parameters to make the performance of the process consistent and reproducible.

In order to conduct process development/optimization studies to efficiently identify the effects of these parameters on cell growth and other CQAs, an appropriate scale-down model system for a specific bioreactor type should be needed so that Design of Experiment (DOE)-based cell culture experiments can be

TABLE 8.1

A List of Categorized Process Parameters for Dynamic Culture in Microcarrier-Based Bioreactors

Category	Parameter	Category	Parameter
Medium type	• Serum-based • Serum-free • Xeno-free • Animal-free • Chemically defined	Bioreactor	• Bioreactor geometry and size • Impeller type and size • Location of impeller and other instruments submerged in culture • Stirring protocol
Medium composition	• Growth factors • Attachment factors • Hormones, lipids, vitamins, minerals, trace elements • Carbohydrates, amino acids, nucleosides vitamins) • Binding proteins • Buffer • Protease inhibitors • Shear protecting agents • Antitoxins • Viscosity modulating agents		• Impeller tip speed • Stirring direction • Continuous vs. Intermittent stirring (cell attachment phase) • Manipulation of stirring speed according to cell growth • Aeration protocol • Head-space aeration • Sparging (use of anti-forming agents) • Material of the inner surface of vessel and impeller (to avoid the attachment of cells and microcarriers and the generation of particulate)

Medium feeding	• Rate
	• Mode
	 • Batch
	 • Fed-batch
	 • Medium exchange
	  • Bolus
	  • Continuous (perfusion)
		Physico- chemical Variables	• Temperature
			• pH
			• Dissolved O_2
			• Dissolved CO_2
			• Viscosity
			• Osmolality

Microcarrier	• Selection of microcarriers • Chemical properties • Physical properties • Geometrical properties (solid, microporous, vs. macroporous) • Microcarrier coating protein (particularly for serum-free culture) • Microcarrier loading density • Particulate generation	Others	• Cell inoculation density • Cell-to-bead ratio • Initial working volume (cell attachment phase) • Preparation of microcarriers and cell inocula • Bead-to-bead transfer • Addition of anti-forming agent • Addition of base

carried out. Some parameters often interact with each other to result in synergic effects; therefore, the DOE approach is particularly important to examine the effect of such parameters in concert during the process optimization studies. Conducting such studies should require an appropriate scale-down model system to enable DOE-based high-throughput experiments. Initial examination of some parameters on cell growth could be performed even using classical small-scale static vessels such as well-plates or petri dishes. As an example, Figure 8.8 shows a synergic effect between medium formulation and oxygen tension in a colony forming unit-fibroblast (CFU-F) assay, which was performed using hMSCs derived from an umbilical cord under two different medium formulations (αMEM + 20%FBS vs PPRF-msc6) and two different oxygen concentrations in the incubator (20% vs 5%). In this assay, a large number of colonies were formed when cells were cultured in PPRF-msc6 under 5% O_2. In contrast, no colonies were generated in the condition using 20% FBS αMEM at 20% O_2, and other combinations led to the generation of a few small colonies only. Although the effect of some parameters could be detected by the use of the classical small-scale cell culture wares, other parameters such as bioreactor- and microcarrier-related parameters in Table 8.1 and those related to high density culture (e.g., medium feeding) might require scale-down models that can mimic the actual manufacturing platform bioreactors. For instance, spinner flasks have been widely

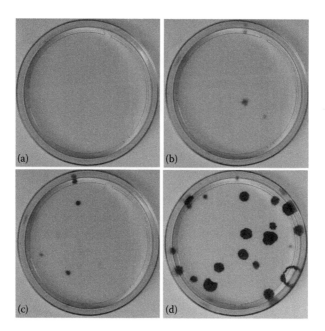

FIGURE 8.8 CFU-F assay for human umbilical cord-derived MSC (UC-MSCs) in PPRF-msc6 versus 20% FBS αMEM using gelatin-coated dishes. hUC-MSCs previously isolated and expanded up to P4 in 20% FBS αMEM were plated at 1.5 cells/cm^2 into 100-mm dishes (i.e., 90 cells per dish) coated with gelatin and cultured in four different conditions: (a) 20% FBS αMEM + 20% O_2, (b) 20% FBS αMEM + 5% O_2, (c) PPRF-msc6 + 20% O_2, and (d) PPRF-msc6 + 5% O_2. The resulting colonies were stained with crystal violet in methanol on day 16.

used as a scale-down model for stirred bioreactors. In addition, miniature disposable bioreactors, such as Sartorius' ambr15 system, have been introduced to run many different culture conditions simultaneously at a very small scale. The ambr15 bioreactor system has been designed to run a maximum of 24 or 48 stirred, miniature disposable bioreactors at a working volume of 10–15 mL for high-throughput cell culture experiments. Moreover, this system allows the monitoring and control of critical physicochemical parameters of individual bioreactors (i.e., DO and pH), and enables programmable, automated liquid addition and sampling using a robot liquid handler. The ambr15 bioreactor system has been proven to support microcarrier-based hMSC growth; however, it has required a significant amount of time and effort to optimize an agitation protocol to run the system properly (Abraham et al., 2017). The rectangular shape and the impeller position of the ambr15 vessel (Nienow et al., 2016a) tended to cause a significant cell/bead clumping in the corners of the vessel when hMSCs were cultured on microcarriers. A unique agitation protocol to support the uniform suspension of microcarrier-mediated cell aggregates needs to be developed to turn possible use of the ambr15 system for hMSC process optimization as a scale-down model for bioreactor culture. Vertical-wheel PBS-mini vessels play a role as a scale-down model for large-scale controlled bioreactors. As described earlier, the MagDrive family of vertical-wheel bioreactors have similar vessel and impeller design, allowing a low shear environment and efficient liquid mixing with minimum dissipation energy gradients across a range of vessel sizes from 0.1 to 80 L. Therefore, the small-size vessels should be useful not only as a seed train (as illustrated in Figure 8.6c) but as a representative scale-down model for process development and optimization, demonstrating similar performance throughout the range of scales.

8.3.4 CURRENT CHALLENGES WITH PROCESS OPTIMIZATION FOR hMSC CULTURE IN BIOREACTORS

Previously we published a review article to summarize the effect of some of the key parameters on hMSCs, including microcarrier type, microcarrier and cell seeding density, medium formulation and feeding, substrate coating material, bioreactor configuration and mixing, initial culture condition, bead-to-bead transfer, and so on (Jung et al., 2012b). Also, toward the accomplishment of high density culture, the importance of developing a well-designed medium feeding regime has already been reviewed (Abraham et al., 2017). Therefore, in the present report we wanted to briefly point out a few additional important issues that often arise in the microcarrier-based culture of hMSCs in dynamic conditions, including (1) the aggregation of hMSCs growing on microcarriers and (2) in-vessel cell harvest at the end of the culture period.

8.3.4.1 Considerations to Reduce the Degree of Cell Clumping in Suspension Bioreactors

It is well known that hMSCs tend to stick to each other during the microcarrier-based bioreactor culture (Ferrari et al., 2012), which could induce uncontrolled agglomeration of cells and microcarriers. This would result in a high degree of heterogeneity of agglomerates in size, sometimes causing a process failure in the worst case with the

TABLE 8.2

Parameters Which Affect Cell Aggregation as Well as Cell Growth in Microcarrier-Based Culture of hMSCs in Bioreactors

Category	Parameter
Mechanical	Stirring speed
	Stirring direction
	• forward vs. reverse or downward vs. upward
	• uni-directional vs. combination of forward and reverse or downward and upward
	Continuous stirring vs. intermittent stirring
Chemical	Use of anti-clumping agents
Medium	Medium compositions
	• serum vs. serum-free
	• serum concentration
	• basal medium formulation
	• concentration of Ca^{++}, and Mg^{++}
	• concentration of ascorbic acid and other components that facilitate the secretion of extracellular matric proteins
Other	Cell seeding density
	Microcarrier loading density

formation of huge masses of cell aggregates being associated with probes or other process instruments submerged in medium and not being suspended. The formation of uncontrolled large aggregates must be avoided because cells in such aggregates are exposed to different microenvironments, which may lead to the generation of cells having different growth and physiological characteristics and even cell death. Table 8.2 points out a list of parameters that are deemed to impact cell aggregation as well as cell growth in microcarrier-based culture of hMSCs in bioreactors, beyond the geometry, size of bioreactor and impeller type, and design that should have a significant impact on the aggregation. The parameters are grouped into four different categories—mechanical, chemical, medium, and others. The manipulation of the impeller stirring regime (speed, direction, and time) is often used to control the size of aggregates. However, as a high agitation rate causes a high shear stress that could be harmful on cells, the manipulation of impeller stirring regime should be considered within a limited range. As a chemical approach, anti-clumping agents could be considered if their use is acceptable in the regulatory perspective. Medium formulation, including the concentration of serum and basal medium components, often also affects cell aggregation. For instance, the addition of ascorbic acid has shown to be beneficial for hMSC growth but increases the secretion of extracellular matrix (ECM) proteins in T-flask cultures (Jung et al., 2010). The secreted ECM molecules may facilitate cellular aggregation in microcarrier culture; therefore, the concentration of ascorbic acid in culture medium may need to be determined considering their dual effect on cell growth and aggregation. Finally, the degree of hMSC aggregation in microcarrier culture was influenced by cell seeding density and microcarrier loading density (Ferrari et al., 2012).

8.3.4.2 Enhancing Cell Harvest Efficiency

Protocols to dissociate cells into single cells at harvest and separate them from micro-carriers with maximum recovery yields while maintaining CQAs of cells should also be developed to make the microcarrier-based production of hMSCs successful. This is particularly important for high density cultures since the extracellular matrix secreted by cells during the culture could increase the size and compactness of aggregates formed with multilayers of cells and thus the recovery of single cells for the large and dense cell-microcarrier aggregates at high efficiency can be technically challenging and time-consuming. To develop an optimized cell harvest protocol, parameters that affect the dissociation of cells from the microcarriers and aggregates and their quality should first be identified. Following is a list of the variables that can be considered:

- Frequency and time of pre-harvest wash
- Formulation and volume of wash buffer
- Type, volume, concentration and pH of proteolytic enzyme
- Concentration of ethylenediaminetetraacetic acid (EDTA), if used
- Temperature
  - Reagents (washing buffer, enzyme)
  - Operational temperature
- Time of enzymatic dissociation
- Impeller type
- Impeller agitation regime (time, speed) and power input
- Criteria to end cell dissociation
- Method for cell separation from microcarriers
- Overall harvest operation time

For instance, Nienow and colleagues demonstrated a cell harvest protocol involving an intense agitation, which led to the reproducible achievement of 95% harvest efficiency of hMSCs from solid microcarriers (Rafiq et al., 2013; Nienow et al., 2014; Nienow et al., 2016b).

8.3.5 Mixing and Aeration—Hydrodynamics and Mass Transfer

There are a number of design aspects that need to be taken into account when growing MSCs in a suspension bioreactor. Two of these key design aspects include (1) oxygen supply and (2) hydrodynamic shear in the liquid medium. These are especially important as cultures are scaled-up from sub-liter volumes to 10s of liter volumes.

Oxygen supply to the reactor is important to maintain cell growth, viability and their therapeutic properties. The specific oxygen consumption rate of mammalian cells has been reported to vary from 0.06 to 0.6 mmol O_2/(L•hour) for a culture at a density 1×10^6 cells/mL (Butler, 2004). In small-scale cultures (i.e., up to 1 L), this oxygen demand can normally be satisfied by gas diffusion from the head space through the culture surface (see Figure 8.9 for diagram of oxygen diffusion to cells in a microcarrier). The rate limiting step in this scenario is the dissolution of oxygen into the culture

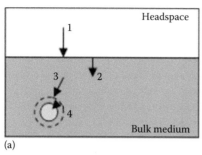
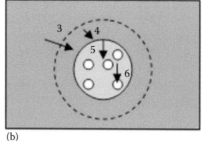

(a)　　　　　　　　　　　　　　　(b)

FIGURE 8.9　Headspace aeration of cell cultures and diffusion of oxygen from headspace to a macroporous microcarrier. (a) Oxygen from the headspace is transferred to the microcarrier in several steps: (1) diffusion from the bulk gas to the gas-liquid interface, (2) dissolution across the gas liquid interface, (3) transport from bulk liquid to the stagnant film at the aggregate surface. (b) Oxygen in the liquid is transferred into the macroporous microcarrier: (5) diffusion from microcarrier surface into the center of the microcarrier, and (6) uptake by the cells. (Adapted from Chiang C. [2008]. With permission.)

medium. Aunins and colleagues (Aunins et al., 1989) developed a correlation for the volumetric mass transfer coefficient ($k_L a$) for surface oxygenation that incorporated the geometry of the impeller, shear stress and medium properties, given by:

$$k_L a = 1.08 \cdot N_{RE}^{0.78} \left[\frac{D_{O_2} a}{D_T} \right] \tag{8.1}$$

where:

N_{RE} is the impeller Reynolds number
D_{O_2} is the oxygen diffusion coefficient
a is the mass transfer interfacial area
D_t is the reactor diameter

Typically, for small bioreactors (less than 1.0 L) the oxygen demands can be satisfied by gas diffusion from the headspace through the surface of the medium. As the working volume of the bioreactor increases, the surface area-to-volume ratio (referred to as the aspect ratio) decreases. As an example, for 10 L and 50 L vessels, the aspect ratio would decrease from 7.1 m^{-1} (10 L) to 1.8 m^{-1} (50 L). Therefore, as bioreactor sizes increase during scaling-up, the area available for oxygen transfer through the interfacial surface of the medium in the bioreactor decreases. Then considering this in light of equation (1), it is clear that as the interfacial area decreases this leads to a decrease in the available mass transfer of oxygen into the liquid medium. In this case, there are three ways to increase the oxygen supply to the culture medium in the bioreactor to meet the oxygen demands of proliferating cells—(1) increasing the oxygen concentration in the headspace, (2) sparging oxygen into the culture, and/or (3) increasing the agitation rate of the impeller. Sparging introduces oxygen by bubbling it through the culture medium. This increases the available surface area for oxygen to be delivered to the culture medium. It is desirable to produce small bubbles that can have less of a chance of entrapping cells and carrying them to the surface where they burst (Shuler and Kargi, 2002). And the effect that sparging has on foaming must be considered, especially when used in serum-containing media. Antifoam can be used but it may have a detrimental effect on cells and needs to be evaluated before its use. As a rule of thumb, it is favorable to increase the oxygen concentration in the headspace first and then supplement the oxygen supply to the culture with sparging. Increasing the agitation rate is another option since this would increase the dissolution rate of oxygen into the medium. However, as the agitation increases, the hydrodynamic shear could be detrimental as it may cause excessive cell damage.

Hydrodynamic shear is also very important to consider when scaling-up bioreactor cultures. Shear in the bioreactor ensures adequate oxygen mass transfer and bulk fluid mixing. However, it is important to ensure that the hydrodynamic shear associated with the level of turbulence is within acceptable limits. In stirred suspension bioreactors, the agitation rate governs the hydrodynamic shear within the vessel and as the agitation increases the hydrodynamic shear rate increases. If the agitation rate is too low, the cultures may not be well-mixed causing problems such as significant aggregation of cells and microcarriers and a non-homogenous cell culture environment. The uncontrolled aggregation may cause limited transfer of oxygen and nutrients to cells inside large aggregates. However, if the agitation rate is too high, this can be detrimental if it causes excessive cell damage. In order to estimate the hydrodynamic shear stress in the bioreactor system, the Kolmogorov's theory of turbulent eddy's (Cherry and Kwon, 1990) and Nagata's correlation (Nagata, 1975) can be used. This is given by the equations:

$$Np = \frac{P}{N^3 D_i^5 \rho} \tag{8.2}$$

$$\varepsilon = \frac{P}{V \rho} \tag{8.3}$$

where:
   Np is the power number
   P is the impeller power consumption
   D_i is the impeller diameter
   N is the impeller speed
   ρ is the density of the medium
   V is the volume of the culture
   ε is the volume average of energy dissipation

ε gives us a relationship of the eddies, which are small vortices that are created in the bioreactor. If these vortices are smaller than the microcarriers, they will shear cells off the surface of the microcarriers. If they are much larger than individual microcarriers, they will allow microcarrier to clump within them and therefore allow the formation of large agglomerations of microcarriers. It will be important to determine the average eddy size that enables consistent hMSC expansion, while maintaining cell quality, in smaller reactors and then scale-up to larger reactors based on these values.

However, one caveat is that this numerical calculation does not consider the flow regime present in the vessel, which may impact the growth of the cells. Therefore, it has also been suggested that suspension experiments and computational fluid dynamics (CFD) studies combined with particle image velocimetry measurements be used to determine optimal scale-up operating parameters (Kaiser et al., 2013).

8.3.6 CELL-BASED VERSUS CELL-FREE PRODUCTS

MSCs are currently being produced for several clinical applications, including GvHD, multiple sclerosis, cardiovascular and neurological disorders, and lung diseases. The mechanism of action of MSCs in treating these indications is not the replacement of damaged/diseased cells, but in the secretion of bioactive molecules, which can modulate the immune/inflammatory system and have a paracrine activity. These secreted bioactive factors have a broad range of therapeutic effects both *in vitro* and *in vivo* (for example, anti-inflammatory, anti-fibrotic, anti-apoptotic, anti-angiogenic, or immunomodulatory) as well as repair/regenerative actions (Meirelles Lda et al., 2009). In addition to the secretion of a number of bioactive factors, hMSCs have been shown to secrete large amounts of exosomes, which are a class of secreted lipid membrane vesicles (Baglio et al., 2012). They are formed by the invagination of endolysosomal vesicles to form multi-vesicular bodies and have a diameter ranging from 40 to 100 nm. Exosomes contain proteins and genetic material (i.e., RNA), which can be transferred to other cells. This transference of material to other cells has been hypothesized to modulate the behavior of these cells and be one of the main contributors to the hMSC therapeutic advantage (Baglio et al., 2012).

8.3.7 IMPACT OF USING A DYNAMIC CULTURE SYSTEM ON THE PHENOTYPIC PROPERTIES OF HUMAN MESENCHYMAL STEM CELLS

The external environment that hMSCs are cultured in can influence their secretory profile and composition of bioactive molecules. Indeed, by modifying the external environment they appear to mimic *in vitro* the disease environment that they might see *in vivo* and can produce, for example, a secretome-derived conditioned medium (CM), which has therapeutic benefits for that disease. There have been several groups that have shown this in a variety of diseases. Oskowitz and colleagues (Oskowitz et al., 2011) showed that, when hMSCs that had been expanded in serum-containing medium were placed in a serum-derived medium, the hMSCs secreted increased levels of pro-angiogenic factors including Vascular Endothelial Growth Factor (VEGF)-A, angiopoietins, Insulin-like Growth Factor (IGF)-1, and Hepatocyte Growth Factor (HGF). Additionally, when they used an *in vivo* modified chick chorioallantoic angiogenesis assay it was observed that the serum-derived hMSCs displayed significantly higher angiogenic potential compared to hMSCs that were cultured in a typical culture condition (i.e., 17% FBS αMEM). Moreover, Chang and colleagues preconditioned bone marrow-derived hMSCs under hypoxic conditions and collected cell secretome (Chang et al., 2013). This was then transplanted into rats with experimental traumatic brain injury (TBI) and resulted in the rats performing significantly better physically in both motor and cognitive function tests. The rats also showed increased neurogenesis and decreased brain damaged compared to TBI rats transplanted with the secretome collected from normoxic-expanded bone marrow-derived hMSCs. Moreover, the hypoxic conditions were able to stimulate the BM-hMSCs to secrete high levels of VEGF and HGF. Conversely, for the treatment of bronchopulmonary dysplasia (BPD), CM from BM-hMSCs preconditioned in hyperoxia (i.e., 95% O_2 in the atmosphere) had the most potent therapeutic benefit compared to CM from BM-hMSCs expanded in normoxic conditions (Waszak et al., 2012). Additionally, hyperoxia was able to induce the higher secretion of stanniocalcin-1. By modifying the external culture environment, they were able to influence the therapeutic properties of hMSCs and their secreted products, which could result in the development of a cell-free therapy.

The use of dynamic, controlled culture systems can allow for an easily scalable expansion of hMSCs while maintaining a homogenous culture environment. However, these culture systems introduce liquid shear and environmental control (i.e., DO, pH and temperature), which can influence the secretion of bioactive molecules from hMSCs. Our group has recently published details on the production of a neutrophic cocktail of factors from hMSCs (i.e., secretome) that could be used for the treatment of diseases of the central nervous system (Teixeira et al., 2015, 2016a, 2016b). In the first study, hMSCs from the Wharton Jelly of the umbilical cord were cultured on microcarriers in PPRF-msc6 medium in computer-controlled stirred tank bioreactors (DASGIP, Germany) in either normoxic (20% O_2) or hypoxic (5% O_2) conditions. The cells were expanded for 72 hours and then the culture medium was removed and replaced with Neurobasal-A medium to be conditioned by hMSCs over 24 hours. We found that, when the secretome from either normoxic or hypoxic-expanded cells was used to culture human neural precursor cells (hNPCs), there was a significant increase in cell number and expression of neural markers, such as doublecortin and Microtubule-Associated Protein (MAP)-2, over hNPCs expanded in Neurobasal-A alone. While no significant difference was observed between the marker expression for hNPCs cultured with the secretome either from normoxic- or hypoxic-expanded cells, we did find that there were significantly more proteins produced in the hypoxic secretome compared to the normoxic secretome (166 proteins in hypoxic secretome, 104 in normoxic secretome of which 81 were co-expressed). Of these proteins with neurotrophic significance, we identified a number of molecules where three of them were significantly increased in the hypoxic secretome: thymosin beta, elongation factor-2, and 14-3-3, all of which have been reported to have important roles in neurite outgrowth and neuroprotective actions (Johnson et al., 1992; van Kesteren et al., 2006; Chen et al., 2007; Sun and Kim, 2007; Ramser et al., 2010; Fraga et al., 2013; Iketani et al., 2013)

In our second secretome study, BM-hMSCs were cultured in similar conditions in the bioreactor and expanded in planar culture under normoxic conditions (Teixeira et al., 2016b). The secretome collected from the bioreactor-expanded cells was found to significantly increase the number of immature neurons (DCX[+]) and mature neurons (MAP2[+]/NeuN[+]) when co-cultured with hNPCs, compared to static-expanded cell secretome. When this secretome was injected into the dentate gyrus of a rat hippocampus, we observed that there was a significant increase in the number of immature neurons generated with the bioreactor-produced secretome (CMd), compared with the static-produced secretome (CMs) (Figure 8.10b and e). There was a higher number of proliferative immature neurons (Ki-67+/DCX+) observed in the

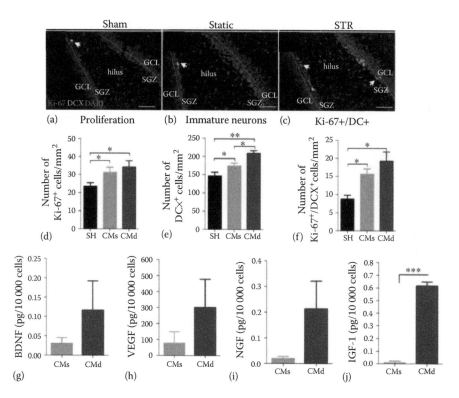

FIGURE 8.10 The dynamic culture of hMSCs in bioreactors can impact the secretion of bioactive molecules by hMSCs and, in our system, has led to a significant increase in neurotrophic molecules. Injection of the hMSC-secretome into the dentate gyrus of an adult rat hippocampus from either (b) static culture (CMs) or (c) dynamic culture (CMd) resulted in an increase in endogenous proliferating immature neuronal cells compared to the sham (a, d–f). The dynamic secretome had increased levels of BDNF, VEGF, NGF, and IGF-1 compared to static secretome (g–j). Data are expressed as mean $\pm$SEM. *p<0.05; **p<0.01. (Adapted from Teixeira, F.G. et al. [2015]. With permission.)

bioreactor-produced secretome (CMd), but this was not significantly higher than those seen for the static-produced secretome (CMs) (Figure 8.10a, c, d, f). Analysis of the secretome revealed many proteins upregulated in the bioreactor-produced secretome (CMd), which include Brain-Derived Neurotrophic Factor (BDNF), VEGF, and Nerve Growth Factor (NGF) (Figure 8.10g–i). There was a significant upregulation in IGF-1 (Figure 8.10j) in the bioreactor-produced secretome, which is a protein that is implicated as being an enhancer of neuronal differentiation and hippocampal neurogenesis, as well as an essential component of the signaling networks regulating neurogenesis (Annenkov, 2009; Carlson et al., 2014).

In our third secretome study, the secretome of BM-hMSCs, collected from our bioreactors, was injected into a rat model of Parkinson' disease (6-OHDA rat model) (Teixeira et al., 2016a). We found that there was a significant increase in the amount and density of thyrosine hydroxylase (a marker of dopamine neurons) in rats that were injected with the secretome compared to our 6-OHDA treated rats (Figure 8.11c, d, f–h),

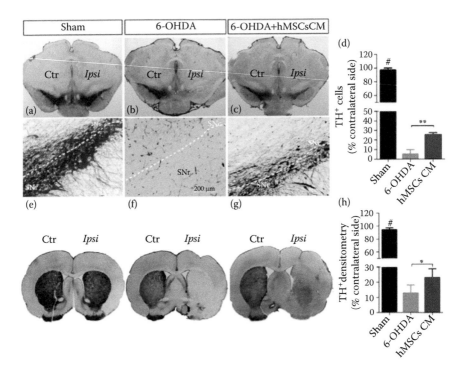

FIGURE 8.11 The dynamic culture of hMSCs in bioreactors produced secreted factors which lead to an increase in the level of TH-positive neurons when injected into the midbrain of animals in a 6-OHDA rat model of Parkinson's disease. Compared to the sham group (a,e), the administration of 6-OHDA exhibited reduced TH staining in both the substantia nigra par compacta and striatum. Injection of the dynamic secretome into a 6-OHDA treated rat-model resulted in an increased number and density of TH-positive cells in the injection site (c,d,g,h) compared to the 6-OHDA non-treated group (b,d,f,h). Data are presented as mean ±SEM (sham = 9, 6-OHDA = 9, dynamic secretome = 8). $*p < 0.05$, $**p < 0.01$, #, sham animals statistically different from 6-OHDA and dynamic secretome injected group. Abbreviations: Ctr, control; lpsi, ipsilateral; 6-OHDA, 6-hydroxidopamine; TH+, tyrosine hydroxylase positive. (Adapted from Teixeira, F.G. et al. [2016a]. With permission.)

compared to controls (Figure 8.11a, b). We also observed that there was a significant improvement in motor functions. In summary, our results demonstrated that the use of dynamic bioreactor culture conditions can not only provide controlled conditions for the expansion of hMSCs but could potentially generate a cocktail of factors (i.e., a secretome) that can be used as a cell-free therapy.

8.4   STREAMLINING THE MANUFACTURING PROCESS

Once each of the unit operations of a cell manufacturing process is optimized to achieve the best performance (i.e., rapid cell growth, high density culture, controlled size of cell-microcarrier aggregates, high cell harvest efficiency, high performance of cells or secretome as final products under an optimal hydrodynamic condition, etc.), the entire process must be carried out in a GMP-compliant manner under a strictly aseptic environment to ensure patients' safety. Operating GMP-compliant cell manufacturing processes in clean rooms that comply with ISO standards could take a prolonged time with long hold steps if the activities are not well organized, particularly when the manufacturing scale is large. The process "time" itself must be considered as a critical factor that could have a significant impact on the quality and potency of final cell products.

Lengthy hold steps should be avoided, and each of unit operations—cell inoculation, medium exchange, cell harvest, and the following of the purification processes—should be quick, efficient, consistent and compliant. For instance, performing a bolus medium exchange to replace spent medium with pre-warmed fresh medium for microcarrier culture involves the settling down of cells loaded on microcarriers to separate them from the spent medium. Once cells are settled down, which should take time, the spent medium is perfused out and the fresh medium is added. In a large-scale cell manufacturing process, the time required to complete the medium exchange could take a long time, during which the agitation of culture is interrupted and thus control process variables, such as temperature, pH and DO, cannot be controlled. As such, streamlining the unit operations should be carried out to reduce process time. In this regard, it is crucial to document (i.e., Standard Operating Procedures) all the steps of cell manufacturing processes in a clear and detailed manner. In addition, to finally achieve a "lean manufacturing process" to reduce the process time of each unit operation, the steps required between unit operations should also be conducted in an efficient way to minimize process hold time. It is important to establish a continuous processing from tissue acquisition or a working cell bank through final products in the form of frozen vials without "holding steps" to maximize the potency and quality of the cell products. In this regard, any constraints to limit the quick and continuous processing of hMSCs need to be identified and resolved.

REFERENCES

Abdelrazik H, Spaggiari GM, Chiossone L et al. 2011. Mesenchymal stem cells expanded in human platelet lysate display a decreased inhibitory capacity on T- and NK-cell proliferation and function. *Eur J Immunol* 41(11):3281–3290.

Abraham E, Gupta S, Jung S et al. 2017. Bioreactor for scale-up: Process control. In: Hematti P, Viswanathan S, (Eds.) *Mesenchymal Stromal Cells Translational Pathways to Clinical Adoption*. London, UK: Academic Press, pp. 139–178.

Annenkov A. 2009. The insulin-like growth factor (IGF) receptor type 1 (IGF1R) as an essential component of the signalling network regulating neurogenesis. *Mol Neurobiol* 40(3):195–215.

Aunins JG, Woodson BA, Hale TK et al. 1989. Effects of paddle impeller geometry on power input and mass transfer in small-scale animal cell culture vessels. *Biotechnol Bioeng* 34(9):1127–1132.

Baghbaderani BA, Mukhida K, Hong M et al. 2011. A review of bioreactor protocols for human neural precursor cell expansion in preparation for clinical trials. *Curr Stem Cell Res Ther* 6(3):229–254.

Baghbaderani BA, Mukhida K, Sen A et al. 2010. Bioreactor expansion of human neural precursor cells in serum-free media retains neurogenic potential. *Biotechnol Bioeng* 105(4):823–833.

Baglio SR, Pegtel DM, Baldini N. 2012. Mesenchymal stem cell secreted vesicles provide novel opportunities in (stem) cell-free therapy. *Front Physiol* 3:359.

Baksh D, Davies JE, Zandstra PW. 2003. Adult human bone marrow-derived mesenchymal progenitor cells are capable of adhesion-independent survival and expansion. *Exp Hematol* 31(8):723–732.

Bartosh TJ, Ylostalo JH, Mohammadipoor A et al. 2010. Aggregation of human mesenchymal stromal cells (MSCs) into 3D spheroids enhances their antiinflammatory properties. *Proc Natl Acad Sci USA* 107(31):13724–13729.

Bieback K, Hecker A, Kocaomer A et al. 2009. Human alternatives to fetal bovine serum for the expansion of mesenchymal stromal cells from bone marrow. *Stem Cells* 27(9):2331–2341.

Brindley DA, Davie NL, Culme-Seymour EJ et al. 2012. Peak serum: Implications of serum supply for cell therapy manufacturing. *Regen Med* 7(1):7–13.

Butler M. 2004. Animal cell culture and technology. London, UK: BIOS Scientific Publishers.

Capelli C, Domenghini M, Borleri G et al. 2007. Human platelet lysate allows expansion and clinical grade production of mesenchymal stromal cells from small samples of bone marrow aspirates or marrow filter washouts. *Bone Marrow Transplant* 40(8):785–791.

Carlson SW, Madathil SK, Sama DM et al. 2014. Conditional overexpression of insulin-like growth factor-1 enhances hippocampal neurogenesis and restores immature neuron dendritic processes after traumatic brain injury. *J Neuropathol Exp Neurol* 73(8):734–746.

Chang CP, Chio CC, Cheong CU et al. 2013. Hypoxic preconditioning enhances the therapeutic potential of the secretome from cultured human mesenchymal stem cells in experimental traumatic brain injury. *Clin Sci (Lond)* 124(3):165–176.

Chen J, Lee CT, Errico SL et al. 2007. Increases in expression of 14-3-3 eta and 14-3-3 zeta transcripts during neuroprotection induced by delta9-tetrahydrocannabinol in AF5 cells. *J Neurosci Res* 85(8):1724–1733.

Cherry RS, Kwon KY. 1990. Transient shear stresses on a suspension cell in turbulence. *Biotechnol Bioeng* 36(6):563–571.

Chiang C. 2008. Expansion of human brain cancer-derived cells [MSc. Thesis]. Calgary, AB: University of Calgary, 188 p.

Croughan MS, Giroux D, Fang D et al. 2016. Novel single-use bioreactors for scale-up of anchorage-dependent cell manufacturing of cell therapies. In: Cabral JMS, De Silva CL, Chase LG, Diogo MM, (Eds.). *Stem Cell Manufacturing*: Atlanta, GA: Elsevier, pp. 105–139.

Croughan MS, Hu WS. 2006. From microcarriers to hydrodynamics: Introducing engineering science into animal cell culture. *Biotechnol Bioeng* 95(2):220–225.

de Soure AM, Fernandes-Platzgummer A, Moreira F et al. 2017. Integrated culture platform based on a human platelet lysate supplement for the isolation and scalable manufacturing of umbilical cord matrix-derived mesenchymal stem/stromal cells. *J Tissue Eng Regen Med* 11(5):1630–1640.

Digirolamo CM, Stokes D, Colter D et al. 1999. Propagation and senescence of human marrow stromal cells in culture: A simple colony-forming assay identifies samples with the greatest potential to propagate and differentiate. *Br J Haematol* 107(2):275–281.

dos Santos F, Andrade PZ, Abecasis MM et al. 2011. Toward a clinical-grade expansion of mesenchymal stem cells from human sources: A microcarrier-based culture system under xeno-free conditions. *Tissue Eng Part C Methods* 17(12):1201–1210.

Doucet C, Ernou I, Zhang Y et al. 2005. Platelet lysates promote mesenchymal stem cell expansion: A safety substitute for animal serum in cell-based therapy applications. *J Cell Physiol* 205(2):228–236.

Dromard C, Bourin P, Andre M et al. 2011. Human adipose derived stroma/stem cells grow in serum-free medium as floating spheres. *Exp Cell Res* 317(6):770–780.

Eibes G, dos Santos F, Andrade PZ et al. 2010. Maximizing the ex vivo expansion of human mesenchymal stem cells using a microcarrier-based stirred culture system. *J Biotechnol* 146(4):194–197.

Federal U.S. Food and Drug Administration, United States. 2007. Pharmaceutical quality for the 21st century a risk-based approach progress report.

Ferrari C, Balandras F, Guedon E et al. 2012. Limiting cell aggregation during mesenchymal stem cell expansion on microcarriers. *Biotechnol Prog* 28(3):780–787.

Fischer UM, Harting MT, Jimenez F et al. 2009. Pulmonary passage is a major obstacle for intravenous stem cell delivery: The pulmonary first-pass effect. *Stem Cells Dev* 18(5):683–692.

Flemming A, Schallmoser K, Strunk D et al. 2011. Immunomodulative efficacy of bone marrow-derived mesenchymal stem cells cultured in human platelet lysate. *J Clin Immunol* 31(6):1143–1156.

Fraga JS, Silva NA, Lourenco AS et al. 2013. Unveiling the effects of the secretome of mesenchymal progenitors from the umbilical cord in different neuronal cell populations. *Biochimie* 95(12):2297–2303.

Frauenschuh S, Reichmann E, Ibold Y et al. 2007. A microcarrier-based cultivation system for expansion of primary mesenchymal stem cells. *Biotechnol Prog* 23(1):187–193.

Freshney RI. 2000. *Culture of Animal Cells: A Manual of Basic Technique.* New York: Wiley-LISS. xxvi, 577 p.,[12]p. of plates p.

Frith JE, Thomson B, Genever PG. 2010. Dynamic three-dimensional culture methods enhance mesenchymal stem cell properties and increase therapeutic potential. *Tissue Eng Part C Methods* 16(4):735–749.

Godara P, McFarland CD, Nordon RE. 2008. Design of bioreactors for mesenchymal stem cell tissue engineering. *J Chem Technol Biotechnol* 83(4):408–420.

Gruber R, Karreth F, Kandler B et al. 2004. Platelet-released supernatants increase migration and proliferation, and decrease osteogenic differentiation of bone marrow-derived mesenchymal progenitor cells under in vitro conditions. *Platelets* 15(1):29–35.

Hanley PJ, Mei Z, Durett AG et al. 2014. Efficient manufacturing of therapeutic mesenchymal stromal cells with the use of the quantum cell expansion system. *Cytotherapy* 16(8):1048–1058.

Hervy M, Weber JL, Pecheul M et al. 2014. Long term expansion of bone marrow-derived hMSCs on novel synthetic microcarriers in xeno-free, defined conditions. *PLoS One* 9(3):e92120.

Iketani M, Iizuka A, Sengoku K et al. 2013. Regulation of neurite outgrowth mediated by localized phosphorylation of protein translational factor eEF2 in growth cones. *Dev Neurobiol* 73(3):230–246.

Johnson G, Gotlib J, Haroutunian V et al. 1992. Increased phosphorylation of elongation factor 2 in Alzheimer's disease. *Brain Res Mol Brain Res* 15(3–4):319–326.

Jung S, Panchalingam KM, Rosenberg L et al. 2012a. Ex vivo expansion of human mesenchymal stem cells in defined serum-free media. *Stem Cells Int* 2012:123030.

Jung S, Panchalingam KM, Wuerth RD et al. 2012b. Large-scale production of human mesenchymal stem cells for clinical applications. *Biotechnol Appl Biochem* 59(2):106–120.

Jung S, Sen A, Kallos MS et al. 2009. Large-scale production of functional beta cells from stem/progenitor cells in suspension bioreactors. In: Halle JP, de Vos P, Rosenberg L, (Eds.). *The Bioartificial Pancreas and Other BioHybrid Therapies.* Kerala, India: Publisher Research Signpost, pp. 529–555.

Jung S, Sen A, Rosenberg L et al. 2010. Identification of growth and attachment factors for the serum-free isolation and expansion of human mesenchymal stromal cells. *Cytotherapy* 12(5):637–657.

Jung S, Sen A, Rosenberg L et al. 2012c. Human mesenchymal stem cell culture: Rapid and efficient isolation and expansion in a defined serum-free medium. *J Tissue Eng Regen Med* 6(5):391–403.

Kaiser S, Jossen V, Schirmaier C et al. 2013. Fluid flow and cell proliferation of mesenchymal adipose-derived stem cells in small-scale, stirred, single-use bioreactors. *Chemie Ingenieur Technik* 85(1–2):95–102.

Karnieli O, Friedner OM, Allickson JG et al. 2017. A consensus introduction to serum replacements and serum-free media for cellular therapies. *Cytotherapy* 19(2):155–169.

Kim J, Ma T. 2012. Perfusion regulation of hMSC microenvironment and osteogenic differentiation in 3D scaffold. *Biotechnol Bioeng* 109(1):252–261.

Kocaoemer A, Kern S, Kluter H et al. 2007. Human AB serum and thrombin-activated platelet-rich plasma are suitable alternatives to fetal calf serum for the expansion of mesenchymal stem cells from adipose tissue. *Stem Cells* 25(5):1270–1278.

Kuznetsov SA, Mankani MH, Robey PG. 2000. Effect of serum on human bone marrow stromal cells: Ex vivo expansion and in vivo bone formation. *Transplantation* 70(12):1780–1787.

Lange C, Cakiroglu F, Spiess AN et al. 2007. Accelerated and safe expansion of human mesenchymal stromal cells in animal serum-free medium for transplantation and regenerative medicine. *J Cell Physiol* 213(1):18–26.

Le Blanc K, Ringden O. 2007. Immunomodulation by mesenchymal stem cells and clinical experience. *J Intern Med* 262(5):509–525.

Lee RH, Pulin AA, Seo MJ et al. 2009. Intravenous hMSCs improve myocardial infarction in mice because cells embolized in lung are activated to secrete the anti-inflammatory protein TSG-6. *Cell Stem Cell* 5(1):54–63.

Liu CH, Wu ML, Hwang SM. 2007. Optimization of serum-free medium for cord blood mesenchymal stem cells. *Biochem Eng J* 33:1–9.

Marshak DR, Holecek JJ. 1999. Chemical defined medium for human mesechymal stem cells. US Patent 5,908,782.

Meirelles Lda S, Fontes AM, Covas DT et al. 2009. Mechanisms involved in the therapeutic properties of mesenchymal stem cells. *Cytokine Growth Factor Rev* 20(5–6):419–427.

Mendicino M, Bailey AM, Wonnacott K et al. 2014. MSC-based product characterization for clinical trials: An FDA perspective. *Cell Stem Cell* 14(2):141–145.

Muller I, Kordowich S, Holzwarth C et al. 2006. Animal serum-free culture conditions for isolation and expansion of multipotent mesenchymal stromal cells from human BM. *Cytotherapy* 8(5):437–444.

Nagata S. 1975. *Mixing: Principles and Applications*. Kodansha, (ED.). New York: Wiley.

Nienow AW, Coopman K, Heathman TRJ et al. 2016a. Bioreactor engineering fundamentals for stem cell manufacturing. In: Cabral JMS, de Silva CL, Chase LG, Diogo MM, editors. *Stem Cell Manufacturing*: Elsevier, pp. 43–75.

Nienow AW, Hewitt CJ, Heathman TRJ et al. 2016b. Agitation conditions for the culture and detachment of hMSCs from microcarriers in multiple bioreactor platforms. *Biochem Eng J* 108:24–29.

Nienow AW, Rafiq QA, Coopman K et al. 2014. A potentially scalable method for the harvesting of hMSCs from microcarriers. *Biochem Eng J* 85:79–88.

Olmer R, Lange A, Selzer S et al. 2012. Suspension culture of human pluripotent stem cells in controlled, stirred bioreactors. *Tissue Eng Part C Methods* 18(10):772–784. United States.

Osiecki MJ, Michl TD, Kul Babur B et al. 2015. Packed bed bioreactor for the isolation and expansion of placental-derived mesenchymal stromal cells. *PLoS One* 10(12):e0144941.

Oskowitz A, McFerrin H, Gutschow M et al. 2011. Serum-deprived human multipotent mesenchymal stromal cells (MSCs) are highly angiogenic. *Stem Cell Res* 6(3):215–225.

Poloni A, Maurizi G, Rosini V et al. 2009. Selection of CD271(+) cells and human AB serum allows a large expansion of mesenchymal stromal cells from human bone marrow. *Cytotherapy* 11(2):153–162.

Potapova IA, Brink PR, Cohen IS et al. 2008. Culturing of human mesenchymal stem cells as three-dimensional aggregates induces functional expression of CXCR4 that regulates adhesion to endothelial cells. *J Biol Chem* 283(19):13100–13107.

Rafiq QA, Brosnan KM, Coopman K et al. 2013. Culture of human mesenchymal stem cells on microcarriers in a 5 l stirred-tank bioreactor. *Biotechnol Lett* 35(8):1233–1245.

Ramser EM, Wolters G, Dityateva G et al. 2010. The 14-3-3zeta protein binds to the cell adhesion molecule L1, promotes L1 phosphorylation by CKII and influences L1-dependent neurite outgrowth. *PLoS One* 5(10):e13462.

Reinhardt J, Stuhler A, Blumel J. 2011. Safety of bovine sera for production of mesenchymal stem cells for therapeutic use. *Hum Gene Ther* 22(6):775–756.

Reinisch A, Bartmann C, Rohde E et al. 2007. Humanized system to propagate cord blood-derived multipotent mesenchymal stromal cells for clinical application. *Regen Med* 2(4):371–382.

Rojewski MT, Fekete N, Baila S et al. 2013. GMP-compliant isolation and expansion of bone marrow-derived MSCs in the closed, automated device quantum cell expansion system. *Cell Trans* 22(11):1981–2000.

Rowley JA. 2010. Developing cell therapy biomanufacturing processes. *Chem Eng Progr* (SBE Stem Cell Engineering Supplement)106(11):50–55.

Sart S, Schneider YJ, Agathos SN. 2009. Ear mesenchymal stem cells: An efficient adult multipotent cell population fit for rapid and scalable expansion. *J Biotechnol* 139(4):291–299.

Sart S, Schneider YJ, Agathos SN. 2010. Influence of culture parameters on ear mesenchymal stem cells expanded on microcarriers. *J Biotechnol* 150(1):149–160.

Schallmoser K, Bartmann C, Rohde E et al. 2007. Human platelet lysate can replace fetal bovine serum for clinical-scale expansion of functional mesenchymal stromal cells. *Transfusion* 47(8):1436–1446.

Schop D, van Dijkhuizen-Radersma R, Borgart E et al. 2010. Expansion of human mesenchymal stromal cells on microcarriers: Growth and metabolism. *J Tissue Eng Regen Med* 4(2):131–140.

Schrepfer S, Deuse T, Reichenspurner H et al. 2007. Stem cell transplantation: The lung barrier. *Transplant Proc* 39(2):573–576.

Sensebé L, Bourin P, Tarte K. 2011. Response to reinhardt et al. *Hum Gene Ther* 22(6):776–777.

Shahdadfar A, Fronsdal K, Haug T et al. 2005. In vitro expansion of human mesenchymal stem cells: Choice of serum is a determinant of cell proliferation, differentiation, gene expression, and transcriptome stability. *Stem Cells* 23(9):1357–1366.

Shuler ML, Kargi F. 2002. *Bioprocess Engineering: Basic Concepts*. Upper Saddle River, NJ: Prentice Hall.

Sousa MF, Silva MM, Giroux D et al. 2015. Production of oncolytic adenovirus and human mesenchymal stem cells in a single-use, Vertical-Wheel bioreactor system: Impact of bioreactor design on performance of microcarrier-based cell culture processes. *Biotechnol Prog* 31(6):1600–1612.

Stute N, Holtz K, Bubenheim M et al. 2004. Autologous serum for isolation and expansion of human mesenchymal stem cells for clinical use. *Exp Hematol* 32(12):1212–1225.

Subramanian K, Park Y, Verfaillie CM et al. 2011. Scalable expansion of multipotent adult progenitor cells as three-dimensional cell aggregates. *Biotechnol Bioeng* 108(2):364–375.

Sun W, Kim H. 2007. Neurotrophic roles of the beta-thymosins in the development and regeneration of the nervous system. *Ann N Y Acad Sci* 1112:210–218.

Tateishi K, Ando W, Higuchi C et al. 2008. Comparison of human serum with fetal bovine serum for expansion and differentiation of human synovial MSC: Potential feasibility for clinical applications. *Cell Transplant* 17(5):549–557.

Teixeira FG, Carvalho MM, Panchalingam KM et al. 2016a. Impact of the secretome of human mesenchymal stem cells on brain structure and animal behavior in a rat model of Parkinson's Disease. *Stem Cells Transl Med.* 6(2):634–646.

Teixeira FG, Panchalingam KM, Anjo SI et al. 2015. Do hypoxia/normoxia culturing conditions change the neuroregulatory profile of Wharton Jelly mesenchymal stem cell
secretome? *Stem Cell Res Ther* 6:133.

Teixeira FG, Panchalingam KM, Assuncao-Silva R et al. 2016b. Modulation of the mesenchymal stem cell secretome using computer-controlled bioreactors: Impact on neuronal
cell proliferation, survival and differentiation. *Sci Rep* 6:27791.

Tekkatte C, Gunasingh GP, Cherian KM et al. 2011. "Humanized" stem cell culture techniques: The animal serum controversy. *Stem Cells Int* 2011:504723.

Tolar J, Le Blanc K, Keating A et al. 2010. Concise review: Hitting the right spot with mesenchymal stromal cells. *Stem Cells* 28(8):1446–1455.

van Kesteren RE, Carter C, Dissel HM et al. 2006. Local synthesis of actin-binding protein
beta-thymosin regulates neurite outgrowth. *J Neurosci* 26(1):152–157.

Waszak P, Alphonse R, Vadivel A et al. 2012. Preconditioning enhances the paracrine effect
of mesenchymal stem cells in preventing oxygen-induced neonatal lung injury in rats.
Stem Cells Dev 21(15):2789–2797.

Weber C, Freimark D, Portner R et al. 2010. Expansion of human mesenchymal stem cells in
a fixed-bed bioreactor system based on non-porous glass carrier—part A: inoculation,
cultivation, and cell harvest procedures. *Int J Artif Organs* 33(8):512–525.

Wu M, Han ZB, Liu JF et al. 2014. Serum-free media and the immunoregulatory properties
of mesenchymal stem cells in vivo and in vitro. *Cell Physiol Biochem* 33(3):569–580.

Yang Y, Rossi FM, Putnins EE. 2007. Ex vivo expansion of rat bone marrow mesenchymal
stromal cells on microcarrier beads in spin culture. *Biomaterials* 28(20):3110–3120.

Zhao F, Ma T. 2005. Perfusion bioreactor system for human mesenchymal stem cell tissue
engineering: Dynamic cell seeding and construct development. *Biotechnol Bioeng*
91(4):482–493.

9 Bioreactor Sensing and Monitoring for Cell Therapy Manufacturing

Ioannis Papantoniou, Toon Lambrechts,
Priyanka Gupta, Sébastien de Bournonville,
Niki Loverdou, Liesbet Geris, and Jean-Marie Aerts

CONTENTS

9.1 MONITORING AND CONTROL OF ATMPs QUALITY ATTRIBUTES THROUGH PROCESS ANALYTICAL TECHNOLOGY

As the field of cell therapy is maturing, the necessity for well-controlled large-scale bioprocesses is imperative. These bioprocesses should allow for the robust expansion of stem/progenitor cell populations, meeting the target number and quality, while at the same time guaranteeing product safety and efficacy (Abraham et al. 2017). In light of the growing need for cell technologies (e.g., for cell therapy and cell-based drug screening), where cost of goods (CoGs), process efficiency, and logistics become critically important for the clinical and commercial translation (Dodson and Levine 2015), a plethora of bioreactor systems has been employed during the last 15 years with increasing scales of operation.

While the introduction of automated bioreactor systems is already a significant step towards the industrialization of cell production processes (Hourd et al. 2014), in a "black box" bioreactor it is still unlikely to efficiently produce cells since it is impossible to precisely define (and later effectively control) the process that is going on inside. Therefore, the incorporation of monitoring technologies in or around these bioreactors systems is a logical continuation of the industrialization of the cell culture process since it is a necessary precondition of providing confidence in the manufacturing process. Additionally, the monitoring results provide the basis for comparability or equivalency of the process (e.g., after the introduction of a process change or the switch to another production location) that is required by the regulators (Williams et al. 2016).

Regulatory bodies such as the Food and Drug Administration (FDA) and European Medicine Agency (EMA) have suggested Process Analytical Technology (PAT) guidelines, which outline a set of tools and methods for making (bio)pharmaceutical manufacturing processes more reliable and efficient (Simaria et al. 2014). Following this direction, regulatory bodies and the academic research institutes have been active in pursuing these tools in the cell therapy field, suggesting that a variety of aspects such as Quality-by-Design, on-line sensors, and statistical experimental design should be incorporated in process development initiatives. Based on the guidelines of the FDA for PAT guidance, there are six aims that should be attained through the implementation of PAT tools: (1) reducing production time; (2) preventing rejection of batches (detrimental for autologous, patient specific, advanced therapy medicinal product (ATMP) manufacturing); (3) enabling real-time release of ATMPs; (4) increasing automation; (5) improving efficiency of material use; and (6) allowing the implementation of continuous processing (important for allogeneic (donor-derived) ATMP manufacturing) (Mandenius and Gustavsson 2015). These goals are expected to be applicable to all biomanufacturing sectors, bio-therapeutic protein-based drugs and gene therapy vectors, as well as cell therapy products with appropriate customization and adaptation in each case.

Cell therapies are either autologous (derived from a single patient, for that patient) or allogeneic (coming from a banked donor source, for many patients). The former does not face the risks associated with cell rejection. However, they are expected to be much more expensive in terms of manufacturing and logistics costs than the latter (Simaria et al. 2014). The differences encountered between the two biomanufacturing paradigms and the high risks associated in their production call for data driven strategies that will allow for the quantification of bioprocess performance metrics and their association with product quality attributes. Specifically, regarding the monitoring of the culture process, the autologous case imposes generally more challenges. For example, while the allogeneic case profits from a clear economies of scale advantage regarding the cost for quality control (QC), the patient-specific autologous production requires an individual QC approach. In most cases, this QC depends at least partially on end-points analysis, thereby increasing the release time of the product significantly (or worse, receiving the final QC data after administration of the cells). While also beneficial for the efficient manufacturing of allogeneic cells, implementing on-line monitoring systems that provide pre-validation data on the release criteria are therefore more critical for autologous processes.

9.2 BASIC CONCEPTS AND DEFINITIONS REGARDING BIOREACTOR MONITORING

The primary goal of bioreactor monitoring is to provide a quantitative description of cell state and fate inside the vessel (e.g., the critical process parameters), which for ATMP manufacturing is closely linked to the resulting cell quality (i.e., the critical quality attributes). There are many modes of operation for sensing (Wendt et al. 2009) that will be discussed in the following paragraphs. Depending on the specific goal of the read-out and considering other factors such as costs, user-friendliness, and required specificity of the results, any of the following strategies can be appropriate to implement in a bioreactor process.

It is important to note that while sensor systems are used to validate the manufacturing process, the sensor systems themselves, and especially (the interpretation of) their results, need validation too. Therefore, it is important to understand the limitations and optimal implementation of the specific sensor under consideration. For example, sensor accuracy (i.e., the magnitude of the error on the reading), sensitivity (i.e., the lowest and higher detectable limit), specificity, response time, and signal drift all must be considered when choosing a specific sensor for an application.

9.2.1 ON-LINE OR IN-LINE *VERSUS* OFF-LINE OR AT-LINE (BIECHELE ET AL. 2015)

1. On-line or in-line sensing strategies generally refer to methods that allow for receiving continuous up-to-date information on the state of the system in real-time or near real-time, respectively. To be able to capture dynamic responses of the stem/progenitor cells in culture, the sample frequency and signal resolution of the on-line or in-line sensor should be at least twice the frequency at which the variable under consideration changes (Nyquist sampling frequency). While on-line or in-line sensing strategies are technically more challenging to implement for certain types of analysis (e.g., secreted protein concentrations), this high-frequency data is ideal for continuous bioreactor process control. pH and dissolved oxygen tension are typical parameters for on-line monitoring in bioreactors.

2. Off-line or at-line sensing strategies do not provide the operator with instant read-outs. On the contrary, most off-line or at-line sensing strategies require

considerable manual preparation and analysis by the operator. While the informational load of the analysis can be extremely high due to the large range of possible analytical systems to be used, these sensing strategies are less optimal for automated control of bioreactors due to the discontinuity between measurement and result (i.e., the result is a retrospect). A frequently used example of off-line monitoring is the analysis of cell surface markers by flow cytometry or high-performance liquid chromatography (HPLC) for medium analysis (e.g., metabolites).

9.2.2 Invasive *versus* Non-Invasive *versus* Indirect Sensing

1. Invasive monitoring is based on *in situ* probes (or labels such as fluorescent tags) that have a direct interface with the cells or culture medium inside the bioreactor. Generally, they can provide direct information on the state of the culture in the vessel in which they are embedded. A typical example of an invasive sensor is an electrochemical sensor where the culture medium must come in close contact with the functionalized membrane of the sensor. Unless these sensor probes can be sterilized in place or installed as a disposable and pre-sterilized component, invasive sensors are generally not preferred for clinical production due to the possible contamination risk, possible interference from the functionalized membrane with the medium and the risk of degradation of the readout quality by sensor fouling.

2. Non-invasive methods can infer information from the culture environment without physical contact to the medium or cells. A typical example is an optical pH or O_2 sensor with a sensor patch with a pH or O_2 sensitive complex that is glued to the (transparent) inside wall of the culture vessel. Plus, it has an optical fiber at the outside of the vessel that allows a contactless interrogation of the bio-environment. While the transparency of the vessel and often relatively lower sensitivity of the sensor must be considered, their ease of use and simpler handlings to maintain sterility make such sensors ideal for bioreactor monitoring.

3. Indirect sensing *(ex-situ)* makes use of sampling methods, either automated via (recirculating) sampling lines, or manual via sampling ports that allow a sterile connection to the culture vessel with a syringe, for example. A very common use of this strategy is found in bioprocesses where samples are drawn from a sample port and analyzed for lactate and glucose concentrations.

9.2.3 Destructive *versus* Non-Destructive

Destructive testing is an end-point analysis that renders the cells unsuitable for further use. This type of analysis potentially results in readouts with a very high sensitivity and specificity. In certain cases, however, at the start of an autologous culture process where cell material is scarce, the resulting cell loss is undesirable. Non-destructive methods do not require using/manipulating the cells at all (e.g., microscopy, analysis of spent medium samples), nor are they able to return the sampled cells to the vessel via a sterile recirculation sampling-loop.

Keeping in mind the end goal of large-scale automated production of clinically relevant cells, this chapter will focus on non-invasive and non-destructive technologies. Special attention will be given to on-line or in-line methods that are essential for the continuous process control, leading to reproducible high-quality products. Additionally, the cost-effectiveness will be discussed since certain monitoring strategies contribute significantly to the CoGs.

9.3 MONITORING OF BIOREACTORS BASED ON THEIR MODE OF OPERATION AND DESIGN

Large-scale expansion and differentiation of different stem cell populations in a monitored and controlled environment are of immense importance in the field of cell therapy and regenerative medicine. The use of bioreactors is considered to be the most efficient method for this and currently we have a plethora of different systems available, differing in design and operation based on user requirements. The following section aims at a concise review of the different bioreactor systems used for stem cell culture and the implications of the vessel design on the monitoring strategy. The reactors are classified according to their mode of operation and according to the method of cell growth within the systems. Most importantly the implication of bioreactor mode of operation with sensor compatibility and readout/data quality will be also discussed.

9.3.1 Modes of Operation

Batch: A batch mode of bioreactor is a closed system, wherein the bioreactor is seeded with a defined volume of medium and cells at the initial time point, with no further addition or withdrawal throughout the culture period. At the end of the culture period, the content of the whole reactor is harvested in one go. Cells in a batch operated reactor follow the well-established cell growth pattern while consuming nutrients and producing waste metabolites until the nutrients are completely depleted and/or maximum cell density for the system is reached. Monitoring of processes in reactors under batch mode allows for continuous and regular on-line monitoring of various parameters like pH, metabolite concentrations, dissolved oxygen, and so on, wherein their absolute values can be obtained at different time points. However, due to the closed nature of the system, off-line analysis of parameters-like secretory products, phenotype markers can only be carried out at the end of the total culture period. This may result in loss of information for the user. The closed nature of the system also does not allow dynamic process changes based on gathered process data. As a result, bioreactors with this mode of operation have been mostly used for preliminary analysis of culture parameters for mammalian cells (Dalili et al. 1990, Merten et al. 1990).

Fed-Batch: In a fed-batch system, an intermittent or continuous feed of nutrient is supplied to the reactor without any withdrawal of cells or supernatant. The nutrient level in such a system is controlled by optimizing the feed rate in a way to reduce the production of waste metabolites without completely depleting the medium of nutrients. Mammalian cells can be cultured for weeks in a fed-batch reactor and results in higher product concentration in comparison to batch reactors. Monitoring tools and methods in a fed-batch system are essentially similar to batch reactors but the change in system volume makes it a relative reading and requires the users to take in to account the feed rate and medium volume of the system for analysis purposes. In the field of mammalian cell culture, fed-batch reactors have been used for different cell types, such as baby hamster kidney (BHK) cells, Chinese hamster ovary (CHO) cells, and so on (Lenas et al. 1997, Andersen et al. 2000).

Perfusion reactor system: A perfusion reactor is a continuous system with constant inflow of fresh medium along with an outflow of cell-free supernatant, keeping the culture volume constant throughout the culture period. This system has a relatively homogenous mixture of nutrients and dissolved oxygen resulting in a high final cell density. Most current day dynamic stem cell culture systems use the perfusion mode of operation for different bioreactor designs. The monitoring system and resulting analysis are more complex in a perfusion system in comparison to batch and fed-batch reactors since the dilutive effect of the influx of fresh medium has to be taken into account during the analytics, and the measurement values of different parameters could have different values at different positions in the system (e.g., at the inlet *versus* the outlet of the system). Experimental data and computational modeling has shown that cell growth and matrix formation are affected by the scaffold position within a perfusion system due to different flow patterns, shear stress, and so on (Papantoniou et al. 2014, Guyot et al. 2016). As a result, it is imperative that measurements and monitoring tools be used at different locations within the system to extract every possible data for the system. As an example, it has been reported that oxygen concentration in a scaffold-based perfusion system is different at the inlet and outlet. While at the inlet, the concentration remains almost constant throughout the experiment, a decrease in oxygen concentration is measured at the outlet. This information can be used to provide an estimation of oxygen consumption by the cells (Simmons et al. 2017). Similar measurements of oxygen concentration at different positions have been used in mechanistic models to link such monitored parameters to cell growth and health (Lambrechts et al. 2014).

9.3.2 Bioreactor Design

Stirred-tank reactor: Stirred-tank reactors are cylindrical vessels with an impeller for stirring motion and are one of the oldest and widely used dynamic cell culture systems. Stirred-tank reactors are easy to operate and scaling-up is a relatively simple task using vessels of different sizes, thus reducing the chances of vessel-to-vessel variability. Other advantages include the possibility of expanding suspension cells, as well as anchorage-dependent cells, feasibility of on-line monitoring, and automation of the system and parameters involved like nutrients, dissolved oxygen, pH, and so on. Although in most cases it is used under fed-batch, continuous feeding is also possible as a perfusion system. One of the earliest known stirred-tank reactor-based expansion of stem cells was published by Zandstra and co-workers in 1994, when they cultured hematopoietic stem cells in a spinner flask. Since then, experiments have been carried out on various other stem cell types under variable growth conditions. Pluripotent stem cell (both embryonic stem cells [ESCs] and induced pluripotent stem cells [iPSCs]) culture has also been reported in bench top spinner flasks either as cell aggregates or on attachment surfaces like microcarriers, scaffolds, and hydrogels (Fok and Zandstra 2005, Cormier et al. 2006, Chen et al. 2010, Azarin and Palecek 2010, Zweigerdt et al. 2011, Gupta et al. 2016, Ashok et al. 2016, Badenes et al. 2016, Abecasis et al. 2017). Unlike pluripotent stem cells (PSCs), mesenchymal stem/stromal cells (MSCs) from different sources have been extensively expanded using stirred-tank reactors of different scales, which range from 15 mL bench top systems to liter scale reactors (Baksh et al. 2003, Sart et al. 2010, Jung et al. 2012, Rafiq et al. 2013, dos Santos et al. 2014, Chen et al. 2015, de Soure et al. 2016, Heathman et al. 2016). Spinner flask systems have also been used for the differentiation of stem cells towards more committed lineages like cardiomyocyte, osteogenic, chondrogenic, and neurogenic. Although shear stress within a stirred-tank reactor is heterogeneous in nature, proper spin rate within stirred-tank reactors is essential for having a homogenous mixing of metabolites, dissolved oxygen, and proper cell growth within the system. The more or less homogenous culture environment allows the users to have sample-based monitoring of the system along with continuous probe-based ones. The suspension method of culture in case of stirred-tank reactors also allows users to use spectroscopic methods for monitoring the process. Near Infrared (NIR) spectroscopic method, Raman spectroscopy, and impedance spectroscopy have been used for measuring and monitoring cell process status, concentration of medium nutrients, and cell concentration in a suspension bioreactor (Zhao et al. 2015)

Parallel plate reactors: Parallel plate bioreactors consist of a vessel wherein medium is perfused parallel to the 2D cell layer resulting in uniform shear stress over the cells. These reactors are straightforward as the required operations resemble the standard flask-based culture, they are easy to use, have automation options, and are mainly used as micro or bench top systems. However, periodic harvest from this type of system is virtually impossible. Hematopoietic, mesenchymal, and pluripotent stem cells have been cultured using parallel plate type systems (Palsson et al. 1993, Dennis et al. 2007, Wolfe and Ahsan 2013, Belair et al. 2015). While most 2D cell cultures are monitored by microscopy, the multi-plate configuration brings additional challenges for visual inspection. Commercial multi-plate bioreactors, such as the Pall Xpansion® Multiplate Bioreactor system, therefore developed a specialized microscope (Ovizio Imaging Systems) for monitoring cell growth in combination with on-line pH and

O_2 sensors (Leferink et al. 2015, Lambrechts et al. 2016). Since mixing in parallel plate bioreactors is limited and nutrient gradients might exist, for example, perpendicular to the plates, care should be taken for sample-based monitoring techniques.

Hollow-fiber reactor: Hollow-fiber bioreactors are double compartment systems, usually consisting of fiber bundles encased in a vessel with ports for intra-capillary and/or extra-capillary medium flow. This type of a system allows for surface areas larger than parallel plate systems but results in more challenging harvesting and scale-up. Despite their disadvantages, embryonic, mesenchymal, and hematopoietic stem cells have all been expanded using hollow-fiber bioreactors by several groups (Housler et al. 2012, Lambrechts et al. 2016). The 3D configurations of the opaque fibers do not allow visual inspection of the cells, and at the same time, due to the relatively large length of the fibers, nutrient gradients are potentially introduced over the length of the fibers. Therefore, precise monitoring of the culture environment is required so the homogeneity of the nutrient availability can be controlled via the interplay between perfusion and recirculation in the extra-capillary environment of the hollow-fiber bioreactor.

Rotating wall vessel: The rotating wall vessel (RWV) bioreactor was designed by NASA's Biotechnology group as a means to have a dynamic culture with relatively low shear and turbulence. The bioreactor simulates a microgravity environment, offers attractive options for stem cell culture, and has been used for culturing different stem cells including embryonic, mesenchymal, hematopoietic and neural stem cells in suspension (Gerecht-Nir et al. 2004, Lin et al. 2004, Liu et al. 2006, Chen et al. 2006). Differentiation of MSCs have been also carried out using the RWV system (Song et al. 2006). The rotating wall makes it challenging to implement probe or patch-based sensors and for visual inspections.

Fixed-bed reactor: Fixed-bed reactors consist of scaffolds or particles fixed within a column. Medium is perfused through the "packed bed" and supplies nutrients, oxygen, and shear stress to the cells. A fixed-bed reactor can provide 3D environment for stem cells, allows for greater cell-cell and cell-extracellular matrix interaction resulting in better mimicking of *in vivo* cell niche. However, due to the high cells density per cubic meter these systems may also give rise to spatial concentration gradient of nutrients oxygen etc. Fixed-bed type reactors with scaffolds have been used extensively for the expansion and/or differentiation of different types of stem cells (Alves da Silva et al. 2011, Sonnaert et al. 2017). They have also been used for the formation of neo-tissues making them extremely important in the field of tissue engineering. These systems are usually used as a perfusion system with in-line and off-line measurements and monitoring carried out in a similar way as a perfusion system (de Peppo et al. 2013, Sonnaert et al. 2015, Sabatino et al. 2015).

9.4　MONITORING BASIC PHYSICOCHEMICAL PARAMETERS

Bioreactor process variables are of a chemical, physical, or biological nature and can be measured in the gas, liquid, and solid phases of a bioprocess. Concentration changes are informative of cell growth, metabolism and productivity. A typical biosensor comprises three parts, a biological detection component that is immobilized on a signal transducer unit, followed by an amplification, and signal conversion unit. The biocomponent recognizes the analyte either through a catalytic mechanism (e.g., cells) or through binding (e.g., antibodies, membrane receptors) (Thevenot et al. 2001). There are also novel components such as thermostable enzymes or aptamers (Song et al. 2008). Monitoring can be performed with in-line sensors or via analytical systems that can be coupled with sampling devices as at-line sensor systems. Biosensors for many different analytes have been developed to date (Mulchandani and Bassi 1995, Bracewell et al. 2002, Rhee et al. 2004, Borisov and Wolfbeis 2008). Typical implementation of these types of sensors are the Clark-type electrode for O_2 or optical readings from O_2 or pH sensitive sensor patches (Sections 9.4.2 and 9.4.3).

9.4.1　GLUCOSE

Most widely implemented biosensor application is that of determining glucose concentration since this consists of the most important energy source for cells during culture. Glucose and lactate levels have been used as in-process control parameters to indicate the active metabolism for differentiation of stem cell populations. Glucose concentration can have an impact on the cell differentiation process and, in high concentration, has been seen to suppresses embryonic stem cell differentiation into cardiomyocytes (Yang et al. 2016) and inhibit the proliferation and migration of bone marrow mesenchymal stem cell (Zhang et al. 2016). In addition, for both PSCs and MSCs, it has been seen that culture under constant low glucose concentration in the bioreactor feed, glucose consumption, and lactate production were limited (Schop et al. 2008, Chen et al. 2010).

In the case of glucose, electrochemical and optical sensing methods have been extensively investigated. However, continuous glucose monitoring is still limited and potentially could be incorporated in the process development stage by the adoption of automated parallelized platforms offering on-line glucose monitoring such as the ambr® system (Rafiq et al. 2017, Xu et al. 2017). Many more electrochemical detection methods (Borisov and Wolfbeis 2008) are in use for glucose monitoring than optical ones. Glucose sensors, according to a comprehensive review of Steiner and co-workers, can be classified in four categories: (a) monitoring of optical properties of enzymes, their cofactors and co-substrates, (b) measurement of enzymatic oxidation products of glucose oxidase, (c) the use of boronic acids, and (d) the use of glucose binding proteins. In the literature, there are also examples of using spectroscopic methods such as NIR and mid infrared (MIR) spectroscopy for measuring glucose (Rhiel et al. 2002, Henriques et al. 2009).

9.4.2   OXYGEN

Oxygen tension is one of the most important parameters in cell culture monitoring. It has been seen that long-term culture of human MSCs under hypoxic conditions (2% O_2) results in a decrease in cell proliferation but not in an increased apoptosis for up to 24 days of culture (Grayson et al. 2007). In addition, it has been reported that hypoxia induced down-regulation of osteoblastic genes in human MSCs *in vitro* (Potier et al. 2007) and low dissolved oxygen values (1% vs 21%) affected osteogenic differentiation with a metabolic switch to anaerobic glycolysis (Hsu et al. 2013). In general, low dissolved oxygen tensions have been seen to maintain progenitor cell populations maintaining stemness of human MSC populations (Mohyeldin et al. 2010).

Semiconducting, electrochemical and paramagnetic sensors are commonly used for oxygen analysis in the gas phase. The dissolved oxygen concentration in the medium (liquid phase) can be measured by electrochemical and optical chemical sensors (chemosensors). Membrane-covered amperometric electrodes, known as Clark-type electrodes, are the most common electrochemical sensor used. Optical measurement of the partial pressure of oxygen is based on the fluorescence quenching of an indicator by molecular oxygen (Harms et al. 2002). In addition, optical sensors can be easily miniaturized (Beutel and Henkel 2011).

9.4.3   pH

pH has been seen to influence osteogenic differentiation and ECM mineralization. In a relevant study by Monfoulet and colleagues, the presence of mineralized nodules in the extracellular matrix of human bone marrow MSCs was seen to be fully inhibited at alkaline (>7.54) pH values. In addition, for a pH range between 7.9 and 8.27, proliferation was not affected while osteogenic differentiation was seen to be inhibited (Monfoulet et al. 2014). In addition, reduction of extracellular pH from 7.8 to 7.0 was seen to negatively affect MSC osteogenic differentiation *in vitro* (Tao et al. 2016). Moving on to glucose, mineral deposition of MSCs during osteogenesis in high-glucose medium is greater than that of MSCs in low-glucose medium (Li et al. 2007).

pH must be controlled in nearly all bioprocesses to keep the cells at optimal cultivation conditions. Electrochemical techniques for pH are based on potentiometric and amperometric measurements, with potentiometric sensors being more commonly used. A working (measurement) electrode (working potential) is correlated to a reference electrode (Henkel and Beutel 2013). Other systems are the ion-sensitive field effect transistors (ISFETs). Apart from the classical electrochemical sensors, also optical chemosensor systems are used. Fiber optic pH sensors are based on absorbing or fluorescing pH indicators. Frequently used pH-sensitive indicators for absorbance are cresol red, phenol red, bromopherol blue, and chlorophenol red (Gupta and Sharma 1998). These indicators are immobilized on solid substrates or in polymers that are mounted directly on a transparent carrier (glass or optical fibers). The pH optodes can easily be miniaturized and have response times in the range of milliseconds. However, they suffer from cross sensitivity to ionic strength, a limited dynamic range, and the loss of sensitivity during sterilization or cleaning procedures (Beutel and Henkel 2011). Nowadays only γ-sterilized optical biosensors are commercially available and are implemented in pre-sterilized disposable bioreactor systems. Finally, pH levels can also be measured by spectroscopic methods, such as MIR and NIR spectroscopy (Schenk et al. 2008).

9.4.4 CARBON DIOXIDE

Monitoring of carbon dioxide in bioprocesses is essential as it is a product of cell respiration and can pass through cell membranes to influence pH inside the cells. Thus, variations of dissolved CO_2 concentration can affect cell morphology, their metabolic products, and consumption/productions rates. Most optical CO_2 sensors operate in the same way as the classical Severinghaus electrodes. They are typically based on the fluorometric or colorimetric changes of pH-sensitive indicators that are added to the buffer system and separated from the analyte solution by a CO_2—permeable membrane.

For monitoring dissolved carbon dioxide, the use of traditional electrochemical sensors can be problematic because they are bulky and invasive (Chatterjee et al. 2015). Disposable optical sensors have a lot of advantages, like high sensitivity and easy miniaturization, and they are only partially invasive. However, there are the disadvantages regarding the toxicity of the patch and the photo-toxicity of illuminating light. Gupta and colleagues (Gupta et al. 2014) used a completely non-invasive approach to monitor dissolved carbon dioxide in disposable small-scale cell culture vessels based on the diffusion through permeable vessel walls, which can be applicable in sensitive cells. A strong limitation of this approach was that it cannot continuously measure the process. A completely non-invasive rate-based technique, measuring the initial diffusion rate, allowed for the measurement of the partial pressure of CO_2. This system has been tested in spinner flasks and the results were comparable to those measured with fluorescence-based patch sensors. The main difficulty regarding the integration and use of carbon dioxide sensors is their limited stability and dependency on the ionic strength of the buffer (Glindcamp et al. 2009). A "credit-card" microsystem for the determination of carbon dioxide for fermentation monitoring was also recently reported (Calvo-Lopez et al. 2016).

9.4.5 SPECTROSCOPIC SENSORS

Spectroscopic analyses are based on the interaction of electromagnetic waves and molecule bonds. Common spectroscopic methods for bioprocess monitoring are focused on the spectral range from UV to MIR, including fluorescence and Raman Spectroscopy. Various process variables can be measured in different spectral ranges (16–21). The advantages of spectroscopic sensors include no sampling is needed (except for calibration), there is no interaction between the sensor and the analytes, and several different process variables can be determined simultaneously. One major drawback is that chemometric data analysis is needed to extract relevant process information from the generated data.

9.5   IMAGE-BASED SINGLE CELL MONITORING DURING BIOPROCESSING

Despite new advances and technological breakthrough in monitoring, visual analysis of cells is still one of the most versatile and powerful tools (Molder et al. 2008). Image-based monitoring of bioprocess parameters is increasingly recognized as a valuable tool in the field, and the use and development of in-situ image acquisition sensors is rising (Bluma et al. 2010). However, today image-based results often depend on manual analysis, which is time consuming and subject to very high variability. Thus, there is a need for non-invasive automated visual cellular analysis that allows objective quantification of critical quality attributes (CQA). In this section, we review the different optical modalities that are currently available and under development in the bioreactor field, as well as the future perspective for imaging of bioreactor-based bioprocesses.

The developments and trends in the field will be reviewed from the most basic optical techniques point of view (i.e., bright field microscope), to the latest high-tech developments (i.e., lens-free imaging), naturally following a quasi-chronological order. In the last decade, different types of imaging modalities have been adapted to bioreactors in *in-situ* setups. Most of these are gathered in Table 9.1.

From this table, one can see that the first innovations in the field consisted in adapting bright field microscopes to bioreactors, using *in-situ* probes in contact with cell suspension and commercially available products (i.e., *in-situ* microscope type III XTF, Sartorius). The probe allows monitoring of a sample volume (assumed to be representative of the batch). By using image processing algorithms, it can provide estimates of the cell concentration, cell size or degree of aggregation (Joeris et al. 2002). Using neural network algorithms trained on manual analyses, Rudolph and co-workers were able to build a model-based monitoring of colonization of microcarriers by cells (Rudolph et al. 2008).

However, bright field systems show inherent limitations for live cell imaging (and processing) because of the low contrast of transparent specimen. Furthermore, the sample volumes imaged with these systems can be quite small compared to the bioreactor volume, making the extrapolated data less representative of the cell suspension. To overcome the contrast limitations, Wei and collaborators adapted a dark field condenser into an existing *in situ* probe (Wei et al. 2007). However, direct phase quantification is not possible with this technique, as it involves a non-linear phase to amplitude conversion, which can result in artefacts in the images (Mann et al. 2005). This applies also to other imaging modalities, like phase contrast or differential interference contrast imaging, which provide good contrast for transparent specimens.

Quantitative phase imaging allows the computation of the optical thickness of a transparent specimen, which gives information on the physical thickness of the samples and their optical index variation (Mann et al. 2005). Lens-free imaging techniques are rising in the field for imaging of 2D planar cell culture as they provide direct phase quantification, great contrast of transparent samples, decoupling of the field-of-view to the resolution, and larger field of views (Greenbaum et al. 2012). Depending on the spatial and temporal coherence of the light source on the image

TABLE 9.1

An Overview of Available Imaging Techniques for Visualization of Cell Features. Advantages and Shortcomings of Each Technology Are Also Listed

Imaging Modality/Setup	Measurements	Pros	Cons	Representative References
Bright field imaging/probe inside stirred system under batch operation.	Cell concentration, cell size, cell volume. As population averages.	Real-time imaging.	Probe is invasive. Relatively small field of view.	Joeris et al. (2002)
	Cell concentration, cell volume. As population averages Cell concentration, cell size distribution, degree of aggregation. With neural network validated on manual analysis → level of microcarrier colonization modelled	Real-time imaging.	Probe is invasive. Computationally heavy methods required. Relatively small field of view.	Rudolph et al. (2008)
Dark field imaging/probe inside stirred system under batch operation.	Cell density. Cell viability reading using a machine-learning algorithm validated against live dead manual analysis.	Real-time imaging.	Probe is invasive. Computationally heavy methods required. Relatively small field of view.	Wei et al. (2007)
Differential interference contrast imaging/probe inside stirred system under batch operation.	Confluence of cells on micro-carriers, cell density, cell distribution.	Real-time imaging.	Probe is invasive. Computationally heavy methods required. Relatively small field of view.	Odeleye et al. (2017)
Phase contrast imaging used at-line	Cell concentration, cell size distribution, 2D attachment morphology,	Non-invasive imaging. Good contrast of transparent cells.	Suited for planar cultures, not for 3D setups. At-line operation.	Jaccard et al. (2015)
Digital holographic imaging	Cell number, cell area, thickness and volume.	Real-time, non-invasive imaging. Excellent contrast. Large field of view, decoupled from resolution. Compact devices used on-line.	Not suited for non-planar cultures.	El-Schish et al. (2010)

reconstruction algorithms, several lens-free imaging (LFI) techniques have emerged (Kim et al. 2012). In general, the interest for digital holography imaging (DHI) has increased in the field of cell monitoring (Carl et al. 2004, Rappaz et al. 2005, Molder et al. 2008, Biechele et al. 2015) to become a powerful quantitative, non-invasive imaging technique for 2D planar cell cultures nowadays. Due to the design simplicity of the technique, DHI allows the development of compact, light, and cost-effective devices. Several commercially available DHI systems (such as iLine S, Ovizio® with Pall®; iLine M, Ovizio®; LUX2, CytoSMART™; and Cytonote, iPRASENSE) exist for the monitoring and controlling of cell bioprocesses with compact designs fitting inside incubators, a wireless connection to computer, wide field of views, and high resolutions. Recently, an integrated optics-based methodology was reported that consistently and accurately assesses viable cell number, confluence, and cell distribution of embryonic stem cells cultured on micro-carriers sampled from suspension bioreactors (Odeleye et al. 2017).

9.6 VISUAL MONITORING OF 3D GROWTH IN SCAFFOLD-BASED PERFUSION BIOREACTORS

Concerning tissue engineered ATMPs, bioreactors are used to enhance tissue proliferation and cell differentiation. The field of tissue engineering is evolving towards the *in vitro* creation of complex 3D tissues and, currently, the gold standard for assessing engineered tissue construct quality is by using destructive end-point assays such as histological sectioning (Doroski et al. 2007, Leferink et al. 2016). The field is developing a growing interest in novel methods for non-destructive 3D construct visualization over time with a decent spatiotemporal resolution (Mather et al. 2007). Most optical microscopy techniques show major constraints for these applications as they have limited penetration depth (Vielreicher et al. 2013). The applicability of a given imaging modality highly depends on the type of carrier structure used for the tissue and the biological components of interest. In this section, we review the currently used or promising imaging technologies for *in vitro* 3D tissue formation in bioreactors.

9.6.1 Micro-Computed Tomography

X-ray computed tomography (CT) exploits the variations in the X-ray absorption properties of a sample to form a 3D reconstructed image. The spatial resolution achieved with this technique depends on the X-ray source spot size, which, more generally, depends on the coherence of the X-ray beam (Appel et al. 2013). To date, by using this technique nanometer scale resolution has been achieved. As the contrast in the reconstructed pictures is dependent on the X-ray absorption properties of the materials, a good contrast can be achieved between denser tissues like bone and softer tissues (i.e., cartilage, fibrous tissue). However, achieving a decent contrast to distinguish between two soft tissues and the background is challenging. Two arising approaches to overcome this limitation in soft tissue engineering consist in using chemical contrast agents that stains the structures of interest (Papantoniou et al. 2014) combining the use of micro-CT with X-ray phase contrast imaging (Zhu et al. 2011, Appel et al. 2015, Bradley et al. 2017). Using synchrotron approaches, even single cell resolution could be achieved without the use of contrast (Voronov et al. 2013). These approaches go one step further in the development of X-ray micro-CT based on-line monitoring tool for tissue engineering.

9.6.2 MAGNETIC RESONANCE IMAGING

Magnetic Resonance Imaging (MRI) has shown to be a promising modality for the imaging of soft tissues (Lalande et al. 2011, Kotecha et al. 2013). Few advances have been achieved in the coupling of MRI devices with bioreactors for longitudinal monitoring of tissue growth (Crowe et al. 2011, Gottwald et al. 2013). The use of contrast agents with MRIs has also allowed cell tracking (Feng et al. 2011, Di Corato et al. 2013). Although this modality inherently provides better contrast than CT for soft tissue 3D visualization, the acquisition of MRI scans with decent resolutions (micro-MRI) require high-field MRI scanners and long scanning times. Such devices capable of applying strong enough magnetic fields for micro-MRI are expensive, and their implementation in tissue engineering laboratories is not yet feasible (Leferink et al. 2016).

9.6.3 IMPEDANCE MONITORING

Radio frequency impedance (RFI) monitoring has long been recognized as an accurate and reliable method for on-line and off-line measurement of live cell biomass as the impedance (measured capacitance) is proportional to the volume of living cells in the bioprocess (Carvell and Dowd 2006). ABER® Instruments' biomass sensor, HP® Colloid Dielectric Probe and CellSine are a few RFI systems currently commercially available. Furthermore, the use of this sensor to scaled-down systems (down to 1 mL) and single-use bioreactors has been reported (Carvell et al. 2014).

9.7 HIGH-THROUGHPUT BIOREACTOR MONITORING THROUGH THE ADOPTION OF OMICS TOOLS

There is a clear need for the development and application of novel methods that enable the comprehensive characterization of cell state, improving the current conventional set of measurements for monitoring cell and tissue quality (Lipsitz et al. 2016). In this context, the development of "omics" technologies could further contribute in capturing intracellular characteristics and, hence, enhance the ability to monitor bioreactor processes.

The term "omics" refers to the quantification and characterization of various biological molecules present at a specific time or condition. These approaches include genomics, transcriptomics, proteomics, and metabolomics, which, respectively, examine genes, messenger RNA (mRNA), proteins, and metabolites present in a particular cellular environment. Each set of biological molecules and their levels can be analyzed as a molecular signature for the cell and condition of interest. This information is a key for understanding cellular physiology and using a system-biology approach to rationally improve cellular performance. Omics technologies, although relatively new, have been rapidly implemented in biotechnology, leading to "revolutions" in the field of cancer biology (Vucic et al. 2012), biofuel production (de Jong et al. 2012), and in the process optimization of industrial mammalian systems (Vernardis et al. 2013). Additionally, major biotechnological-biopharmaceutical companies like Bristol-Myers Squibb, Bayer, Pfizer, and Life Technologies (part of ThermoFisher Scientific) had already conducted studies with the goal of integrating omics technologies in the bioprocessing field (Xu et al. 2011, Slade et al. 2012).

Recently, the adoption of omics technologies in the field of stem cell bioprocessing has been initiated. More specifically, Silva and colleagues (Silva et al. 2015) used genome-transcriptome profiling as a critical evaluation tool for the robustness of the expansion process of human ESCs in stirred-tank bioreactor cultures. Furthermore, they explored the use of omics tools and, more specifically, metabolomics to investigate the physiological and metabolic changes during human ESC expansion in the aforementioned bioprocess. They claimed that this new information can guide bioprocess design and medium optimization. In the same context, Alves and co-workers (Gomes-Alves et al. 2016) explored the use of omics tools for the characterization of the *in vitro* expansion of human cardiac progenitor cells (CPC). They used gene expression microarrays and mass spectrometry-based approaches to compare transcriptome, proteome, surface markers, and secretion profiles of human CPC cultured in static monolayers and in stirred microcarrier-based systems. In this work, the authors used omics tools as a quality control for the developed bioprocess, and they underlined the importance of this methodology as a future potency assay of their stem cell population. In another study, Abecasis and colleagues (Abecasis et al. 2017) used stirred-tank bioreactors for the expansion of human iPSCs as aggregates. Extensive cell characterization was conducted by using whole proteomic analysis as a quality control tool for the maintenance of cells' pluripotent character during the culture.

Although bioreactor-based ATMP manufacturing still relies on empirical techniques, it is clear that this is a rapidly evolving landscape. The use of omics represents a unique opportunity to develop improved methods to characterize and understand cell culture state, and rationally optimize bioreactor cultivation processes as illustrated in the comprehensive review of Lewis and co-workers (Lewis et al. 2016). The information gained from omics approaches can be used for the critical evaluation of the impact of cell culture conditions but also to define potency criteria. All the above can help even to redesign operating conditions, coupled with medium formulations and bioreactor design approaches (Clarke et al. 2011). In addition, the identification of sets of efficacy-specific biomarkers, which could be easily monitored on-line with existing sensors, is another important aspect of the use of omics strategies.

9.7.1 Limitations of Omics Technologies in Stem Cell Bioprocessing

Although they possess a great potential for the rational improvement of stem cell bioprocessing, omics technologies face some critical limitations that should be addressed before this technology could be adopted in industrial practice. These limitations should be mostly identified in the data analysis cost required for such large and complex data sets. Despite great advancements in the bioinformatics field, a strong limitation of omics methodologies is the limited ability to meaningfully interpret the large amount of data that are generated by these methods. For that purpose, there is continuing need to develop more efficient computational tools that would facilitate the extraction of conclusions from such large data sets at a higher frequency. Furthermore, the identification of meaningful links between omics datasets and desired phenotypes is a major challenge. Large data sets can be used to identify specific biomarkers of cell physiology. However, validation of these biomarkers is a critical and often neglected step for monitoring purposes. Once validated, these biomarkers can be routinely monitored. However, biomarkers can be cell line dependent, and, as a result, omics strategies need to be developed to incorporate this methodology into the rapid timelines of the industrial bioprocesses. As biomarkers can be cell line specific, there is also the need to develop omics strategies for integrating this methodology in the rapid timelines of the industrial bioprocesses.

9.8 MODEL-ENHANCED MONITORING AND PROCESS CONTROL

In automated bioreactors, it is standard practice that measurements of physicochemical or environmental variables (i.e., critical process parameters (CPPs) such as pH and O_2) are used as input for a controller that, in turn, takes care of providing suitable conditions for cell culture inside the bioreactor. It is then presumed, based also on the mapping of a design space, that these suitable environmental conditions will deliver cells with the proper quality attributes for clinical application. While currently these CQAs are verified by end-point and destructive analysis, it would be more efficient to have continuous feedback on the CQAs during the culture process and implement controllers that directly optimize the CQAs (instead of indirectly via the CPPs), ultimately leading to higher process certainty and more potent cell-based products.

As it can be seen from the previous overview of monitoring techniques, it is not necessarily a problem of measuring the CQAs, but the limiting factor, to date, is the lack of non-invasive sensor systems with a measurement rate that allows on-line monitoring of the CQAs. Furthermore, sensor hardware development is required to have cost-effective on-line and non-invasive CQA measurements from inside the culture vessel (e.g., replace classic off-line methods for immunophenotyping or quantitative polymerase chain reaction [qPCR]). Alternatively, it is a promising approach to estimate the unmeasurable CQAs in real-time, based on a combination of CPP measurements that can be linked to the CQA of interest by means of a model. This model-enhanced monitoring strategy is also sometimes referred to as a "soft(ware) sensor" or "inferential measurement" methods (de Assis and Maciel 2000, Kadlec et al. 2009). In practice, the models are used to translate sensor data into more informative process read-outs that can be used to make informed process decisions.

This "model-based monitoring" strategy, combined with a desired output trajectory can be used as the basis for "model-based control" in which the measured process data and a model are used to determine or predict the most appropriate controller setting to reach the desired state of the process. Although more often used in bioreactors for mammalian cell culture (Kovarova-Kovar et al. 2000, Ramaswamy et al. 2005, Aehle et al. 2012), only limited examples of model-based control can be found in literature for stem cell bioreactors (Csaszar et al. 2012, Lambrechts et al. 2014). While the clinical use of these adaptive or predictive manufacturing strategies might require adjustments to the current regulatory guidelines, they provide a possible robust and objective strategy to deal with heterogenic input material (e.g., donor-related variability in autologous processes).

9.9 CONCLUDING REMARKS

The steady increase in the production scale of cell-based therapeutics highlights the continuous maturation of the field. However, there is still considerable room for the adoption of additional sensors and monitoring strategies that will further aid the understanding of bioprocesses. This is necessary to face the challenges encountered during the transition from the early preclinical stage to the late commercial stage of manufacturing. In addition to the adoption of novel biosensors, it is crucial to develop model-based strategies for making most of the existing sensor-derived data sets. There is a need for determining additional cost-effective metrics allowing for the definition of process efficiency and how model-based analysis of bioreactor processes could significantly contribute to this. Moreover, the identification of the risks associated with batch failure through model-based monitoring could also serve as an excellent tool for addressing important bottlenecks in the current field of stem cell bioprocessing, which is growing into a personalized biomanufacturing sector.

REFERENCES

Abecasis, B., T. Aguiar, E. Arnault, R. Costa, P. Gomes-Alves, A. Aspegren, M. Serra, and P. M. Alves. 2017. Expansion of 3D human induced pluripotent stem cell aggregates in bioreactors: Bioprocess intensification and scaling-up approaches. *J Biotechnol* 246:81–93. doi:10.1016/j.jbiotec.2017.01.004.
Abraham, E., B. B. Ahmadian, K. Holderness, Y. Levinson, and E. McAfee. 2017. Platforms for manufacturing allogeneic, autologous and iPSC cell therapy products: An industry perspective. *Adv Biochem Eng Biotechnol.* doi:10.1007/10_2017_14.
Aehle, M., K. Bork, S. Schaepe, A. Kuprijanov, R. Horstkorte, R. Simutis, and A. Lubbert. 2012. Increasing batch-to-batch reproducibility of CHO-cell cultures using a model predictive control approach. *Cytotechnology* 64 (6):623–634. doi:10.1007/s10616-012-9438-1.
Alves da Silva, M. L., A. Martins, A. R. Costa-Pinto, V. M. Correlo, P. Sol, M. Bhattacharya, S. Faria, R. L. Reis, and N. M. Neves. 2011. Chondrogenic differentiation of human bone marrow mesenchymal stem cells in chitosan-based scaffolds using a flow-perfusion bioreactor. *J Tissue Eng Regen Med* 5 (9):722–732. doi:10.1002/term.372.
Andersen, D. C., T. Bridges, M. Gawlitzek, and C. Hoy. 2000. Multiple cell culture factors can affect the glycosylation of Asn-184 in CHO-produced tissue-type plasminogen activator. *Biotechnology and Bioengineering* 70 (1):25–31. doi:10.1002/1097-0290(20001005)70:1<25::AID-BIT4>3.0.CO;2-Q.
Appel, A. A., M. A. Anastasio, J. C. Larson, and E. M. Brey. 2013. Imaging challenges in biomaterials and tissue engineering. *Biomaterials* 34 (28):6615–6630. doi:10.1016/j.biomaterials.2013.05.033.

Appel, A. A., J. C. Larson, A. B. Garson, H. F. Guan, Z. Zhong, B. N. B. Nguyen, J. P. Fisher, M. A. Anastasio, and E. M. Brey. 2015. X-ray phase contrast imaging of calcified tissue and biomaterial structure in bioreactor engineered tissues. *Biotechnol Bioeng* 112 (3):612–620. doi:10.1002/bit.25467.

Ashok, P., Y. Fan, M. R. Rostami, and E. S. Tzanakakis. 2016. Aggregate and microcarrier cultures of human pluripotent stem cells in stirred-suspension systems. *Methods Mol Biol* 1502:35–52. doi:10.1007/7651_2015_312.

Azarin, S. M., and S. P. Palecek. 2010. Development of scalable culture systems for human embryonic stem cells. *Biochem Eng J* 48 (3):378. doi:10.1016/j.bej.2009.10.020.

Badenes, S. M., T. G. Fernandes, C. S. Cordeiro, S. Boucher, D. Kuninger, M. C. Vemuri, M. M. Diogo, and J. M. Cabral. 2016. Defined essential 8 medium and vitronectin efficiently support scalable xeno-free expansion of human induced pluripotent stem cells in stirred microcarrier culture systems. *PLoS One* 11 (3):e0151264. doi:10.1371/journal.pone.0151264.

Baksh, D., J. E. Davies, and P. W. Zandstra. 2003. Adult human bone marrow–derived mesenchymal progenitor cells are capable of adhesion-independent survival and expansion. *Exp Hematol* 31 (8):723–732. doi:10.1016/S0301-472X(03)00106-1.

Belair, D. G., J. A. Whisler, J. Valdez, J. Velazquez, J. A. Molenda, V. Vickerman, R. Lewis et al. 2015. Human vascular tissue models formed from human induced pluripotent stem cell derived endothelial cells. *Stem Cell Rev Rep* 11 (3):511–525. doi:10.1007/s12015-014-9549-5.

Beutel, S., and S. Henkel. 2011. In situ sensor techniques in modern bioprocess monitoring. *App Microbiol Biotechnol* 91 (6):1493–1505. doi:10.1007/s00253-011-3470-5.

Biechele, P., C. Busse, D. Solle, T. Scheper, and K. Reardon. 2015. Sensor systems for bioprocess monitoring. *Eng Life Sci* 15 (5):469–488. doi:10.1002/elsc.201500014.

Bluma, A., T. Hopfner, P. Lindner, C. Rehbock, S. Beutel, D. Riechers, B. Hitzmann, and T. Scheper. 2010. In-situ imaging sensors for bioprocess monitoring: State of the art. *Anal Bioanal Chem* 398 (6):2429–2438. doi:10.1007/s00216-010-4181-y.

Borisov, S. M., and O. S. Wolfbeis. 2008. Optical biosensors. *Chem Rev* 108 (2):423–461. doi:10.1021/cr068105t.

Bracewell, D. G., A. Gill, and M. Hoare. 2002. An in-line flow injection optical biosensor for real-time bioprocess monitoring. *Food Bioprod Proc* 80 (C2):71–77. doi:10.1205/09603080252938690.

Bradley, R. S., I. K. Robinson, and M. Yusuf. 2017. 3D X-Ray nanotomography of cells grown on electrospun scaffolds. *Macromol Biosci* 17 (2). doi:10.1002/mabi.201600236.

Calvo-Lopez, A., O. Ymbern, D. Izquierdo, and J. Alonso-Chamarro. 2016. Low cost and compact analytical microsystem for carbon dioxide determination in production processes of wine and beer. *Anal Chim Acta* 931:64–69. doi:10.1016/j.aca.2016.05.010.

Carl, D., B. Kemper, G. Wernicke, and G. von Bally. 2004. Parameter-optimized digital holographic microscope for high-resolution living-cell analysis. *Appl Opt* 43 (36):6536–6544. doi:10.1364/Ao.43.006536.

Carvell, J., M. Lee, and P. Sandhar. 2014. Recent developments in scaling down and using single use probes for measuring the live cell concentration by dielectric spectroscopy. *New Biotechnol* 31:S50. doi:10.1016/j.nbt.2014.05.1725.

Carvell, J. P., and J. E. Dowd. 2006. On-line measurements and control of viable cell density in cell culture manufacturing processes using radio-frequency impedance. *Cytotechnology* 50 (1–3):35–48. doi:10.1007/s10616-005-3974-x.

Chatterjee, M., X. D. Ge, S. Uplekar, Y. Kostov, L. Croucher, M. Pilli, and G. Rao. 2015. A unique noninvasive approach to monitoring dissolved O-2 and CO_2 in cell culture. *Biotechnol Bioeng* 112 (1):104–110. doi:10.1002/bit.25348.

Chen, A. K., X. Chen, A. B. Choo, S. Reuveny, and S. K. Oh. 2010. Expansion of human embryonic stem cells on cellulose microcarriers. *Curr Protoc Stem Cell Biol* Chapter 1: Unit 1C 11. doi:10.1002/9780470151808.sc01c11s14.

Chen, A. K.-L., Y. K. Chew, H. Y. Tan, S. Reuveny, and S. Kah Weng Oh. 2015. Increasing efficiency of human mesenchymal stromal cell culture by optimization of microcarrier concentration and design of medium feed. *Cytotherapy* 17 (2):163–173. doi:10.1016/j.jcyt.2014.08.011.

Chen, X., A. Chen, T. L. Woo, A. B. Choo, S. Reuveny, and S. K. Oh. 2010. Investigations into the metabolism of two-dimensional colony and suspended microcarrier cultures of human embryonic stem cells in serum-free media. *Stem Cells Dev* 19 (11):1781–1792. doi:10.1089/scd.2010.0077.

Chen, Xi, H. Xu, C. Wan, M. McCaigue, and G. Li. 2006. Bioreactor expansion of human adult bone marrow-derived mesenchymal stem cells. *Stem Cells* 24 (9):2052–2059. doi:10.1634/stemcells.2005-0591.

Clarke, C., P. Doolan, N. Barron, P. Meleady, F. O'Sullivan, P. Gammell, M. Melville, M. Leonard, and M. Clynes. 2011. Large scale microarray profiling and coexpression network analysis of CHO cells identifies transcriptional modules associated with growth and productivity. *J Biotechnol* 155 (3):350–359. doi:10.1016/j.jbiotec.2011.07.011.

Cormier, J. T., N. I. Zur Nieden, D. E. Rancourt, and M. S. Kallos. 2006. Expansion of undifferentiated murine embryonic stem cells as aggregates in suspension culture bioreactors. *Tissue Eng* 12 (11):3233–3245. doi:10.1089/ten.2006.12.3233.

Crowe, J. J., S. C. Grant, T. M. Logan, and T. Ma. 2011. A magnetic resonance-compatible perfusion bioreactor system for three-dimensional human mesenchymal stem cell construct development. *Chem Eng Sci* 66 (18):4138–4147. doi:10.1016/j.ces.2011.05.046.

Csaszar, E., D. C. Kirouac, M. Yu, W. J. Wang, W. L. Qiao, M. P. Cooke, A. E. Boitano, C. Ito, and P. W. Zandstra. 2012. Rapid expansion of human hematopoietic stem cells by automated control of inhibitory feedback signaling. *Cell Stem Cell* 10 (2):218–229. doi:10.1016/j.stem.2012.01.003.

Dalili, M., G. D. Sayles, and D. F. Ollis. 1990. Glutamine-limited batch hybridoma growth and antibody production: Experiment and model. *Biotechnol Bioeng* 36 (1):74–82. doi:10.1002/bit.260360110.

de Assis, A. J., and R. Maciel. 2000. Soft sensors development for on-line bioreactor state estimation. *Comp Chem Eng* 24 (2–7):1099–1103. doi:10.1016/S0098-1354(00)00489-0.

de Jong, B., V. Siewers, and J. Nielsen. 2012. Systems biology of yeast: Enabling technology for development of cell factories for production of advanced biofuels. *Cur Opin Biotechnol* 23 (4):624–630. doi:10.1016/j.copbio.2011.11.021.

de Peppo, G. M., I. Marcos-Campos, D. J. Kahler, D. Alsalman, L. Shang, G. Vunjak-Novakovic, and D. Marolt. 2013. Engineering bone tissue substitutes from human induced pluripotent stem cells. *Proc Natl Acad Sci* 110 (21):8680–8685. doi:10.1073/pnas.1301190110.

de Soure, A. M., A. Fernandes-Platzgummer, F. Moreira, C. Lilaia, S.-H. Liu, C.-P. Ku, Y.-F. Huang, W. Milligan, J. M. S. Cabral, and C. L. da Silva. 2016. Integrated culture platform based on a human platelet lysate supplement for the isolation and scalable manufacturing of umbilical cord matrix-derived mesenchymal stem/stromal cells. *J Tissue Eng Reg Med*. doi:10.1002/term.2200.

Dennis, J. E., K. Esterly, A. Awadallah, C. R. Parrish, G. M. Poynter, and K. L. Goltry. 2007. Clinical-scale expansion of a mixed population of bone marrow-derived stem and progenitor cells for potential use in bone tissue regeneration. *Stem Cells* 25 (10):2575–2582. doi:10.1634/stemcells.2007-0204.

Di Corato, R., F. Gazeau, C. Le Visage, D. Fayol, P. Levitz, F. Lux, D. Letourneur, N. Luciani, O. Tillement, and C. Wilhelm. 2013. High-resolution cellular MRI: Gadolinium and iron oxide nanoparticles for in-depth dual-cell imaging of engineered tissue constructs. *Acs Nano* 7 (9):7500–7512. doi:10.1021/nn401095p.

Dodson, B. P., and A. D. Levine. 2015. Challenges in the translation and commercialization of cell therapies. *BMC Biotechnol* 15:70. doi:10.1186/s12896-015-0190-4.

Doroski, D. M., K. S. Brink, and J. S. Temenoff. 2007. Techniques for biological characterization of tissue-engineered tendon and ligament. *Biomaterials* 28 (2):187–202. doi:10.1016/j.biomaterials.2006.08.040.

dos Santos, F., A. Campbell, A. Fernandes-Platzgummer, P. Z. Andrade, J. M. Gimble, Y. Wen, S. Boucher, M. C. Vemuri, C. L. da Silva, and J. M. S. Cabral. 2014. A xenogeneic-free bioreactor system for the clinical-scale expansion of human mesenchymal stem/stromal cells. *Biotechnol Bioeng* 111 (6):1116–1127. doi:10.1002/bit.25187.

El-Schish, Z., A. Mölder, M. Sebesta, L. Gisselsson, K. Alm, and A. Gjörloff Wingren. 2010. Digital holographic microscopy—Innovative and non-destructive analysis of living cells. In *Microscopy: Science, Technology, Applications and Education*, A. Mendez-Vilas, and J. Diaz Álvarez (Eds.), pp. 1055–1062. Badajoz, Spain: Formatex Research Center.

Feng, Y., X. H. Jin, G. Dai, J. Liu, J. R. Chen, and L. Yang. 2011. *In vitro* targeted magnetic delivery and tracking of superparamagnetic iron oxide particles labeled stem cells for articular cartilage defect repair. *J Huazhong Uni Sci Technol-Med Sci* 31 (2):204–209. doi:10.1007/s11596-011-0253-2.

Fok, E. Y. L., and P. W. Zandstra. 2005. Shear-controlled single-step mouse embryonic stem cell expansion and embryoid body–based differentiation. *Stem Cells* 23 (9):1333–1342. doi:10.1634/stemcells.2005-0112.

Gerecht-Nir, S., S. Cohen, and J. Itskovitz-Eldor. 2004. Bioreactor cultivation enhances the efficiency of human embryoid body (hEB) formation and differentiation. *Biotechnol Bioeng* 86 (5):493–502. doi:10.1002/bit.20045.

Glindkamp, A., D. Riechers, C. Rehbock, B. Hitzmann, T. Scheper, K. F. Reardon. 2009. Sensors in disposable bioreactors status and trends. *Adv Biochem Eng Biotech* 115.

Gomes-Alves, P., M. Serra, C. Brito, C. P. Ricardo, R. Cunha, M. F. Sousa, B. Sanchez, A et al. 2016. *In vitro* expansion of human cardiac progenitor cells: Exploring "omics tools for characterization of cell-based allogeneic products." *Trans Res* 171:96–110. doi:10.1016/j.trsl.2016.02.001.

Gottwald, E., T. Kleintschek, S. Giselbrecht, R. Truckenmuller, B. Altmann, M. Worgull, J. Dopfert, L. Schad, and M. Heilmann. 2013. Characterization of a chip-based bioreactor for three-dimensional cell cultivation via magnetic resonance imaging. *Zeitschrift Fur Medizinische Physik* 23 (2):102–110. doi:10.1016/j.zemedi.2013.01.003.

Grayson, W. L., F. Zhao, B. Bunnell, and T. Ma. 2007. Hypoxia enhances proliferation and tissue formation of human mesenchymal stem cells. *Biochem Biophys Res Commun* 358 (3):948–953. doi:10.1016/j.bbrc.2007.05.054.

Greenbaum, A., W. Luo, T. W. Su, Z. Gorocs, L. Xue, S. O. Isikman, A. F. Coskun, O. Mudanyali, and A. Ozcan. 2012. Imaging without lenses: Achievements and remaining challenges of wide-field on-chip microscopy. *Nat Meth* 9 (9):889–895. doi:10.1038/nmeth.2114.

Gupta, B. D., and S. Sharma. 1998. A long-range fiber optic pH sensor prepared by dye doped sol-gel immobilization technique. *Optics Commun* 154 (5–6):282–284. doi:10.1016/S0030-4018(98)00321-6.

Gupta, P. A., X. Ge, Y. Kostov, and G. Rao. 2014. A completely noninvasive method of dissolved oxygen monitoring in disposable small-scale cell culture vessels based on diffusion through permeable vessel walls. *Biotechnol Prog* 30 (1):172–177. doi:10.1002/btpr.1838.

Gupta, P., M. Z. Ismadi, P. J. Verma, A. Fouras, S. Jadhav, J. Bellare, and K. Hourigan. 2016. Optimization of agitation speed in spinner flask for microcarrier structural integrity and expansion of induced pluripotent stem cells. *Cytotechnology* 68 (1):45–59. doi:10.1007/s10616-014-9750-z.

Guyot, Y., I. Papantoniou, F. P. Luyten, and L. Geris. 2016. Coupling curvature-dependent and shear stress-stimulated neotissue growth in dynamic bioreactor cultures: A 3D computational model of a complete scaffold. *Biomech Mod Mechanobiol* 15 (1):169–180. doi:10.1007/s10237-015-0753-2.

Harms, P., Y. Kostov, and G. Rao. 2002. Bioprocess monitoring. *Curr Opin Biotechnol* 13 (2):124–127. doi:10.1016/S0958-1669(02)00295-1.

Heathman, T. R. J., A. Stolzing, C. Fabian, Q. A. Rafiq, K. Coopman, A. W. Nienow, B. Kara, and C. J. Hewitt. 2016. Scalability and process transfer of mesenchymal stromal cell production from monolayer to microcarrier culture using human platelet lysate. *Cytotherapy* 18 (4):523–535. doi:10.1016/j.jcyt.2016.01.007.

Henkel, S., and S. Beutel. 2013. Determination of pH Value in biotechnology. *Chemie Ingenieur Technik* 85 (6):872–885. doi:10.1002/cite.201200099.

Henriques, J. G., S. Buziol, E. Stocker, A. Voogd, and J. C. Menezes. 2009. Monitoring mammalian cell cultivations for monoclonal antibody production using near-infrared spectroscopy. *Opt Sen Sys Biotechnol* 116:73–97. doi:10.1007/10_2009_11.

Hourd, P., A. Chandra, D. Alvey, P. Ginty, M. McCall, E. Ratcliffe, E. Rayment, and D. J. Williams. 2014. Qualification of academic facilities for small-scale automated manufacture of autologous cell-based products. *Regen Med* 9 (6):799–815. doi:10.2217/rme.14.47.

Housler, G. J., T. Miki, E. Schmelzer, C. Pekor, X. Zhang, L. Kang, V. Voskinarian-Berse, S. Abbot, K. Zeilinger, and J. C. Gerlach. 2012. Compartmental hollow fiber capillary membrane-based bioreactor technology for in vitro studies on red blood cell lineage direction of hematopoietic stem cells. *Tissue Eng Part C Methods* 18 (2):133–142. doi:10.1089/ten. TEC.2011.0305.

Hsu, S. H., C. T. Chen, and Y. H. Wei. 2013. Inhibitory effects of hypoxia on metabolic switch and osteogenic differentiation of human mesenchymal stem cells. *Stem Cells* 31 (12):2779–2788. doi:10.1002/stem.1441.

Jaccard, N., N. Szita, and L. D. Griffin. 2015. Segmentation of phase contrast microscopy images based on multi-scale local basic image features histograms. *Comput Methods Biomech Biomed Eng Imaging Vis* 5:359–367.

Joeris, K., J. G. Frerichs, K. Konstantinov, and T. Scheper. 2002. In-situ microscopy: Online process monitoring of mammalian cell cultures. *Cytotechnology* 38 (1–2):129–134. doi:10.1023/A:1021170502775.

Jung, S., K. M. Panchalingam, R. D. Wuerth, L. Rosenberg, and L. A. Behie. 2012. Large-scale production of human mesenchymal stem cells for clinical applications. *Biotechnol Appl Biochem* 59 (2):106–120. doi:10.1002/bab.1006.

Kadlec, P., B. Gabrys, and S. Strandt. 2009. Data-driven soft sensors in the process industry. *Comp Chem Eng* 33 (4):795–814. doi:10.1016/j.compchemeng.2008.12.012.

Kim, S. B., H. Bae, K. I. Koo, M. R. Dokmeci, A. Ozcan, and A. Khademhosseini. 2012. Lens-free imaging for biological applications. *Jala* 17 (1):43–49. doi:10.1177/2211068211426695.

Kotecha, M., D. Klatt, and R. L. Magin. 2013. Monitoring cartilage tissue engineering using magnetic resonance spectroscopy, imaging, and elastography. *Tissue Eng Part B-Rev* 19 (6):470–484. doi:10.1089/ten.teb.2012.0755.

Kovarova-Kovar, K., S. Gehlen, A. Kunze, T. Keller, R. von Daniken, M. Kolb, and A. P. G. M. van Loon. 2000. Application of model-predictive control based on artificial neural networks to optimize the fed-batch process for riboflavin production. *J Biotechnol* 79 (1):39–52. doi:10.1016/S0168-1656(00)00211-X.

Lalande, C., S. Miraux, S. M. Derkaoui, S. Mornet, R. Bareille, J. C. Fricain, J. M. Franconi et al. 2011. Magnetic resonance imaging tracking of human adipose derived stromal cells within three-dimensional scaffolds for bone tissue engineering. *Europ Cells Mat* 21:341–354. doi:10.22203/eCM.v021a25.

Lambrechts, T., I. Papantoniou, M. Sonnaert, J. Schrooten, and J. M. Aerts. 2014. Model-based cell number quantification using online single-oxygen sensor data for tissue engineering perfusion bioreactors. *Biotechnol Bioeng* 111 (10):1982–1992. doi:10.1002/bit.25274.

Lambrechts, T., I. Papantoniou, S. Viazzi, T. Bovy, J. Schrooten, F. P. Luyten, and J. M. Aerts. 2016. Evaluation of a monitored multiplate bioreactor for large-scale expansion of human periosteum derived stem cells for bone tissue engineering applications. *Biochem Eng J* 108:58–68. doi:10.1016/j.bej.2015.07.015.

Lambrechts, T., I. Papantoniou, B. Rice, J. Schrooten, F. P. Luyten, and J.-M. Aerts. 2016. Large-scale progenitor cell expansion for multiple donors in a monitored hollow fibre bioreactor. *Cytotherapy* 18 (9):1219–1233. doi:10.1016/j.jcyt.2016.05.013.

Leferink, A. M., C. A. van Blitterswijk, and L. Moroni. 2016. Methods of monitoring cell fate and tissue growth in three-dimensional scaffold-based strategies for *in vitro* tissue engineering. *Tissue Eng Part B-Rev* 22 (4):265–283. doi:10.1089/ten.teb.2015.0340.

Leferink, A. M., Y.-C. Chng, C. A. van Blitterswijk, and L. Moroni. 2015. Distribution and viability of fetal and adult human bone marrow stromal cells in a biaxial rotating vessel bioreactor after seeding on polymeric 3D additive manufactured scaffolds. *Front Bioeng Biotechnol* 3 (169). doi:10.3389/fbioe.2015.00169.

Lenas, P., T. Kitade, H. Watanabe, H. Honda, and T. Kobayashi. 1997. Adaptive fuzzy control of nutrients concentration in fed-batch culture of mammalian cells. *Cytotechnology* 25 (1):9–15. doi:10.1023/a:1007950002663.

Lewis, A. M., N. R. Abu-Absi, M. C. Borys, and Z. J. Li. 2016. The use of "Omics technology to rationally improve industrial mammalian cell line performance." *Biotechnol Bioeng* 113 (1):26–38. doi:10.1002/bit.25673.

Li, Y. M., T. Schilling, P. Benisch, S. Zeck, J. Meissner-Weigl, D. Schneider, C. Limbert et al. 2007. Effects of high glucose on mesenchymal stem cell proliferation and differentiation. *Biochem Biophys Res Commun* 363 (1):209–215. doi:10.1016/j.bbrc.2007.08.161.

Lin, H. J., T. J. O'Shaughnessy, J. Kelly, and W. Ma. 2004. Neural stem cell differentiation in a cell–collagen–bioreactor culture system. *Devel Brain Res* 153 (2):163–173. doi:10.1016/j.devbrainres.2004.08.010.

Lipsitz, Y. Y., N. E. Timmins, and P. W. Zandstra. 2016. Quality cell therapy manufacturing by design. *Nat Biotechnol* 34 (4):393–400. doi:10.1038/nbt.3525.

Liu, Y., T. Liu, X. Fan, X. Ma, and Z. Cui. 2006. Ex vivo expansion of hematopoietic stem cells derived from umbilical cord blood in rotating wall vessel. *J Biotechnol* 124 (3):592–601. doi:10.1016/j.jbiotec.2006.01.020.

Mandenius, C. F., and R. Gustavsson. 2015. Mini-review: Soft sensors as means for PAT in the manufacture of bio-therapeutics. *J Chem Technol Biotechnol* 90 (2):215–227. doi:10.1002/jctb.4477.

Mann, C. J., L. F. Yu, C. M. Lo, and M. K. Kim. 2005. High-resolution quantitative phase-contrast microscopy by digital holography. *Opt Express* 13 (22):8693–8698. doi:10.1364/Opex.13.008693.

Mather, M. L., S. P. Morgan, and J. A. Crowe. 2007. Meeting the needs of monitoring in tissue engineering. *Reg Med* 2 (2):145–160. doi:10.2217/174607451.2.2.145.

Merten, O.-W., H. Keller, L. Cabanié, M. Leno, and M. Hardefelt. 1990. Batch production and secretion kinetics of hybridomas: Pulse-chase experiments. *Cytotechnology* 4 (1):77–89. doi:10.1007/bf00148813.

Mohyeldin, A., T. Garzon-Muvdi, and A. Quinones-Hinojosa. 2010. Oxygen in stem cell biology: A critical component of the stem cell niche. *Cell Stem Cell* 7 (2):150–161. doi:10.1016/j.stem.2010.07.007.

Molder, A., M. Sebesta, M. Gustafsson, L. Gisselson, A. G. Wingren, and K. Alm. 2008. Non-invasive, label-free cell counting and quantitative analysis of adherent cells using digital holography. *J Micro* 232 (2):240–247.

Monfoulet, L. E., P. Becquart, D. Marchat, K. Vandamme, M. Bourguignon, E. Pacard, V. Viateau, H. Petite, and D. Logeart-Avramoglou. 2014. The pH in the microenvironment of human mesenchymal stem cells is a critical factor for optimal osteogenesis in tissue-engineered constructs. *Tissue Eng Part A* 20 (13–14):1827–1840. doi:10.1089/ten.tea.2013.0500.

Mulchandani, A., and A. S. Bassi. 1995. Principles and applications of biosensors for bioprocess monitoring and control. *Crit Rev Biotechnol* 15 (2):105–124. doi:10.3109/07388559509147402.

Odeleye, A. O. O., S. Castillo-Avila, M. Boon, H. Martin, and K. Coopman. 2017. Development of an optical system for the non-invasive tracking of stem cell growth on microcarriers. *Biotechnol Bioeng* 114 (9):2032–2042. doi:10.1002/bit.26328.

Palsson, B. O., S. H. Paek, R. M. Schwartz, M. Palsson, G. M. Lee, S. Silver, and S. G. Emerson. 1993. Expansion of human bone marrow progenitor cells in a high cell density continuous perfusion system. *Biotechnology (N Y)* 11 (3):368–372.

Papantoniou, I., M. Sonnaert, L. Geris, F. P. Luyten, J. Schrooten, and G. Kerckhofs. 2014. Three-dimensional characterization of tissue-engineered constructs by contrast-enhanced nanofocus computed tomography. *Tissue Eng Part C-Meth* 20 (3):177–187. doi:10.1089/ten.tec.2013.0041.

Papantoniou, I., Y. Guyot, M. Sonnaert, G. Kerckhofs, F. P. Luyten, L. Geris, and J. Schrooten. 2014. Spatial optimization in perfusion bioreactors improves bone tissue-engineered construct quality attributes. *Biotechnol Bioeng* 111 (12):2560–2570. doi:10.1002/bit.25303.

Potier, E., E. Ferreira, R. Andriamanalijaona, J. P. Pujol, K. Oudina, D. Logeart-Avramoglou, and H. Petite. 2007. Hypoxia affects mesenchymal stromal cell osteogenic differentiation and angiogenic factor expression. *Bone* 40 (4):1078–1087. doi:10.1016/j.bone.2006.11.024.

Rafiq, Q. A., M. P. Hanga, T. R. J. Heathman, K. Coopman, A. W. Nienow, D. J. Williams, and C. J. Hewitt. 2017. Process development of human multipotent stromal cell microcarrier culture using an automated high-throughput microbioreactor. *Biotechnol Bioeng* doi:10.1002/bit.26359.

Rafiq, Q. A., K. M. Brosnan, K. Coopman, A. W. Nienow, and C. J. Hewitt. 2013. Culture of human mesenchymal stem cells on microcarriers in a 5 l stirred-tank bioreactor. *Biotechnol Lett* 35 (8):1233–1245. doi:10.1007/s10529-013-1211-9.

Ramaswamy, S., T. J. Cutright, and H. K. Qammar. 2005. Control of a continuous bioreactor using model predictive control. *Proc Biochem* 40 (8):2763–2770. doi:10.1016/j.procbio.2004.12.019.

Rappaz, B., P. Marquet, E. Cuche, Y. Emery, C. Depeursinge, and P. J. Magistretti. 2005. Measurement of the integral refractive index and dynamic cell morphometry of living cells with digital holographic microscopy. *Opt Express* 13 (23):9361–9373. doi:10.1364/Opex.13.009361.

Rhee, I. L., A. Ritzka, and T. Scheper. 2004. On-line monitoring and control of substrate concentrations in biological processes by flow injection analysis systems. *Biotechnol Bioproc Eng* 9 (3):156–165.

Rhiel, M., P. Ducommun, I. Bolzonella, I. Marison, and U. von Stockar. 2002. Real-time in situ monitoring of freely suspended and immobilized cell cultures based on mid-infrared spectroscopic measurements. *Biotechnol Bioeng* 77 (2):174–185.

Rudolph, G., P. Lindner, A. Gierse, A. Bluma, G. Martinez, B. Hitzmann, and T. Scheper. 2008. Online monitoring of microcarrier based fibroblast cultivations with in situ microscopy. *Biotechnol Bioeng* 99 (1):136–145. doi:10.1002/bit.21523.

Sabatino, M. A., R. Santoro, S. Gueven, C. Jaquiery, D. J. Wendt, I. Martin, M. Moretti, and A. Barbero. 2015. Cartilage graft engineering by co-culturing primary human articular chondrocytes with human bone marrow stromal cells. *J Tissue Eng Regen Med* 9 (12):1394–1403. doi:10.1002/term.1661.

Sart, S., Y.-J. Schneider, and S. N. Agathos. 2010. Influence of culture parameters on ear mesenchymal stem cells expanded on microcarriers. *J Biotechnol* 150 (1):149–160. doi:10.1016/j.jbiotec.2010.08.003.

Schenk, J., I. W. Marison, and U. von Stockarl. 2008. pH prediction and control in bioprocesses using mid-infrared spectroscopy. *Biotechnol Bioeng* 100 (1):82–93. doi:10.1002/bit.21719.

Schop, D., F. W. Janssen, E. Borgart, J. D. de Bruijn, and R. van Dijkhuizen-Radersma. 2008. Expansion of mesenchymal stem cells using a microcarrier-based cultivation system: Growth and metabolism. *J Tissue Eng Regen Med* 2 (2–3):126–135. doi:10.1002/term.73.

Silva, M. M., A. F. Rodrigues, C. Correia, M. F. Q. Sousa, C. Brito, A. S. Coroadinha, M. Serra, and P. M. Alves. 2015. Robust expansion of human pluripotent stem cells: Integration of bioprocess design with transcriptomic and metabolomic characterization. *Stem Cells Trans Med* 4 (7):731–742. doi:10.5966/sctm.2014-0270.

Simaria, A. S., S. Hassan, H. Varadaraju, J. Rowley, K. Warren, P. Vanek, and S. S. Farid. 2014. Allogeneic cell therapy bioprocess economics and optimization: Single-use cell expansion technologies. *Biotechnol Bioeng* 111 (1):69–83. doi:10.1002/bit.25008.

Simmons, A. D., C. Williams III, A. Degoix, and V. I. Sikavitsas. 2017. Sensing metabolites for the monitoring of tissue engineered construct cellularity in perfusion bioreactors. *Biosens Bioelectron* 90:443–449. doi:10.1016/j.bios.2016.09.094.

Slade, P. G., M. Hajivandi, C. M. Bartel, and S. F. Gorfien. 2012. Identifying the CHO secretome using mucin-type O-linked glycosylation and click-chemistry. *J Prot Res* 11 (12):6175–6186. doi:10.1021/pr300810f.

Song, K., Z. Yang, T. Liu, W. Zhi, X. Li, L. Deng, Z. Cui, and X. Ma. 2006. Fabrication and detection of tissue-engineered bones with bio-derived scaffolds in a rotating bioreactor. *Biotechnol Appl Biochem* 45 (2):65–74. doi:10.1042/BA20060045.

Song, S. P., L. H. Wang, J. Li, J. L. Zhao, and C. H. Fan. 2008. Aptamer-based biosensors. *Trac-Trends Anal Chem* 27 (2):108–117. doi:10.1016/j.trac.2007.12.004.

Sonnaert, M., I. Papantoniou, V. Bloemen, G. Kerckhofs, F. P. Luyten, and J. Schrooten. 2017. Human periosteal-derived cell expansion in a perfusion bioreactor system: Proliferation, differentiation and extracellular matrix formation. *J Tissue Eng Regen Med* 11 (2):519–530. doi:10.1002/term.1951.

Sonnaert, M., G. Kerckhofs, I. Papantoniou, S. Van Vlierberghe, V. Boterberg, P. Dubruel, F. P. Luyten et al. 2015. Multifactorial optimization of contrast-enhanced nanofocus computed tomography for quantitative analysis of neo-tissue formation in tissue engineering constructs. *PLoS One* 10 (6):e0130227. doi:10.1371/journal.pone.0130227.

Tao, S. C., Y. S. Gao, H. Y. Zhu, J. H. Yin, Y. X. Chen, Y. L. Zhang, S. C. Guo, and C. Q. Zhang. 2016. Decreased extracellular pH inhibits osteogenesis through proton-sensing GPR4-mediated suppression of yes-associated protein. *Sci Rep* 6. doi:10.1038/Srep26835.

Thevenot, D. R., K. Toth, R. A. Durst, and G. S. Wilson. 2001. Electrochemical biosensors: Recommended definitions and classification. *Anal Lett* 34 (5):635–659. doi:10.1081/AI-100103209.

Vernardis, S. I., C. T. Goudar, and M. I. Klapa. 2013. Metabolic profiling reveals that time related physiological changes in mammalian cell perfusion cultures are bioreactor scale independent. *Metab Eng* 19:1–9. doi:10.1016/j.ymben.2013.04.005.

Vielreicher, M., S. Schurmann, R. Detsch, M. A. Schmidt, A. Buttgereit, A. Boccaccini, and O. Friedrich. 2013. Taking a deep look: Modern microscopy technologies to optimize the design and functionality of biocompatible scaffolds for tissue engineering in regenerative medicine. *J R Soc Interface* 10 (86). doi:10.1098/rsif.2013.0263.

Voronov, R. S., S. B. VanGordon, R. L. Shambaugh, D. V. Papavassiliou, and V. I. Sikavitsas. 2013. 3D tissue-engineered construct analysis via conventional high-resolution micro-computed tomography without X-ray contrast. *Tissue Eng Part C-Meth* 19 (5):327–335. doi:10.1089/ten.tec.2011.0612.

Vucic, E. A., K. L. Thu, K. Robison, L. A. Rybaczyk, R. Chari, C. E. Alvarez, and W. L. Lam. 2012. Translating cancer "omics" to improved outcomes. *Genome Res* 22 (2):188–195. doi:10.1101/gr.124354.111.

Wei, N., J. You, K. Friehs, E. Flaschel, and T. W. Nattkemper. 2007. An in situ probe for on-line monitoring of cell density and viability on the basis of dark field microscopy in conjunction with image processing and supervised machine learning. *Biotechnol Bioeng* 97 (6):1489–1500. doi:10.1002/bit.21368.

Wendt, D., S. A. Riboldi, M. Cioffi, and I. Martin. 2009. Bioreactors in tissue engineering: Scientific challenges and clinical perspectives. *Bioreact Sys Tissue Eng* 112:1–27. doi:10.1007/10_2008_1.

Williams, D. J., R. Archer, P. Archibald, I. Bantounas, R. Baptista, R. Barker, J. Barry et al. 2016. Comparability: Manufacturing, characterization and controls, report of a *UK Regenerative Medicine Platform Pluripotent Stem Cell Platform Workshop*, Trinity Hall, Cambridge, September 14–15, 2015. *Regen Med* 11 (5):483–492. doi:10.2217/rme-2016-0053.

Wolfe, R. P., and T. Ahsan. 2013. Shear stress during early embryonic stem cell differentiation promotes hematopoietic and endothelial phenotypes. *Biotechnol Bioeng* 110 (4):1231–1242. doi:10.1002/bit.24782.

Xu, P., C. Clark, T. Ryder, C. Sparks, J. Zhou, M. Wang, R. Russell, and C. Scott. 2017. Characterization of TAP Ambr 250 disposable bioreactors, as a reliable scale-down model for biologics process development. *Biotechnol Prog* 33 (2):478–489. doi:10.1002/btpr.2417.

Xu, X., H. Nagarajan, N. E. Lewis, S. K. Pan, Z. M. Cai, X. Liu, W. B. Chen et al. The genomic sequence of the Chinese hamster ovary (CHO)-K1 cell line. *Nat Biotechnol* 29 (8):U735–U131. doi:10.1038/nbt.1932.

Yang, P. H., X. Chen, S. Kaushal, E. A. Reece, and P. X. Yang. 2016. High glucose suppresses embryonic stem cell differentiation into cardiomyocytes High glucose inhibits ES cell cardiogenesis. *Stem Cell Res Ther* 7:187. doi:10.1186/s13287-016-0446-5.

Zandstra, P. W., C. J. Eaves, and J. M. Piret. 1994. Expansion of hematopoietic progenitor cell populations in stirred suspension bioreactors of normal human bone marrow cells. *Nat Biotech* 12 (9):909–914.

Zhang, B., N. Liu, H. G. Shi, H. Wu, Y. X. Gao, H. X. He, B. Gu, and H. C. Liu. 2016. High glucose microenvironments inhibit the proliferation and migration of bone mesenchymal stem cells by activating GSK3 beta. *J Bone Miner Metab* 34 (2):140–150. doi:10.1007/s00774-015-0662-6.

Zhao, L., H.-Y. Fu, W. Zhou, and W.-S. Hu. 2015. Advances in process monitoring tools for cell culture bioprocesses. *Eng Life Sci* 15 (5):459–468. doi:10.1002/elsc.201500006.

Zhu, N., D. Chapman, D. Cooper, D. J. Schreyer, and X. B. Chen. 2011. X-ray diffraction enhanced imaging as a novel method to visualize low-density scaffolds in soft tissue engineering. *Tissue Eng Part C-Meth* 17 (11):1071–1080. doi:10.1089/ten.tec.2011.0102.

Zweigerdt, R., R. Olmer, H. Singh, A. Haverich, and U. Martin. 2011. Scalable expansion of human pluripotent stem cells in suspension culture. *Nat Protocols* 6 (5):689–700.

10 Bioreactors for Tendon Tissue Engineering
Challenging Mechanical Demands Towards Tendon Regeneration

*Ana I. Gonçalves, Dominika Berdecka,
Márcia T. Rodrigues, Rui L. Reis,
and Manuela E. Gomes*

CONTENTS

10.1 INTRODUCTION

Tendons are unique connective tissues with the vital role to store and return elastic energy, resist damage, provide mechanical feedback and amplify or attenuate muscle power, and transmit forces from muscle to bone. Although tendon relevance in joint biomechanics and overall human body is often misunderstood and disregarded to other tissues, such as bone or cartilage, recent growing interest in tendon mechanical properties has highlighted potential studies working towards improved therapeutic strategies in the orthopedic field.

The term "tendon" comes from the Latin word, *tendere*, meaning to stretch (Liu et al. 2011). Despite the high tensile strength, tendons have limited intrinsic healing capabilities. It is estimated that approximately 50% of all musculoskeletal injuries are tendon-related (Praemer et al. 1992). Upon injury, the tendon undergoes degeneration and morpho-histological misalignment of the collagen fibers. Ultimately, severe damage will result in pain and disability. Thus, a major challenge in tissue engineering (TE) and regenerative medicine is to recreate the tendon niche and replicate biomechanical forces involved in tendon functionality including stretching, loading, compression, and torsion.

One potential approach to artificially generating the biomechanical demands of tendons is using complex advanced systems such as bioreactors. Bioreactors are designed considering the specific parameters of the replacing tissue or organ, especially in what concerns to tissue biomechanics and maintenance of a desired phenotype prior to implantation. Since the birth of TE in the early 1990s, bioreactors have been primarily used for studying basic pathways, expand and grow tissue/organ substitutes, maintain *ex vivo* organ vitality and priming therapeutic cells before implantation. The use of suitable biomechanical and biophysical environments in which cells could synthesize a functional matrix results in a closer mimicry of tendon tissue, leading to maturation of cell-laden constructs prior to implantation *in vivo*. This way, the utilization of bioreactors as *in vitro* models is expected to minimize the number of animal experiments as the implantation step only occurs when the morphological, biological and biomechanical properties of the engineered construct match those of the natural tissue.

In the tendon scenario, bioreactors have been used to culture tendon engineered substitutes and to investigate suitable *in vitro* conditions for establishing benchmarks and protocols for effective cell programming toward the tenogenic phenotype. Thus, bioreactors are promising tools for developing and culturing *in vitro* generated tendon substitutes, as potential alternatives to pharmacological therapies and to fulfill the current need for tissue substitutes to treat tendinopathies.

This chapter will outline state of the art TE strategies on cell culture or cell laden 3D matrices using mechanically active environments provided by bioreactor systems for tendon regeneration as a potential means to obtain functional tendon substitutes. For a better understanding on the performance of these systems and their role in strategies applied to tendon TE, the intrinsic properties and requirements of tendons will be explored with an emphasis on the role of biomechanical stimulation in tendon development and maturation, as well as the biomechanics-tissue functionality relationship.

10.2 TENDON STRUCTURE-FUNCTION AND MECHANOBIOLOGY BEHAVIOR

10.2.1 ROLE OF MECHANICAL STIMULATION IN TENDON DEVELOPMENT AND FUNCTIONALITY

Tendons are specialized connective tissues that serve to transmit forces between muscles and bones, and thus allow body motion. Their crucial role in musculoskeletal functionality implies distinct mechanical properties, which are assured by tissue-specific structure and molecular composition, namely highly organized collagen fibers arranged parallel to the tendon axis. The smallest structural unit of tendon is fibril composed of collagen molecules assembled in a quarter-staggered D-periodic pattern (Kastelic et al. 1978). Fibrils form fibers, which group together to form fiber bundles or fascicles, enveloped by thin layer of connective tissue called endotenon. Fascicles bundles are, in turn, enclosed by another layer of loose connective tissue sheath, the epitenon, that provides vasculature, lymphatics, and innervation to the tendon unit. Tendons may be eventually enveloped by paratenon sheath of paratenon that serves to reduce friction with adjacent tissues, thus enabling free tendon movement against its surroundings (Kastelic et al. 1978).

Tendon fibroblasts tenocytes are found longitudinally aligned in the rows between collagen fibers, and are mainly responsible for the synthesis and maintenance of the extracellular matrix (ECM). Interestingly, a population of cells with universal stem cell characteristics, named *tendon stem/progenitor cells* (TSPCs), has been identified in both human and murine tendons (Bi et al. 2007).

The major component of tendons is collagen type I that represents approximately 95% of the total collagen content and around 60% of the tissue dry mass (Riley et al. 1994). Tendon-specific mechanical integrity and function is acquired through a multistep process of collagen fibrillogenesis during tendon development (Birk and Zycband 1994, Silver et al. 2003, Zhang et al. 2004). In the first stage, collagen molecules assemble in the extracellular space to form immature fibril intermediates. Fibril intermediates associate subsequently end-to-end forming longer and mechanically functional fibrils. The linear growth is then followed by a lateral growth step, where fibrils associate laterally generating fully mature fibrils with larger diameters (Zhang et al. 2004). The process of fibril assembly is regulated by heterotypic interactions between fibril-forming and fibril-associated collagens, and fibril-associated proteoglycans. For example, the interaction between two fibrillar collagen type I and type III plays a role in initial fibril assembly and control of fibril diameters (Banos et al. 2008). The ratio of collagen type III to collagen type I exerts spatial and temporal variations throughout tendon development. On the other hand, in the mature tendon, collagen type III is present mainly in the endotenon and epitenon. Similarly, collagen type V may assemble with collagen type I and has been implicated in fibril nucleation and diameter regulation (Birk et al. 1990, Wenstrup et al. 2004). Type XII and type XIV collagens, which represent Fibril-Associated Collagens with Interrupted Triple-helices (FACITs), are localized near the surface of fibrils and may contribute to fibrillogenesis regulation by providing molecular bridges between collagen fibrils and other components of the ECM. Although their role in fibril assembly is not well understood yet, it was hypothesized that collagen type XII may stabilize fibril structure during tendon development (Chiquet 1999), while type XIV limits fibril diameter (Young et al. 2002). Beside the collagen class, molecules that belong to the family of small-leucine-rich proteoglycans (SLRPs), such as decorin, biglycan, fibromodulin, and lumican, are believed to actively regulate tendon fibrillogenesis, since their targeted disruption in mouse models lead to abnormal fibril phenotypes (Vogel and Heinegård 1985, Yoon and Halper 2005, Subramanian and Schilling 2015). Interestingly, biglycan and fibromodulin have been recognized as critical components of tendon stem cell niche regulating TSPCs differentiation and tendon formation *in vivo* (Bi et al. 2007).

Noteworthy, the cellular composition and collagen organization are not uniform along the tendon length and demonstrate regional differences in the myotendinous junction, which is the interface between tendon and muscle and in the tendon to bone attachment site, called enthesis (Thomopoulos et al. 2003). These molecular and cellular variations are translated in different mechanical properties of specific tendon regions that reflect nonhomogeneous mechanical loading requirements in different anatomical sites (Genin et al. 2009).

The molecular mechanism governing the synthesis and spatial organization of collagen in developing tendons has not been fully elucidated. Since collagen is the main component of various connective tissues, it is impossible to trace tendon development by mapping its expression. In fact, no marker unique for tendons has been identified to date. The basic helix-loop-helix transcription factor scleraxis (Scx) has been described as an early tendon marker, whereas a type II transmembrane glycoprotein tenomodulin (Tnmd) is considered a late tendon marker (Shukunami et al. 2006). Though not specific to tendon, two other transcription factors are involved in tendon development, the homeobox protein Mohawk (*Mkx*) and a member of zinc finger transcription factor family, early growth response factor 1 (Egr1). *Mkx*-null mice presented a wavy-tail phenotype and hypoplastic and less vibrant tendons throughout the body with reduced fibril diameters and down-regulation of type I collagen expression, when compared to the wild type (WT) counterparts. In addition to disruption of postnatal collagen fibrillogenesis, mutant mice exhibited abnormal tendon sheaths (Ito et al. 2010, Liu, Watson et al. 2010). Similarly, Egr1 has been shown to positively regulate collagen transcription in postnatal tendons. Egr1$^{-/-}$ mutant mice demonstrated a deficiency in expression of *Scx*, *Col1a1* and *Col1a2* genes, reduced fibril diameter and packing density, resulting in mechanically weaker tendons, compared with their WT littermates (Lejard et al. 2011, Guerquin et al. 2013). Interestingly, ectopic expression of either *Mkx* or Egr1 promoted tenogenic differentiation of mesenchymal stem/stromal cells (MSCs) via activation of TGF-β (Transforming growth factor beta) signaling pathway (Guerquin et al. 2013, Liu et al. 2014).

Tendon embryogenesis has not been fully investigated, and most of the data comes from developmental studies in invertebrates and chick and mouse models. The vertebrate tendons originate from mesoderm or mesectoderm, more specifically, the craniofacial tendons are derived from neural crest cells, axial tendons originate from syndetome, whereas limb tendons come from the limb lateral plate. Notably, tendons share same embryological origins with cartilage and bone, but not with skeletal muscles, which originate from dermomyotome (axial), mesoderm (head), or somites (limb). Despite these distinctions, the development of various components of the musculoskeletal system progresses in their close spatial and temporal association. It has been demonstrated that depending on the anatomic location, tendon development requires the presence of muscle. In chick somites, surgical ablation of dermomyotome prior to myotome formation results in the absence of *Scx* expression, indicating that muscle is required for initiation of development of axial tendons (Brent et al. 2003). Similarly, in *Myf5$^{-/-}$; MyoD$^{-/-}$* double-mutant embryos *Scx* expression is undetectable in mouse somites, and further supports the fact that myotome specification is indispensable for axial tendon progenitor formation (Brent et al. 2005). Contrarily, limb and head tendon development are initiated independently of muscles in mouse, chick, and zebrafish embryos. *Scx* expression is induced normally in muscleless limbs of *Pax3* mutant mice (Schweitzer et al. 2001) and *Myod1-Myf5* deficient zebrafish embryos, as well as in murine and zebrafish craniofacial tendons (Berthet et al. 2013). However, eventually the absence of muscles results in tendon development arrest and loss of *Scx* expression (Schweitzer et al. 2001, Bonnin et al. 2005). Hence, muscle is crucial for *Scx* induction in axial tendons, as well as for the maintenance of its expression in cranial and limb tendons. Since this pattern has been conserved across different species, it may indicate a requirement for mechanical forces provided by muscle during tendon morphogenesis. The two main signaling pathways identified as being involved in tendon development are TGFβ-SMAD2/3 and Fibroblast growth factor (FGF)-ERK/MAPK pathways (Havis et al. 2014). FGF signaling from the myotome was first associated with the induction of *Scx*-expressing tendon progenitors in adjacent somatic subcompartment of developing axial tendons in chicks (Brent et al. 2003). Disruption of TGFβ signaling in *Tgfb2$^{-/-}$* and *Tgfb3$^{-/-}$* double-mutant embryos leads to the loss of most tendons and ligaments (Pryce et al. 2009). Since *Scx* expression is disrupted only at E12.5, it has been suggested that TGFβ is required for tendon progenitor maintenance. Interestingly, in pharmacologically immobilized chick embryos both FGF and TGFβ signaling cascades were downregulated, suggesting that FGF and TGFβ ligands regulate tendon differentiation acting downstream to mechanical forces present in developing embryos (Havis et al. 2016). Additionally, growth differentiation factors (GDFs) that belong to the bone morphogenetic protein (BMP) family have been implicated in tendon development. *GDF-5* deficient mice exhibited altered ultrastructure and composition and inferior mechanical properties of Achilles tendon, when compared with control littermates (Mikic et al. 2001). Similarly, *GDF-6* deficiency in mice was associated with reduction in tail tendon collagen content and compromised tail material properties (Mikic et al. 2009). Beside those mentioned thus far, some other signaling pathways, such as the highly conserved Wnt pathway or calcium signaling might be involved in tendon morphogenesis.

10.3 BIOMECHANICS IN HOMEOSTASIS AND TENDINOPATHIC TENDONS

Being subjected to dynamic mechanical forces *in vivo*, tendons exhibit a unique crimp pattern and viscoelastic properties akin to a spring that enable tendon to effectively store and subsequently release mechanical energy. In a typical tendon stress-strain curve, four different regions can be distinguished. The initial toe region, where tendon is strained up to 2%, corresponds to the stretching-out of the characteristic crimp pattern. In the linear region of the curve, where stretching does not exceed 4%, collagen fibers lose their crimp pattern; the slope of the linear region defines the Young's modulus (i.e., elasticity) of the tendon. Stretching over 4% results in microscopic tearing, whereas strain beyond 8%–10% leads to macroscopic failure and tendon rupture (Wang 2006). Studies of force-length relationship revealed that with increasing forces, tendons lengthen to a certain degree (ascending limb)- after a certain point application of force results in tendon failure (descending limb) (Maganaris et al. 2004). Viscoelastic properties of tendon are defined by creep, that indicates increasing deformation under constant load, stress relaxation upon deformation, as well as hysteresis, or energy dissipation, which implies that an amount of energy is lost during loading. Consequently, the loading and unloading curves look differently. Mechanical properties vary depending on tendon anatomical site and specific function and are therefore dictated by the level of mechanical load to which a particular tendon is subjected (Bennett et al. 1986). These mechanical forces placed on tendons are, in turn, determined by the type of activity, passive or active mobilization, joint position, level of muscle contraction, tendon relative size, and so on. Additionally, variations in the rate and frequency of mechanical loading would result in different tendon forces (Wang 2006).

10.3.1 LOADING AND OVERUSE

Tendons are metabolically active tissues and tendon-resident fibroblasts respond to dynamic mechanical loading by alterations in the synthesis of ECM components and matrix degrading enzymes. A growing body of evidence supports the key role of mechanical stress in promotion and maintenance of tendon-specific phenotype. While mechanical forces are essential for tendon development and homeostasis, both complete unloading and contrarily excessive loading beyond a physiological range might have detrimental effects on tendon functionality.

Hannafin and colleagues investigated the effect of stress deprivation and cyclic tensile loading on histological and mechanical characteristics of the canine flexor *digitorum profundus* tendon (Hannafin et al. 1995). Stress deprivation resulted in significant changes in cell morphology and number, collagen fiber alignment, and progressive decrease in the tensile modulus over an eight-week period. However, tendons subjected to cyclic tensile loading for four weeks demonstrated increased Young's modulus (93% of the control) when compared to stress-deprived tendons (68% of the control), as well as maintained normal histological patterns (Hannafin et al. 1995). Surgical release of tensile strain in an engineered human tendon model resulted in disruption of tendon architecture, downregulation of tendon-related markers and induction of pro-inflammatory mediators (Bayer et al. 2014). To determine if the loss of tensile tension could induce apoptosis in tendon cells, Egerbacher and co-workers cultured rat tail tendons for 24 hours under cyclic loading or stress-deprived conditions. Upregulated caspase-3 expression and the increased number of apoptotic cells in stress-deprived tendons, when compared with the loaded group, indicated that loss of homeostatic tension induces programmed cell death (Egerbacher et al. 2008). Employing a transgenic mouse model, where green fluorescent protein (GFP) expression is driven by the *Scx* promoter, Maeda and colleagues demonstrated that gradual and temporary loss of transmittal forces from skeletal muscles by application of botulinum toxin A resulting in reversible loss of *Scx* expression and a decline in tendon mechanical properties. Acute loss of tensile loading by complete tendon transection led, in turn, to destabilization of the ECM structure, excessive release of active TGF-β and massive tenocyte death (Maeda et al. 2011).

Wang and collaborators investigated the effect of different mechanical stimulation regimes on rabbit Achilles tendon integrity in a bioreactor system (Wang et al. 2013). In the absence of loading, gradual loss of collagen fiber organization, increased cellularity, and cell roundness were observed, indicating a progressive divergence from the native tendon phenotype. Tendons stimulated with 3% cyclic tensile strain demonstrated moderate ECM disruption and elevated expression levels of matrix metalloproteinases (MMPs), MMP-1, -3 and -12, whereas excessive loading of 9% resulted in partial tendon ruptures. However, tendons stimulated with 6% cyclic tensile strain maintained their structural integrity and cellular function, suggesting that there is a narrow range of tensile loading promoting an anabolic effect and tendon tissue homeostasis (Wang et al. 2013). In a follow-up study, the model was extended to characterization of degenerative changes observed in tendons under loading-deprived conditions. When unloaded for 6 and 12 days, tendons exhibited progressive degenerative alterations, abnormal collagen type III production, increased cell apoptosis, and impaired mechanical properties. However, the application of a 6% cyclic tensile strain at day 7 for another six days was able to reverse morphological degenerative changes and partially restore mechanical properties of the unloaded tendon to the levels characteristic for the healthy tissue (Wang et al. 2015). Although mechanical stimulation is crucial for tendon-specific phenotype maintenance, excessive mechanical loading has been implicated as the major causative factor of tendon overuse injuries, collectively referred to as tendinopathies.

Histopathological presentation of painful tendons may comprise increased or decreased cellularity, cell rounding, increased vascularity and innervation, increased collagen type III expression and proteoglycans content, and collagen fibril disorganization. Molecular changes in tendinopathy include elevated expression of collagen type I and III, biglycan, fibromodulin, aggrecan, fibronectin, tenascin C (TNC-C) and alterations in expression levels (both upregulation and downregulation) of MMPs, and tissue inhibitor of metalloproteinases (TIMPs) that regulate ECM turnover (Corps et al. 2006, Jones et al. 2006). However, the etiology of tendon injuries has not been fully elucidated yet, and especially the role of inflammation in tendon pathology and healing process remains the subject of debate and ongoing controversy (Riley 2008, Dakin et al. 2015, Dean et al. 2016, Millar et al. 2017, Fredberg and Stengaard-Pedersen 2008). Several factors have been postulated to be implicated in tendon disease occurrence, including age, gender, body weight, vascular perfusion, nutrition, joint laxity, systemic diseases, muscle weakness, physical load, repetitive loading, abnormal movement, poor technique and training errors, fast progression and high intensity, environmental conditions, running surface, and more. Moreover, genetic predisposition (e.g., variants within COL5A1, TNC-C, and MMP3 genes in Achilles tendinopathy), treatment with corticosteroids or fluoroquinolones, oral contraceptives uptake, as well as existing comorbidities, such as obesity, diabetes, or hyperlipidemia, have been proposed as risk factors in tendon pathology development (Fredberg and Stengaard-Pedersen 2008, Magra and Maffulli 2005, 2008, Maffulli et al. 2013).

Soslowsy and colleagues employed an intensive running regime for 4, 8, and 16 weeks to induce an overuse injury in a rat model. Compared to the control group, which was allowed normal cage activity, the supraspinatus tendons in the exercised animals demonstrated increased cellularity and collagen fiber disorganization, the features that are normally observed in human tendinopathy. The tendons from the running group exhibited enlarged cross-sectional area and decreased mechanical properties, when compared to the control group (Soslowsky et al. 2000). In an *in vitro* study by Thorpe and co-workers, the application of cyclic loading mimicking high intensity exercise resulted in matrix damage and cell rounding in equine superficial digital flexor tendon explants. Those morphological changes were accompanied by increased expression of inflammatory mediators and MMPs in the loaded samples, when compared to the control group (Thorpe et al. 2015). Similar ECM damage and inflammatory response was observed in bovine flexor tendon overloading model (Spiesz et al. 2015).

Due to low cellularity, poor vascularization and innervation, tendons demonstrate restricted intrinsic healing capacity. A repaired tendon never regains the mechanical properties and hence full functionality of the pre-injured tissue, indeed, final tensile strength of healed tendon might be reduced by up to 30% (Majewski et al. 2008) or not restored two years after surgical repair (Geremia et al. 2015). After an acute injury, the tendon healing process normally follows a course of distinct, overlapping stages of early inflammation, proliferation, and remodeling, each orchestrated by a specific set of cellular and biochemical components. In the initial inflammatory phase, erythrocytes, platelets, and inflammatory cells (e.g., neutrophils, monocytes, macrophages) infiltrate the wound site and release vasoactive and chemotactic agents to promote angiogenesis and fibroblasts recruitment. During the proliferative phase, tendon fibroblasts synthesize collagen and other ECM components leading to granulation tissue formation around the wound site. After six to eight weeks, final remodeling phase commences. This stage is characterized by decreasing cellularity and reorganization of collagen architecture where collagen type III is replaced by collagen type I. As the scar matures, covalent bonding between collagen fibers increases, which leads to higher tendon stiffness and tensile strength. Yet, the healed tendon never matches characteristics of intact tissue. During tendon healing upregulation of several growth factors and cytokines, such as insulin-like growth factor-1 (IGF-1), platelet derived growth factor (PDGF), basic fibroblast growth factor (bFGF), vascular endothelial growth factor (VEGF) and transforming growth factor beta (TGF-β), stimulate cell migration, proliferation, angiogenesis and synthesis of collagen and other ECM components (Voleti et al. 2012, Millar et al. 2017).

Tendon stem/progenitor cells role in tendon homeostasis and disease is not well understood yet, however it was suggested that TSPC malfunction may contribute to impaired healing and repair, or tendon pathology development. Especially, age-related depletion of the stem cell pool and/or a decline in stem cell function associated with entrance of senescence state might be implicated in pathology onset and progression. The mechanoresponse of TSPCs have been studied both *in vitro* and *in vivo* indicating a critical role of mechanical loading in tendon stem cell fate and function. Mechanical loading at physiological level (4% stretching) promoted TSPCs proliferation and differentiation into tenocytes, whereas at excessive stretching (8%) TSPCs differentiated in non-tenocytes such as adipocytes, chondrocytes and osteocytes, in addition to tenocytes (Zhang and Wang 2010a). In a follow-up *in vivo* study employing mouse treadmill running model it was found that tendons subjected to repetitive, strenuous mechanical loading produced high levels of PGE_2, which in turn was associated with decreased proliferation of TSPCs and TSPC differentiation into adipocytes and osteocytes (Zhang and Wang 2010b). Such non-tenocyte differentiation of TSPCs under abnormal mechanical loading may explain some pathological features of late tendinopathy such as lipid accumulation, mucoid formation and tissue calcification.

Management of tendon injuries includes surgical procedures and nonsurgical modalities such as physiotherapy. Application of mechanical stimulation may be beneficial for the proper organization of collagen fibers and prevention of adhesion formation during tendon healing. In injured canine flexor tendons, active mobilization increased their tensile strength and restored gliding surfaces while reducing intrasynovial adhesion formation (Gelberman et al. 1986, Wada et al. 2001). In a rabbit model of Achilles tendon healing, early mobilization after tenotomy favored a more rapid restoration of tissue functionality, when compared to the group subjected to continuous immobilization (Pneumaticos et al. 2000). A study of 64 human patients with Achilles tendon ruptures treated surgically and with early mobilization indicated that application of an early mobilization rehabilitation program reduces the range of motion loss and muscle atrophy, increases blood supply, as well as improves strength of calf the muscles and ankle movement (Sorrenti 2006). 12 weeks of eccentric resistance training in elite soccer players increased peritendinous type I collagen synthesis in individuals suffering from Achilles tendinosis, whereas collagen metabolism was not affected in the healthy control group (Langberg et al. 2007). However, a 10-year follow up study of postoperative regimes of Achilles tendon ruptures showed that early mobilization and immobilization in tension resulted in similar clinical outcomes and isokinetic strengths (Lantto et al. 2015). Although some conflicting data exists and the optimal rehabilitation protocol and precise molecular mechanism underlying the beneficial effects of mobilization remain to be determined, controlled tendon-loading and motion plays crucial role in tendinopathy management (Rees et al. 2006).

10.4   CURRENT THERAPIES FOR THE MANAGEMENT OF TENDINOPATHIES

Conservative treatments and/or grafting surgeries are the gold standards for the treatment of tendon injuries. The treatment of choice is influenced by tendon location, type and severity of lesion as well as on the symptoms and clinical evidence of injury.

Independently of the treatment selected, the mid to long-term outcomes are not completely satisfactory with a risk of recurrence of symptoms that include pain, instability and degradation of mechanical function.

In the case of tissue grafting, besides the morbidity of the donor tissue and the risk of (re)rupture of the inflicted tendon, both tendons may experience long-term consequences as loss of mechanical competence, functional disability and degeneration that may progress into nearby tissues.

Alternatively, tissue grafting from autologous or cadaveric sources, biological augmentation matrices of decellularized mammalian-origin tissues mainly human (GraftJacket®), porcine (Restore™), equine (OrthADAPT®) or bovine (TissueMend®) have been investigated and presented to the clinical field, revised by Chen et al (Chen et al. 2009). The main reasons for a lack of compliance on the medical use of these devices may be caused by the decellularization process that may be insufficient to remove all the resident cells and there is a potential risk of immune-rejection and for zoonoses transmission.

Artificial augmentation devices constitute an alternative to tissue grafting procedures and to biological augmentation devices (Liu et al. 2010). Commercially available devices as LARS™, Kennedy ligament augmentation device, Dacron®, Gore-Tex and Trevira, revised by Batty and colleagues (Batty et al. 2015), were described to avoid and provide improved knee stability (Liu et al. 2010) and full weight bearing. However, artificial devices have shown controversial outcomes on the long-term follow up, concerning mechanical failure or mechanical mismatch with native tissues, instability, synovitis, chronic effusions and progression to early osteoarthritis.

10.5   TENDON TISSUE ENGINEERING STRATEGIES

Tendons require a unique combination of cells within an abundant, hierarchically organized ECM coordinated by mechanical, biochemical, and architectural sensing and signaling. A failure to this balance results in significant non-functional modifications and/or disease.

A traditional TE strategy is inspired by the natural elements within a tissue niche, namely cells, a 3D structure and their highly orchestrated biochemical signaling in different combinatorial approaches with the final goal of stimulating and inducing new tendon formation with restored function (Figure 10.1). As mechano-responsive tissues, mechanical conditioning of tendons is essential and a critical parameter of the native environment for tissue development and maturation, which ultimately will translate into successful 3D tissue equivalents and improved clinical therapies (Figure 10.1).

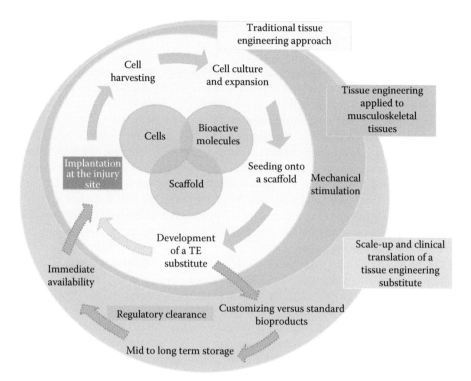

FIGURE 10.1 Diagram on the development of a successful tissue engineered substitute towards clinical translation using a traditional TE approach by employing bioreactors.

10.5.1 THE ROLE OF CELLS IN TENDON TISSUE ENGINEERING

Tendon resident cells, mainly tenoblasts and tenocytes, are scarce and are responsible for the maintenance of ECM. These elongated cells stretched between the collagen fibers of the tendon, synthesize collagen and other macromolecules, assemble these molecules into a cohesive unit, and organize the fiber phase in parallel with the direction of tensile load, which is capable of transducing and responding to mechanical stimuli (Koob 2002). This organization makes it extremely difficult to repair damage to tendinous structures. Natural tendon healing usually leads to the formation of fibrotic tissue, impairing tendon mechanical properties, and resulting in a poor quality of life. More recently, a tendon stem/progenitor cell population was identified (Bi et al. 2007) and described to participate in the endogenous regeneration process of tendons.

The hypocellular and hypovascular nature of tendons can relate to significant limitations in the process of repair and regeneration, especially upon tears and ruptures. Cells, especially the resident cell populations, have a critical role in the regulation of the tendon matrix in both normal and pathological conditions (Clegg et al. 2007) being an important parameter when considering engineering the functional tendon inspired substitutes.

Several cell sources have been investigated for tendon strategies (Gonçalves et al. 2015). Tendon cells are often the first choice despite the donor tissue morbidity and the infliction of a secondary defect to heal the damaged tendon. However, they are still a great source to study mechanisms and predict native biological responses. Other sources, such as dermal fibroblasts or muscle-derived cells (Chen et al. 2016), have been also exploited as cell alternatives for tendon approaches.

Stem cells are also a promising source due to their high self-renewal potential and ability to differentiate into tissue-oriented lineages, with evidence to commit toward a tenogenic fate (Gonçalves et al. 2013, 2017). The secretion of a broad range of bioactive molecules with paracrine effects is believed to be the main mechanism by which MSCs achieve their therapeutic effect (Meirelles et al. 2009). The fact that tendons have their own local stem cell population may indicate that stem cells are likely to participate and even mediate the renewal and remodeling of tissues, but also influence fibrous scarring due to abnormal or irregular signaling mechanisms and influence tissue recovery upon injury.

More recently, embryonic stem cells (ESCs) (Cohen et al. 2010, Chen et al. 2012) and induced pluripotent stem cells (iPSCs) technology (Xu et al. 2013, Bavin et al. 2015) have also been proposed for tendon tissue engineering (TTE). Ethical considerations associated with human embryos research and the risk of *in vivo* teratoma formation presents significant drawbacks for the clinical application of ESCs. A potential strategy to overcome many current ethical concerns in ESCs-based therapy is the use of iPSCs (Docheva et al. 2015), which are highly available cells obtained from multiple sources of autologous cells and can be reprogrammed into a wide range of cell types. The very few studies on human iPSCs differentiated for tendon applications (Xu et al. 2013, Bavin et al. 2015) show controversial outcomes and their potential for tendon fundamental studies and clinical approaches should be more deeply investigated.

Apart from cell sources, cell-based therapies offer the potential to induce a regenerative response by stimulating local cells and inspiring the synthesis of a structural matrix to ensure remodeling of damaged tendon tissues.

10.5.2 BIOACTIVE MOLECULES IN TENDON ENGINEERED SUBSTITUTES

The lack of understanding on cell-to-cell and cell-matrix communication in tendon niches results in limited knowledge to guide biological responses for effective treatment. Moreover, cell secretome and signaling interactions are envisioned to represent the most biologically significant cell role towards repair mechanisms.

Several biomolecules, some identified in developmental biology studies and discussed earlier in this chapter, have been described to participate in tenogenic commitment, namely EGF, FGF, PDGF, IGF-1, GDF-5, TGF-β (Gonçalves et al. 2013, Barsby and Guest 2013, Holladay et al. 2014, Bottagisio et al. 2017); however, their precise spatio-temporal distribution requires further research to promote *in vitro* tenogenic differentiation. Thus, progress towards the recognition of the molecular players favoring homeostasis and regeneration will assist the establishment of tendon benchmarks and methodologies paving the way for clinical translation of cell-based therapies.

10.5.3 RECAPITULATING TENDON MATRIX IN FUNCTIONAL SUBSTITUTES

The matrix of tendon tissues is quite unique and complex: it complies with a supportive function of cells as its organization and structure relates to tendon mechanical properties and function adapted to the anatomical location.

Thus, another main challenge in the successful design of a tendon substitute is to mimic the intrinsic alignment of tendons and recapitulate their complex hierarchical architecture, while remaining mechanically competent towards achieving proper biomechanical functionality to support a complete regeneration of damaged tissues.

Leading edge biomaterial advances have featured fiber fabrication technologies to produce architecturally aligned scaffolds aiming at tendon replacement strategies. These technologies include 3D-printing (Goncalves et al. 2016), a combination of polyelectrolyte complexation and microfluidics (Costa-Almeida et al. 2016, Shilpa 2017), electrospinning (Shuakat and Lin 2014, Wang et al. 2015), and an electrochemical alignment technique (Gurkan et al. 2010, Younesi et al. 2014) combined with textile techniques.

A magnetic responsive scaffold based on a polymeric blend of starch and polycaprolactone (PCL) fibers (SPCL) with aligned structural features incorporating magnetic nanoparticles by 3D-printing was shown to assist the tenogenic differentiation of human adipose-derived stem cells (hASCs) upon the actuation of an external magnetic field and evidenced good biocompatibility and integration within the surrounding tissues upon implantation in an ectopic rat model (Goncalves et al. 2016).

Instructive tenogenic matrices with microscaled parallel aligned fibrils were developed by combining polyelectrolyte complexation and microfluidics (Costa-Almeida et al. 2016, Shilpa 2017). These fibrous photocrosslinkable hydrogels were fabricated with chitosan (positively charged) and methacrylated gellan gum (negatively charged) (Sant et al. 2016), a combination of alginate (ALG) with methacrylated hyaluronic acid (MeHA, negatively charged), or chondroitin sulfate (Costa-Almeida et al. 2016). The MeHA-ALG fibers could be manipulated using textile technologies, allowing the fabrication of 3D constructs with increasing complexity and functionality.

Electrospinning is a familiar technique to TE approaches. It allows the production of long and continuous fibers with controlled diameter mimicking the ECM of the tendon at the nanoscale. Nano-scaled fibers are expected to provide topographical cues at the cell level and stimulate cell response by contact. Accordingly, several studies report that aligned nanofibers can stimulate different cell sources, including dermal fibroblasts (Wang et al. 2016) and iPSCs (Zhang et al. 2015) to commit towards the tenogenic phenotype (Wang et al. 2016) in both synthetic and natural based scaffolds (Teh et al. 2013, Zhang et al. 2015, Wang et al. 2016, Sensini et al. 2017) and to enhance tendon regeneration *in vivo* (Wang et al. 2016). However, nanofiber scaffolds produced by electrospinning are mainly 2D systems and their scaling up into 3D scaffolds is limited unless combined with other scaffolding fabrication techniques (for instance rapid prototyping or textile approaches such as braiding and weaving).

The advancement of medical textiles has created a new generation of biomimetic scaffolding fabrics ranging from simple gauze or bandage materials to scaffolds for tissue culturing and a large variety of prostheses for permanent body implants.

Textile technologies are powerful tools for producing complex and hierarchical 3D constructs using bio-inspired fiber-based materials as building blocks. Textile platforms (weaving, twisting, braiding, and knitting) offer unique advantages, such as the versatility to fine-tune the properties of a scaffold size, shape, porosity, and mechanical properties by varying the assemble parameters—namely fiber diameter, fiber number, or braiding angle (Czaplewski et al. 2014)—with the potential to develop improved 3D constructs with biomimetic properties for achieving tenogenic differentiation (Czaplewski et al. 2014). Moreover, with textile technologies it is expected to grow in 3D and in a hierarchical architecture mimicking the native structure of the tendon.

A novel biofabrication modality, electrochemically aligned collagen (ELAC), allows the continuous production of aligned collagen threads through a pH gradient process between two parallel electrodes (Cheng et al. 2008) and combines textile techniques to fabricate complex 3D scaffolds. Since the development of this technique in 2008, several works report their potential for tendon and ligament-based approaches by mimicking the native tendon's structure and mechanical properties. ELAC scaffolds were shown to induce MSCs tenogenic differentiation (Kishore et al. 2012, Younesi et al. 2014) with the production of a tenomodulin positive matrix. Moreover, in a rabbit patellar tendon model, ELAC braided scaffold was shown to be biocompatible and biodegradable and assisted the increase in volume fraction of the tendon fascicle compared with the control (Kishore et al. 2012).

Although these fabrication technologies assist the alignment of collagen fibers and mimic tendon structure, the topographical and mechanical stimulation of cells is still limited in these scaffolds. Thus, the application of bioreactors to culture and stimulate biological processes in cells laden scaffolds may more effectively recreate tissue dynamic environment, improving the biofunctionality of these constructs aiming to mimic *in vivo* physiological conditions.

10.6 BIOREACTORS IN TISSUE ENGINEERING

10.6.1 DESIGNING BIOREACTORS FOR TENDON TISSUE ENGINEERING

Bioreactors have been widely investigated as advanced tools for tissue engineering of musculoskeletal tissues. These dynamic systems are designed to provide different mechanical conditioning to cells or cell laden 3D matrices in order to resemble the physical forces experienced in the native tissue environment, such as shear stress, hydrostatic pressure, flow perfusion, microgravity or mechano-magnetic stimulation. For example, in bone tissues, nutrients and wastes are transported within a lacunocanalicular network, whose circulation has been recreated by shear stress and flow perfusion systems (Gonçalves et al. 2011, Gardel et al. 2013). Bioreactors applying perfusion have also been shown to have promising results in cartilage studies (Gonçalves et al. 2011), as well as hydrostatic pressure bioreactors described to simulate the main forces to which cartilage is subjected to in the articular joints (Correia et al. 2012). Bioreactors offer controllable and reproducible dynamic environments with enhanced access of the cells to nutrient supply from the culture medium, improved oxygen diffusion and a more efficient metabolic waste removal (Table 10.1). Moreover, bioreactors allow homogeneous and long-term cultures with the possibility

TABLE 10.1
Summary of the Main Characteristics of a Bioreactor System

Cultures in Bioreactors

- Control over *in vitro* environment:
  - pH
  - temperature
  - humidity
  - oxygen tension
  - nutrient supply / waste removal
  - cell metabolite quantification
- Improved oxygen diffusion
- Homogeneous and long-term cultures (non-stop up to several months)
- Standardization of protocols
- Biochemical conditioning:
  - Single or multi dosage without interrupting with the experimental setup
- Biomechanical conditioning:
  - Stretching (mostly cyclic)
  - Loading
  - Tension/compression
  - Mechano-magnetic
- Closer environment to a tissue niche than static 3D cultures

Bioreactor Components

- Non-toxic, especially the parts in direct contact with the cultures
- Suitable for aseptic conditions
- Quick and easily assembled
- Sterilizable if reusable
- Preferably low-cost
- Easy to clean and store
- Portable, to fit in cell culture incubators if necessary
- Multiparameter/Tunability of parameters
- Computer control of parameters
- Real time monitoring using imaging techniques as microscopy, MRI or micro computed tomography
- Possibility for computer-assisted automation
- Possibility for scaling-up strategies

for standardization and automation procedures. Depending on the system, external factors such as pH, temperature, and cell metabolite concentration can be monitored and adjusted, enabling higher cell proliferation rates and decreasing the number of cells that must be initially seeded, while favoring desirable cell responses. With the appropriate stimulus, it is also envisioned a reduction/elimination of medium supplementation including serum requirements. It is also expected that cultured cells subjected to the mechanical conditioning will be able to synthetize native tissue-like ECM in shorter periods of time and following a more controlled and organized distribution as the mechanical forces will likely better mimic the native niches. Thus, applying bioreactors in tissue engineering strategies offers great advantages over 2D cultures and cell laden 3D matrices cultured in static conditions. These include (1) biomechanical conditioning of 3D constructs, providing adequate loading regimes according to the type of stimulation required; (2) increase of mass transport, as the supply of oxygen and soluble nutrients is a critical concern when culturing 3D constructs *in vitro*; (3) controlled culture conditions enabling the systematic study of tissue-specific physiological requirements; (4) computer monitoring and programmable options to control and adjust environmental parameters, reducing the limitations associated to a human operator; and (5) reproducible cycles of stimulation/standardization.

The development of bioengineered products is a time-consuming task and, thus, approaches to potentially accelerate their clinical use are needed. Bioreactor design can be more or less complex depending on the final application and monitoring parameters. However, all bioreactors are composed of a driving system, a control box and connection cables. The driving system is the motor or pump responsible for impelling mechanical stimulation or medium circulation through perfusion forces into the samples, often located in culture chambers. The control box or computer-aided software allows controlling the system, including fluid velocity and biochemical parameters. The tubing cables, often made of materials permeable to gases, are necessary to connect the different parts of the bioreactor to the power socket.

Most of the bioreactors for tissue engineering settings are designed to operate under aseptic conditions and inside a standard CO_2 incubator at 37°C. Thus, the assembly of the bioreactor and the positioning of the samples in the beginning of the experiment, as well as the handling of the system for medium exchange or collection of the samples for analyses are performed within biosafety hoods. The bioreactor materials should be non-toxic, especially the ones in contact with the constructs or tendon tissue samples, and bioreactor parts should be sterilizable if re-usable. Connecting parts such as tubes, nuts, o-rings, lids, and luer adaptors keep the system closed, and must be well tightened to prevent malfunctioning, fluid leakages and consequent contaminations. In order to assure gas interchange and pressure compensation, a 0.22 µm filter is normally used.

10.6.2 The Role of Bioreactors in Tendon/Ligament Tissue Engineering

The combination of multiple factors known to exist in tendon niches in a tridimensional and complex environment may enable the generation of predictive models relating to cellular responses towards scaffold design parameters and ultimately to recapitulate the alignment, the hierarchical architecture and tendon tissue formation. Commercial bioreactor systems for TTE have been designed to meet these requirements proposing a solution to the limitations of static cultures. Generally, these systems allow the researcher to control and manipulate the deformation cycles as well as the strain and rate levels applied on the sample. Hereafter, we will address some of these systems with different complexity that may be used for biologic tendon samples or TTE constructs.

10.6.2.1 Mechanical Stimulation

The bioreactors that have been used so far for TTE greatly differ in terms of complexity and multiparameter analysis (Figure 10.2).

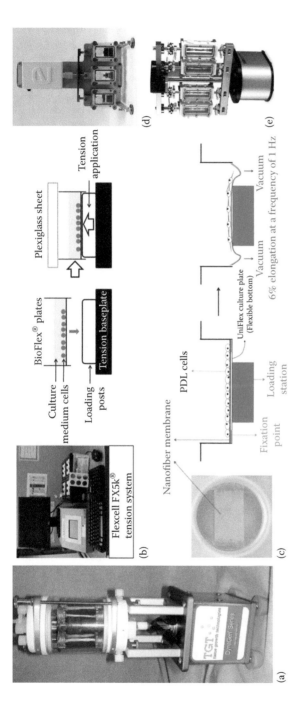

FIGURE 10.2 Images of some commercially available bioreactors used in TTE strategies. (a) LigaGen model L30-4C, DynaGen series, Tissue Growth Technologies. (Reprinted from *J. Hand Surg.*, 35A, Angelidis, I.K. et al., Tissue engineering of flexor tendons: The effect of a tissue bioreactor on adipoderived stem cell-seeded and fibroblast-seeded tendon constructs, 1466–1472, Copyright 2010, with permission from Elsevier.); (b) Schematic diagram of the FX5K® Tension System. (With kind permission from Springer Science+Business Media: *Knee Surg. Sports Traumatol. Arthros.*, Impact of cyclic mechanical stimulation on the expression of extracellular matrix proteins in human primary rotator cuff fibroblasts, 24, 2016, 3884–3891, Lohberger, B. et al.); (c) The Flexcell system equipped with the PDL cells supported on nanofiber matrix, where the dynamic mechanical tensional force was applied to the matrix/cell through equipment vacuum. (Reprinted with permission from Kim, J.H. et al., *PLoS One*, 11(3), e0149967, 2016); (d) TC-3 load bioreactor system, Reprinted with permission from EBERS Medical Technology SL, Cartuja Baja, Zaragoza); (e) ElectroForce® multi-specimen BioDynamic 5200. (Reprinted with permission from TA Instruments, New Castle, DE.)

One of the simplest systems is the Cell Stretching System from STREX, which was designed for stretching cells in culture and applying a stress load to cells up to 20% of stretching ratio. Morita and co-workers have been working with this device to investigate the optimal uniaxial cyclic stretching stimulation to bone marrow MSCs (Morita et al. 2013, 2014, 2015, 2017) towards tenogenic differentiation. This work suggest optimal normal strains between 7.9% and 8.5% for assisting the production of Col I and TNC proteins, respectively.

Flexcell International Corporation developed Flexcell® Tension Systems, which are computer-regulated bioreactors that use vacuum pressure to apply cyclic or static strain to cells growing *in vitro* with control over the magnitude and frequency of the stretching. Depending on the model, some systems can check and analyze real-time cellular biochemical changes in response to strain. In a work developed by Kim and colleagues (Kim et al. 2016), the FX-5000 tension system from Flexcell® was used to apply mechanical stress (strain of 6% elongation at a frequency of 1Hz) to rat periodontal ligament (PDL) cells seeded onto nanofiber-equipped culture plates with random or aligned topography. The cells cultured on the oriented nanofibers combined with the mechanical stress produced PDL specific markers, including periostin and TNC, undergoing ligamentogenesis with simultaneous down-regulation of osteogenesis. Moreover, the cell/nanofiber constructs engineered under mechanical stress showed sound integration into tissue defects with significantly enhanced new bone volume and area, in a rat premaxillary periodontal defect model (Kim et al. 2016). In another recent study from Sun and co-workers (Sun et al. 2016), rabbit fibroblasts from ligament tissues and bone marrow MSCs were mechanically tested under Uniflex/Bioflex culture system from Flexcell®, as a mean to mimic mechanical strain in ligament tissue. Results showed that uniaxial stretch (15% at 0.5 Hz; 10% at 1.0 Hz) stimulated fibroblast proliferation and collagen production, while uniaxial strains (5%, 10%, and 15%) at 0.5 Hz and 10% strain at 1.0 Hz were favorable for MSCs. Similar results on the increment production of total collagen by human fibroblasts from the rotator cuff with cyclic strain (Flexcell FX5K™ Tension System; 10% elongation and 0.5 Hz frequency) were achieved by Lohberger and colleagues (Lohberger et al. 2016). Also, increasing levels of the matrix metalloproteinases MMP1, MMP3, MMP13, and MMP14, analyzed by RT-qPCR, were observed in stimulated conditions as well as tenascin-C and scleraxis.

The LigaGen® Ligament and Tendon Bioreactor from BISS TGT was also designed to provide mechanical stimulation, imposing axial stress or strain to 3D tissue engineered constructs or decellularized tissues to recreate physiological conditions *in vitro*, with studies aiming at hand tendons and anterior cruciate ligament (ACL) regenerative medicine. The chambers of this bioreactor deliver oscillatory axial stimulation to the samples. The stress/strain profiles are defined by the operator, which can be in the form of a simple harmonic (sinusoidal) or a physiological waveform. The bioreactor can be complemented with a perfusion system to provide convective media transport around the samples.

The TC-3 load bioreactor from EBERS Medical Technology SL is a computer-controlled system designed to enclose tissue samples or cell laden scaffolds under mechanical tension and compression axial loading.

Herein, the tension grips apply tension loads on samples as different as sheet-like, membrane substrates or 3D-like samples. This specific bioreactor can operate in two different working modes, horizontal or vertical configuration, depending on the type of experiment to be developed as the requirement for immersion or air liquid interface, for example. TC-3 load bioreactor allows simultaneous flow and deformation conditions, but hydrostatic pressure conditions can also be simulated.

ElectroForce® BioDynamic® systems from TA Instruments can be used to simulate *in vivo* conditions and provide accurate characterization of biomaterials and biological specimens under tension and perfusion flow regimes. Also, an integrated digital video extensometer can be added to the system for primary, secondary, and shear strain measurements. Ligaments, tendons or other thin and elongated specimens are attached to the tensile grips while the chamber is perfused with nutrients. The great advantage of this bioreactor in comparison to all of the above described bioreactors is the possibility of integrating a Dynamic Mechanical Analysis (DMA) software, allowing determination of the viscoelastic properties at the same time of culture/stimulation.

Apart from the model used, the significant cost of commercial bioreactors and the limited number of samples the operator can handle per experimental setup are the main disadvantages pointed out. Therefore, several research groups have custom designed bioreactors, developing new systems to meet more accurately the specific parameters of a tissue or tissue substitute to be screened and evaluated.

One of the most relevant parameters for tendon substitute development is the application of cyclic strain (Screen et al. 2005, Doroski et al. 2010, Andarawis-Puri et al. 2012, Legerlotz et al. 2013), and thus a major consideration to the customization of bioreactors.

Wang and colleagues (Wang et al. 2013) developed a bioreactor system, which applies pre-programmed uniaxial stimulation, to study different cyclic tensile strain (0.25 Hz for 8 hours/day, 0%–9% for six days) on rabbit Achilles tendons. Overall results showed that 3% cyclic tensile strain did not prevent matrix deterioration (gene expression of MMP1, 3, and 12 were highly upregulated by 3% strain stimulation compared to the other groups), whilst at 6% cyclic tensile strain the structural integrity and cellular function of the tendons were maintained. Moreover, at 9%, massive rupture of the collagen bundles was also verified.

Youngstrom and co-workers (Youngstrom et al. 2015) studied the influence of cyclic mechanical conditioning (0%, 3%, or 5% strain at 0.33 Hz for up to one hour daily for 11 days) provided by a custom bioreactor on the maturation and cellular phenotype of decellularized tendons obtained from four equine sources seeded with bone marrow-derived MSC. Cultured cells at 3% and 0.33 Hz integrated within these tissue-derived scaffolds, exhibited higher elastic modulus and higher expression of tenogenic genes.

10.6.2.2 Magnetic Stimulation

In recent years, magnetic driven actuation has been investigated as an alternative form of bio-stimulation in TE strategies. It is known that magnetic forces influence biological processes, and magnetotherapy protocols have been proposed in tissue regeneration and inflammation control after injury. Furthermore, magnetic stimulus may act synergistically with magnetizable nanoparticles internalized by cells in culture or embedded within 3D scaffolds creating local forces, which can be physically sensed by cells assisting mechanotransduction processes that will ultimately lead to an *in vitro* maturation of the cell-laden constructs prior to implantation. This approach has been previously hypothesized and reported by our group on the use of magnetic bioreactors in the stimulation of stem cells towards tenogenic, osteogenic, or chondrogenic differentiation (Lima et al. 2015, Goncalves et al. 2016, Santos et al. 2016) and by others for osteogenesis (Meng et al. 2013, Kang et al. 2013), cardiac TE (Sapir et al. 2014), and neuronal regeneration (Antman-Passig and Shefi 2016).

3D-printed magnetic scaffolds cultured with hASCs exposed to oscillation frequency conditions provided by a magnefect-nano transfection device (nanoTherics Ltd, UK) showed that magnetic stimulation tend to accelerate the production of collagen and noncollagenous proteins by cells after seven days (Goncalves et al. 2016). This device was initially set up for magnetofection purposes, but the magnetic properties of the system showed potential for applications in magnetic force-based TE.

On the other hand, magnetic responsive membranes, which were implanted subcutaneously in rats exposed to a pulsed electro-magnetic field (PEMF) waveform with a magnetic field intensity peak of 0.01 T, a duty cycle of 6.3 ms and a frequency of 75 Hz for two hours a day, five days a week (Magnum XL Pro, Globus), showed to modulate tissue inflammatory response, translated by a decrease in the number of mast cells infiltration and reduction of the thickness of the fibrous capsule (Santos et al. 2016). The coils that provided the mechano-magnetic stimulation within a therapeutic mat were placed under the animals' cage. Magnum devices from Globus are commercially available magnetotherapy devices that provide low-frequency pulsed magnetic fields, being composed of solenoids that permit both the superimposition and the opposition of the magnetic field to treat surface or deep pathologies. These instruments are generally used in human clinical procedures, mostly physiotherapy centers, for applications in muscle, bone-tendon, and anti-ageing treatments. The use of magnetic forces in tissue healing is quite recent and some pioneer studies suggest the influence of magnetic fields in modulating tendon injury recovery after rat Achilles transection (Strauch et al. 2006). Besides pain relief (Nelson et al. 2013) and stimulation of blood circulation, magnetotherapy has been reported to stimulate tendon cell proliferation (Seeliger et al. 2014, Randelli et al. 2016) in the promotion of the healing process.

Bioreactors that generate PEMF have also been investigated. Recently, Liu and colleagues (Liu et al. 2016) and Tucker and colleagues (Tucker et al. 2017) described the use of a commercial device, Physio-Stim® PEMF system from Orthofix Inc, to promote gene expression of human tenocytes (collagen I, TGFβ-1, PDGFβ, BMP12, and TIMP4) and to improve early tendon healing in a rat rotator cuff model, respectively. The FDA approved Orthofix stimulators claim to generate a uniform, low-level PEMF shown to be safe in clinical studies for the healing of nonunion fractures (Garland et al. 1991).

The portable SomaPulse® is another non-invasive PEMF system. It applies a sequence of magnetic pulses programmed to introduce a magnetic field into musculoskeletal tissues. Despite the multiple devices available for magnetic stimulation, the application of electromagnetic fields is still not properly understood, nor how the exposure to PEMF influences tendon resident cells or tendon tissue responses.

Despite the recent scientific interest on the magnetic force impact over biological tissues, the wide range of magnetic properties, such as intensity, time of exposure or frequency, has to be more deeply explored and optimized to individual conditions, tendon anatomical location, and associated pathologies. Electromagnetic fields are expected to influence cells response at the molecular levels or to act as mediators of inflammation. Girolamo and co-workers (Girolamo et al. 2013) reported that a PEMF (1.5mT, 75 Hz) enhanced tendon cell proliferation and the release of anti-inflammatory cytokines and angiogenic factors (IL-1B, IL-6, IL-8, and TGF-β). Herein, the PEMF was generated by a pair of rectangular horizontal coils placed opposite to each other.

10.7 MECHANOREGULATION MECHANISMS

Physiological responses to mechanical loading are initiated by a process called mech-anotransduction, in which cells detect physical changes in their microenvironment through specialized machinery and then translate that information into an appropriate biological response (Santos et al. 2015). This mechanosensitive feedback mechanism modulates cellular functions as proliferation, differentiation, migration, and apopto-sis, and is crucial for organ development and homeostasis (DuFort et al. 2011).

Tendon tissues, physiologically adapted to transmit mechanical forces in a daily basis, are the perfect model to study the mechanisms involved in the translation of mechanical forces into a functional response.

Growing evidence suggests that mechanical forces regulate the expression of the basic helix-loop-helix (bHLH) transcription factor Scx through activation of the TGF-β/Smad2/3 pathway in adult tenocyte cultures, which, in turn, is required for maintenance of tendon-specific ECM (Maeda et al. 2011, Havis et al. 2016, Gaut and Duprez 2016).

Furthermore, it is accepted that these forces can be at least partially mimicked by the stimuli provided by bioreactors.

Mohawk (Mkx) and the downstream tendon-associated genes Tenomodulin (Tnmd), Collagen type I ($Col1a1$ and $Col1a2$), but not Scleraxis (Scx), showed an increased expression in Achilles tendon-derived rat tenocytes subjected to stretching at 2% and 0.25 Hz for six hours in a FX-5000 tension system (Flexcell International) (Kayama et al. 2016).

Moreover, magnetic-mechano actuation directed to cell surface receptors is a good example to remotely deliver mechanical stimuli into individual cells. Studies reported that a magnetic field of variable frequency may influence cellular response and intracellular signaling favoring the differentiation into desired phenotypes and higher proliferation rates in a shorter culture time and in a more reproducible man-ner (Girolamo et al. 2013, Henstock et al. 2014, Rotherham and El Haj 2015, Sapir-Lekhovitser et al. 2016).

3D scaffolds may also be used for tendon mechanobiology studies, in which the actuation of mechanical loads provided by bioreactors may be combined with the stimulation from the topographical and physicochemical properties of the scaffold to the seeded cells. An example is the fiber composite hydrogels developed by Screen and co-workers (Screen et al. 2010) who envisioned to be a mechanotransduction research platform. Collagen type I gene expression was upregulated in NIH/3T3 fibro-blasts laden in the hydrogels and subjected to cyclic tensile loading of 5% dynamic tensile strain, at 1 Hz for 24 hours.

Also, Jones and colleagues (Jones et al. 2013) showed that TGFβ activation plays an important role in mechanotransduction, specifically in the regulation of MMP genes of human tenocytes isolated from tissues with tendinopathic conditions. These tenocytes were seeded onto 3D collagen gels and a 5% cyclic uniaxial strain at 1 Hz for 48 hours was applied over these constructs using the Flexcell FX-4000™ device. Treatments with TGFβ1/TGFβRI inhibitor were compared to mechanical strain regimes, and the outcomes with strain or TGFβ treatment were similar. Overall, there was a decrease in MMP1, -3, -11, -13, and -17 and an increase in collagen type I at the mRNA level (Jones et al. 2013).

10.8 CONCLUDING REMARKS

Since the first investigation of tendon/ligament bioreactors published in the 1990s (Hannafin et al. 1995), these devices have evolved into more complex systems able to test more specimens simultaneously and control/program several parameters. Despite the advances in recent years and the awareness for mimicking the different fundamental aspects of tendons, which are intrinsically associated to tissue function and activity, currently available tendon substitutes are not biomechanically competent as artificial replacements of tendons. Nevertheless, bioreactors can fulfill this functional gap offering a powerful solution for improving and assisting the development of new tissue engineering equivalents as they provide a controlled, dynamic and monitorable environment that more closely resembles native tissues, with potential toward scale up strategies.

Moreover, bioreactors provide the possibility for testing a variety of different cell laden 3D structures, including scaffolds, membranes, tissue explants, and more, that can be investigated and assessed in systematic and reproducible conditions as predictive tools of tissue substitute performance in similar physiological conditions, resembling the native environment. However, the optimal *in vitro* conditions and the optimal 3D tissue substitute have not been established, and the challenge stands for an accurate time spatial recapitulation of physical and biochemical signals that cells may experience in tendon niches, as well as cell response to such potential stimuli, providing important insights into the long-term capability of engineered constructs to maintain tissue proper functionality. Cellular mechano-sensing mechanisms and the information exchange in biomechanical regulatory signals between the cell and its surroundings also have an important role in determining the potential outcomes of bioreactor microenvironments towards pre-clinical models. These issues need to be thoroughly addressed in forthcoming years in order to achieve bioreactor designs that fully comprise biological and biomechanical demands of tendon tissue.

LIST OF ABBREVIATIONS

2D   Two dimensional
3D   Three dimensional
ACL   Anterior cruciate ligament

BGN	Byglican
bHLH	basic helix-loop-helix
BMP	Bone morphogenetic protein
Col I	Collagen type I
DCN	Decorin
DMA	Dynamic Mechanical Analysis
ECM	Extracellular matrix
EGF	Epidermal growth factor
ELAC	Electrochemically aligned collagen
FACITs	Fibril-Associated Collagens with Interrupted Triple-helices
FGF	Fibroblast growth factor
GDF-5	Growth and differentiation factor 5
GFP	Green fluorescent protein
hASCs	Human adipose-derived stem cells
Hz	Hertz
IGF-1	Insulin-like growth factor 1
MRI	Magnetic resonance imaging
MSCs	Mesenchymal stem/stromal cells
MMPs	Matrix metalloproteinases
Mkx	Mohawk
PCL	Poly(ε-caprolactone)
PDGF	Platelet-derived growth factor
PDL	Periodontal ligament
PEMF	Pulsed Electromagnetic Fields
SPCL	Starch poly(ε-caprolactone)
TE	Tissue engineering
TIMP	Tissue inhibitor of metalloproteinases
TSPCs	Tendon stem/progenitor cells
TTE	Tendon Tissue Engineering
TNC-C	Tenascin C
TNC	Tenascin
Tnmd	Tenomodulin
TGF-β	Transforming growth factor beta
Scx	Scleraxis

REFERENCES

Andarawis-Puri, N., J. B. Sereysky, K. J. Jepsen, and E. L. Flatow. 2012. The relationships between cyclic fatigue loading, changes in initial mechanical properties, and the in vivo temporal mechanical response of the rat patellar tendon. *J Biomech* 45 (1):59–65. doi:10.1016/j.jbiomech.2011.10.008.

Angelidis, I. K., J. Thorfinn, I. D. Connolly, D. Lindsey, H. M. Pham, and J. Chang. 2010. Tissue engineering of flexor tendons: The effect of a tissue bioreactor on adipoderived stem cell-seeded and fibroblast-seeded tendon constructs. *J Hand Surg* 35A:1466–1472.

Antman-Passig, M., and O. Shefi. 2016. Remote magnetic orientation of 3D collagen hydrogels for directed neuronal regeneration. *Nano Lett* 16 (4):2567–2573. doi:10.1021/acs. nanolett.6b00131.

Banos, C. C., A. H. Thomas, and C. K. Kuo. 2008. Collagen fibrillogenesis in tendon development: Current models and regulation of fibril assembly. *Birth Defects Res Part C* 84:228–244.

Barsby, T., and D. Guest. 2013. Transforming growth factor beta3 promotes tendon differentiation of equine embryo-derived stem cells. *Tissue Eng Part A* 19 (19–20):2156–2165. doi:10.1089/ten. TEA.2012.0372.

Batty, L. M., C. J. Norsworthy, N. J. Lash, J. Wasiak, A. K. Richmond, and J. A. Feller. 2015. Synthetic devices for reconstructive surgery of the cruciate ligaments: A systematic review. *Arthroscopy* 31 (5):957–968. doi:10.1016/j.arthro.2014.11.032.

Bavin, E. P., O. Smith, A. E. Baird, L. C. Smith, and D. J. Guest. 2015. Equine induced pluripotent stem cells have a reduced tendon differentiation capacity compared to embryonic stem cells. *Front Vet Sci* 2:55. doi:10.3389/fvets.2015.00055.

Bayer, M. L., P. Schjerling, A. Herchenhan, C. Zeltz, K. M. Heinemeier, L. Christensen, M. Krogsgaard, D. Gullberg, and M. Kjaer. 2014. Release of tensile strain on engineered human tendon tissue disturbs cell adhesions, changes matrix architecture, and induces an inflammatory phenotype. *PLoS One* 9 (1):e86078. doi:10.1371/journal.pone.0086078.

Bennett, M. B., R. F. Ker, N. J. Dimery, and R. M. Alexander. 1986. Mechanical-properties of various mammalian tendons. *J Zoology* 209:537–548.

Berthet, E., C. Chen, K. Butcher, R. A. Schneider, T. Alliston, and M. Amirtharajah. 2013. Smad3 binds scleraxis and mohawk and regulates tendon matrix organization. *J Orthop Res* 31 (9):1475–1483. doi:10.1002/jor.22382.

Bi, Y., D. Ehirchiou, T. M. Kilts, C. A. Inkson, M. C. Embree, W. Sonoyama, L. Li et al. 2007. Identification of tendon stem/progenitor cells and the role of the extracellular matrix in their niche. *Nat Med* 13 (10):1219–1227. doi:10.1038/nm1630.

Birk, D. E., and E. Zycband. 1994. Assembly of the tendon extracellular matrix during development. *J Anat* 184 (Pt 3):457–463.

Birk, D. E., J. M. Fitch, J. P. Babiarz, K. J. Doane, and T. F. Linsenmayer. 1990. Collagen fibrillogenesis invitro—Interaction of Type-I and Type-V collagen regulates fibril diameter. *J Cell Sci* 95:649–657.

Bonnin, M.-A., C. Laclef, R. Blaise, S. Eloy-Trinquet, F. Relaix, P. Maire, and D. Duprez. 2005. Six1 is not involved in limb tendon development, but is expressed in limb connective tissue under Shh regulation. *Mechan Develop* 122 (4):573–585.

Bottagisio, M., S. Lopa, V. Granata, G. Talo, C. Bazzocchi, M. Moretti, and A. Barbara Lovati. 2017. Different combinations of growth factors for the tenogenic differentiation of bone marrow mesenchymal stem cells in monolayer culture and in fibrin-based three-dimensional constructs. *Differentiation* 95:44–53. doi:10.1016/j. diff.2017.03.001.

Brent, A. E., R. Schweitzer, and C. J. Tabin. 2003. A somitic compartment of tendon progenitors. *Cell* 113 (2):235–248. doi:10.1016/S0092-8674(03)00268-X.

Brent, A. E., T. Braun, and C. J. Tabin. 2005. Genetic analysis of interactions between the somitic muscle, cartilage and tendon cell lineages during mouse development. *Development* 132 (3):515–528. doi:10.1242/dev.01605.

Chen, B., J. Ding, W. Zhang, G. Zhou, Y. Cao, W. Liu, and B. Wang. 2016. Tissue engineering of tendons: A comparison of muscle-derived cells, tenocytes, and dermal fibroblasts as cell sources. *Plast Reconstr Surg* 137 (3):536e–544e. doi:10.1097/01. prs.0000479980.83169.31.

Chen, J., J. Xu, A. Wang, and M. Zheng. 2009. Scaffolds for tendon and ligament repair: Review of the efficacy of commercial products. *Expert Rev Med Devices* 6 (1):61–73. doi:10.1586/17434440.6.1.61.

Chen, X., Z. Yin, J.-L. Chen, W.-L. Shen, H.-H. Liu, Q.-M. Tang, Z. Fang, L.-R. Lu, J. Ji, and H.-W. Ouyang. 2012. Force and scleraxis synergistically promote the commitment of human ES cells derived MSCs to tenocytes. *Sci Rep* 2:977.

Cheng, X., U. A. Gurkan, C. J. Dehen, M. P. Tate, H. W. Hillhouse, G. J. Simpson, and O. Akkus. 2008. An electrochemical fabrication process for the assembly of aniso-tropically oriented collagen bundles. *Biomaterials* 29 (22):3278–3288. doi:10.1016/j. biomaterials.2008.04.028.

Chiquet, M. 1999. Regulation of extracellular matrix gene expression by mechanical stress. *Matrix Biol* 18 (5):417–426.

Clegg, P. D., S. Strassburg, and R. K. Smith. 2007. Cell phenotypic variation in normal and damaged tendons. *Int J Exp Pathol* 88 (4):227–235.

Cohen, S., L. Leshansky, E. Zussman, M. Burman, S. Srouji, E. Livne, N. Abramov, and J. Itskovitz-Eldor. 2010. Repair of full-thickness tendon injury using connective tis-sue progenitors efficiently derived from human embryonic stem cells and fetal tissues. *Tissue Eng Part A* 16 (10):3119-37. doi:10.1089/ten.TEA.2009.0716.

Corps, A. N., A. H. Robinson, T. Movin, M. L. Costa, B. L. Hazleman, and G. P. Riley. 2006. Increased expression of aggrecan and biglycan mRNA in Achilles tendinopathy. *Rheumatol (Oxford)* 45 (3):291–294. doi:10.1093/rheumatology/kei152.

Correia, C., A. L. Pereira, A. R. C. Duarte, A. M. Frias, A. J. Pedro, J. T. Oliveira, R. A Sousa, and R. L. Reis. 2012. Dynamic culturing of cartilage tissue: The significance of hydro-static pressure. *Tissue Eng Part A* 18 (19–20):1979–1991.

Costa-Almeida, R., L. Gasperini, J. Borges, P. S. Babo, M. T. Rodrigues, J. F. Mano, R. L. Reis, and M. E. Gomes. 2016. Microengineered multicomponent hydrogel fibers: Combining polyelectrolyte complexation and microfluidics. *ACS Biomat Sci Eng.*

Czaplewski, S. K., T. L. Tsai, S. E. Duenwald-Kuehl, R. Vanderby, and W. J. Li. 2014. Tenogenic differentiation of human induced pluripotent stem cell-derived mesenchymal stem cells dictated by properties of braided submicron fibrous scaffolds. *Biomaterials* 35 (25):6907–6917. doi:10.1016/j.biomaterials.2014.05.006.

Dakin, S. G., F. O. Martinez, C. Yapp, G. Wells, U. Oppermann, B. J. Dean, R. D. Smith et al. 2015. Inflammation activation and resolution in human tendon disease. *Sci Transl Med* 7 (311):311ra173. doi:10.1126/scitranslmed.aac4269.

Dean, B. J., P. Gettings, S. G. Dakin, and A. J. Carr. 2016. Are inflammatory cells increased in painful human tendinopathy? A systematic review. *Br J Sports Med* 50 (4):216–220. doi:10.1136/bjsports-2015-094754.

Docheva, D., S. A. Muller, M. Majewski, and C. H. Evans. 2015. Biologics for tendon repair. *Adv Drug Deliv Rev* 84:222–239. doi:10.1016/j.addr.2014.11.015.

Doroski, D. M., M. E. Levenston, and J. S. Temenoff. 2010. Cyclic tensile culture promotes fibroblastic differentiation of marrow stromal cells encapsulated in poly(ethylene glycol)-based hydrogels. *Tissue Eng Part A* 16 (11):3457–3466. doi:10.1089/ten.tea.2010.0233.

DuFort, C. C., M. J. Paszek, and V. M. Weaver. 2011. Balancing forces: Architectural control of mechanotransduction. *Nat Rev Mol Cell Biol* 12 (5):308–319. doi:10.1038/nrm3112.

Egerbacher, M., S. P. Arnoczky, O. Caballero, M. Lavagnino, and K. L. Gardner. 2008. Loss of homeostatic tension induces apoptosis in tendon cells: An in vitro study. *Clin Ortho Rel Res* 466 (7):1562–1568. doi:10.1007/s11999-008-0274-8.

Fredberg, U., and K. Stengaard-Pedersen. 2008. Chronic tendinopathy tissue pathology, pain mechanisms, and etiology with a special focus on inflammation. *Scand J Med Sci Sports* 18 (1):3–15. doi:10.1111/j.1600-0838.2007.00746.x.

Gardel, L. S., C. Correia-Gomes, L. A. Serra, M. E. Gomes, and R. L. Reis. 2013. A novel bidirectional continuous perfusion bioreactor for the culture of large-sized bone tissue-engineered constructs. *J Biomed Mater Res B Appl Biomater* 101 (8):1377–1386. doi:10.1002/jbm.b.32955.

Garland, D. E., B. Moses, and W. Salyer. 1991. Long-term follow-up of fracture nonunions treated with PEMFs. *Contemp Orthop* 22 (3):295–302.

Gaut, L., and D. Duprez. 2016. Tendon development and diseases. *Wiley Interdiscip Rev Dev Biol* 5 (1):5–23. doi:10.1002/wdev.201.

Gelberman, R. H., P. R. Manske, W. H. Akeson, S. L. Woo, G. Lundborg, and D. Amiel. 1986. Flexor tendon repair. *J Orthop Res* 4 (1):119–128. doi:10.1002/jor.1100040116.

Genin, G. M., A. Kent, V. Birman, B. Wopenka, J. D. Pasteris, P. J. Marquez, and S. Thomopoulos. 2009. Functional grading of mineral and collagen in the attachment of tendon to bone. *Biophys J* 97 (4):976–985. doi:10.1016/j.bpj.2009.05.043.

Geremia, J. M., M. F. Bobbert, M. Casa Nova, R. D. Ott, A. Lemos Fde, O. Lupion Rde, V. B. Frasson, and M. A. Vaz. 2015. The structural and mechanical properties of the Achilles tendon 2 years after surgical repair. *Clin Biomech (Bristol, Avon)* 30 (5):485–492. doi:10.1016/j.clinbiomech.2015.03.005.

Girolamo, L. de, D. Stanco, E. Galliera, M. Viganò, A. Colombini, S. Setti, E. Vianello, M. M. Corsi Romanelli, and V. Sansone. 2013. Low frequency pulsed electromagnetic field affects proliferation, tissue-specific gene expression, and cytokines release of human tendon cells. *Cell Biochem Biophys* 66 (3):697–708.

Gonçalves, A., P. Costa, M. T. Rodrigues, I. R. Dias, R. L. Reis, and M. E. Gomes. 2011. Effect of flow perfusion conditions in the chondrogenic differentiation of bone marrow stromal cells cultured onto starch based biodegradable scaffolds. *Acta biomater* 7 (4):1644–1652.

Gonçalves, A. I., M. T. Rodrigues, P. P. Carvalho, M. Banobre-Lopez, E. Paz, P. Freitas, and M. E. Gomes. 2016. Exploring the potential of starch/polycaprolactone aligned magnetic responsive scaffolds for tendon regeneration. *Adv Healthc Mater* 5 (2):213–222. doi:10.1002/adhm.201500623.

Gonçalves, A. I., M. T. Rodrigues, S. J. Lee, A. Atala, J. J. Yoo, R. L. Reis, and M. E. Gomes. 2013. Understanding the role of growth factors in modulating stem cell tenogenesis. *PLoS One* 8 (12):e83734. doi:10.1371/journal.pone.0083734.

Gonçalves, A. I, Gershovich, P. M, Rodrigues, M. T, Reis, R. L, Gomes, M. E. 2017. Human adipose tissue-derived tenomodulin positive subpopulation of stem cells: A promising source of tendon progenitor cells. *J Tissue Eng Regen Med* 1–13. doi:10.1002/term.2495.

Gonçalves, A. I., R. Costa-Almeida, P. Gershovich, M. T. Rodrigues, R. L. Reis, and M. E. Gomes. 2015. Chapter 6 - Cell-based approaches for tendon regeneration. In *Tendon Regeneration*, edited by Manuela E. Gomes, Rui L. Reis and Márcia T. Rodrigues, pp. 187–203. Boston, MA: Academic Press.

Guerquin, M. J., B. Charvet, G. Nourissat, E. Havis, O. Ronsin, M. A. Bonnin, M. Ruggiu et al. 2013. Transcription factor EGR1 directs tendon differentiation and promotes tendon repair. *J Clin Invest* 123 (8):3564–3576. doi:10.1172/JCI67521.

Gurkan, U. A., X. Cheng, V. Kishore, J. A. Uquillas, and O. Akkus. 2010. Comparison of morphology, orientation, and migration of tendon derived fibroblasts and bone marrow stromal cells on electrochemically aligned collagen constructs. *J Biomed Mater Res A* 94 (4):1070–1079. doi:10.1002/jbm.a.32783.

Hannafin, J. A., S. P. Arnoczky, A. Hoonjan, and P. A. Torzilli. 1995. Effect of stress deprivation and cyclic tensile loading on the material and morphologic properties of canine flexor digitorum profundus tendon: an in vitro study. *J Orthopaedic Res* 13 (6):907–914.

Havis, E., M. A. Bonnin, J. Esteves de Lima, B. Charvet, C. Milet, and D. Duprez. 2016. TGFbeta and FGF promote tendon progenitor fate and act downstream of muscle contraction to regulate tendon differentiation during chick limb development. *Development* 143 (20):3839–3851. doi:10.1242/dev.136242.

Havis, E., M.-A. Bonnin, I. Olivera-Martinez, N. Nazaret, M. Ruggiu, J. Weibel, C. Durand, M.-J. Guerquin, C. Bonod-Bidaud, and F. Ruggiero. 2014. Transcriptomic analysis of mouse limb tendon cells during development. *Development* 141 (19):3683–3696.

Henstock, J. R., M. Rotherham, H. Rashidi, K. M. Shakesheff, and A. J. El Haj. 2014. Remotely activated mechanotransduction via magnetic nanoparticles promotes mineralization synergistically with bone morphogenetic protein 2: Applications for injectable cell therapy. *Stem Cells Transl Med* 3 (11):1363–1374. doi:10.5966/sctm.2014-0017.

Holladay, C., S.-A. Abbah, C. O'Dowd, A. Pandit, and D. I. Zeugolis. 2014. Preferential tendon stem cell response to growth supplementation. *J Tissue Eng Rege Med*. doi:10.1002/term.1852.

Ito, Y., N. Toriuchi, T. Yoshitaka, H. Ueno-Kudoh, T. Sato, S. Yokoyama, K. Nishida, T et al. 2010. The mohawk homeobox gene is a critical regulator of tendon differentiation. *Proc Natl Acad Sci U S A* 107 (23):10538–10542. doi:10.1073/pnas.1000525107.

Jones, E. R., G. C. Jones, K. Legerlotz, and G. P. Riley. 2013. Cyclical strain modulates metalloprotease and matrix gene expression in human tenocytes via activation of TGFβ. *Biochimica et Biophysica Acta* 1833:2596–2607.

Jones, G. C., A. N. Corps, C. J. Pennington, I. M. Clark, D. R. Edwards, M. M. Bradley, B. L. Hazleman, and G. P. Riley. 2006. Expression profiling of metalloproteinases and tissue inhibitors of metalloproteinases in normal and degenerate human achilles tendon. *Arthritis Rheum* 54 (3):832–842. doi:10.1002/art.21672.

Kang, K. S., J. Min Hong, J. A. Kang, J.-W. Rhie, Y. Hun Jeong, and D.-W. Cho. 2013. Regulation of osteogenic differentiation of human adipose-derived stem cells by controlling electromagnetic field conditions. *Exp Mol Med* 45:e6.

Kastelic, J., A. Galeski, and E. Baer. 1978. The multicomposite structure of tendon. *Conn Tissue Res* 6 (1):11–23.

Kayama, T., M. Mori, Y. Ito, T. Matsushima, R. Nakamichi, H. Suzuki, S. Ichinose, M. Saito, K. Marumo, and H. Asahara. 2016. Gtf2ird1-dependent mohawk expression regulates mechanosensing properties of the tendon. *Mol Cell Biol* 36 (8):1297–1309. doi:10.1128/MCB.00950-15.

Kim, J. H., M. S. Kang, M. Eltohamy, T. H. Kim, and H. W. Kim. 2016. Dynamic mechanical and nanofibrous topological combinatory cues designed for periodontal ligament engineering. *PLoS One* 11 (3):e0149967. doi:10.1371/journal.pone.0149967.

Kishore, V., J. A. Uquillas, A. Dubikovsky, M. A. Alshehabat, P. W. Snyder, G. J. Breur, and O. Akkus. 2012. In vivo response to electrochemically aligned collagen bioscaffolds. *J Biomed Mater Res B Appl Biomater* 100 (2):400–408. doi:10.1002/jbm.b.31962.

Kishore, V., W. Bullock, X. Sun, W. Scott Van Dyke, and O. Akkus. 2012. Tenogenic differentiation of human MSCs induced by the topography of electrochemically aligned collagen threads. *Biomaterials* 33 (7):2137–2144. doi:10.1016/j.biomaterials.2011.11.066.

Koob, T. J. 2002. Biomimetic approaches to tendon repair. *Comp Biochem Physiol A Mol Integr Physiol* 133 (4):1171–1192.

Langberg, H., H. Ellingsgaard, T. Madsen, J. Jansson, S. P. Magnusson, P. Aagaard, and M. Kjaer. 2007. Eccentric rehabilitation exercise increases peritendinous type I collagen synthesis in humans with Achilles tendinosis. *Scand J Med Sci Sports* 17 (1):61–66. doi:10.1111/j.1600-0838.2006.00522.x.

Lantto, I., J. Heikkinen, T. Flinkkila, P. Ohtonen, J. Kangas, P. Siira, and J. Leppilahti. 2015. Early functional treatment versus cast immobilization in tension after achilles rupture repair: Results of a prospective randomized trial with 10 or more years of follow-up. *Am J Sports Med* 43 (9):2302–2309. doi:10.1177/0363546515591267.

Legerlotz, K., G. C. Jones, H. R. Screen, and G. P. Riley. 2013. Cyclic loading of tendon fascicles using a novel fatigue loading system increases interleukin-6 expression by tenocytes. *Scand J Med Sci Sports* 23 (1):31–37. doi:10.1111/j.1600-0838.2011.01410.x.

Lejard, V., F. Blais, M. J. Guerquin, A. Bonnet, M. A. Bonnin, E. Havis, M. Malbouyres et al. 2011. EGR1 and EGR2 involvement in vertebrate tendon differentiation. *J Biol Chem* 286 (7):5855–5867. doi:10.1074/jbc. M110.153106.

Lima, J., A. I. Goncalves, M. T. Rodrigues, R. L. Reis, and M. E. Gomes. 2015. The effect of magnetic stimulation on the osteogenic and chondrogenic differentiation of human stem cells derived from the adipose tissue (hASCs). *J Mag Mag Mat* 393:526–536. doi:10.1016/j.jmmm.2015.05.087.

Liu, C. F., L. Aschbacher-Smith, N. J. Barthelery, N. Dyment, D. Butler, and C. Wylie. 2011. What we should know before using tissue engineering techniques to repair injured tendons: A developmental biology perspective. *Tissue Eng Part B Rev* 17 (3):165–176. doi:10.1089/ten.TEB.2010.0662.

Liu, H., C. Zhang, S. Zhu, P. Lu, T. Zhu, X. Gong, Z. Zhang, J. Hu, Z. Yin, B. C. Heng, X. Chen, and H. W. Ouyang. 2014. Mohawk promotes the tenogenesis of mesenchymal stem cells through activation of the TGFbeta signaling pathway. *Stem Cells*. doi:10.1002/stem.1866.

Liu, M., C. Lee, D. Laron, N. Zhang, E. I. Waldorff, J. T. Ryaby, B. Feeley, and X. Liu. 2016. Role of pulsed electromagnetic fields (PEMF) on tenocytes and myoblasts-Potential application for treating rotator cuff tears. *J Orthop Res*. doi:10.1002/jor.23278.

Liu, W., S. S. Watson, Y. Lan, D. R. Keene, C. E. Ovitt, H. Liu, R. Schweitzer, and R. Jiang. 2010. The atypical homeodomain transcription factor Mohawk controls tendon morphogenesis. *Mol Cell Biol* 30 (20):4797–4807. doi:10.1128/MCB.00207-10.

Liu, Z. T., X. L. Zhang, Y. Jiang, and B. F. Zeng. 2010. Four-strand hamstring tendon autograft versus LARS artificial ligament for anterior cruciate ligament reconstruction. *Int Orthop* 34 (1):45–49. doi:10.1007/s00264-009-0768-3.

Lohberger, B., H. Kaltenegger, N. Stuendl, B. Rinner, A. Leithner, and P. Sadoghi. 2016. Impact of cyclic mechanical stimulation on the expression of extracellular matrix proteins in human primary rotator cuff fibroblasts. *Knee Surg Sports Traumatol Arthros* 24 (12):3884–3891. doi:10.1007/s00167-015-3790-6.

Maeda, T., T. Sakabe, A. Sunaga, K. Sakai, A. L. Rivera, D. R. Keene, T. Sasaki et al. 2011. Conversion of mechanical force into TGF-β mediated biochemical signals. *Curr Biol* 21 (11):933–941.

Maffulli, N., K. Margiotti, U. G. Longo, M. Loppini, V. M. Fazio, and V. Denaro. 2013. The genetics of sports injuries and athletic performance. *Mus Ligaments Tendons J* 3 (3):173–189.

Maganaris, C. N., M. V. Narici, L. C. Almekinders, and N. Maffulli. 2004. Biomechanics and pathophysiology of overuse tendon injuries: Ideas on insertional tendinopathy. *Sports Med* 34 (14):1005–1017.

Magra, M., and N. Maffulli. 2005. Molecular events in tendinopathy: A role for metalloproteases. *Foot Ankle Clin* 10 (2):267–277. doi:10.1016/j.fcl.2005.01.012.

Magra, M., and N. Maffulli. 2008. Genetic aspects of tendinopathy. *J Sci Med Sport* 11 (3):243–247. doi:10.1016/j.jsams.2007.04.007.

Majewski, M., S. Schaeren, U. Kohlhaas, and P. E. Ochsner. 2008. Postoperative rehabilitation after percutaneous Achilles tendon repair: Early functional therapy versus cast immobilization. *Disabil Rehabil* 30 (20–22):1726–1732. doi:10.1080/09638280701786831.

Meirelles, L. da S., A. Maria Fontes, D. Tadeu Covas, and A. I. Caplan. 2009. Mechanisms involved in the therapeutic properties of mesenchymal stem cells. *Cytokine Growth Factor Rev* 20:419–427.

Meng, J., B. Xiao, Y. Zhang, J. Liu, H. D. Xue, J. Lei, H. Kong, Y. G. Huang, Z. Y. Jin, N. Gu, and H. Y. Xu. 2013. Super-paramagnetic responsive nanofibrous scaffolds under static magnetic field enhance osteogenesis for bone repair in vivo. *Sci Rep* 3:1–7. doi:10.1038/Srep02655.

Mikic, B., B. J. Schalet, R. T. Clark, V. Gaschen, and E. B. Hunziker. 2001. GDF-5 deficiency in mice alters the ultrastructure, mechanical properties and composition of the Achilles tendon. *J Orthop Res* 19 (3):365–371. doi:10.1016/S0736-0266(00)90018-4.

Mikic, B., K. Rossmeier, and L. Bierwert. 2009. Identification of a tendon phenotype in GDF6 deficient mice. *Anat Rec (Hoboken)* 292 (3):396–400. doi:10.1002/ar.20852.

Millar, N. L., G. A. Murrell, and I. B. McInnes. 2017. Inflammatory mechanisms in tendinopathy—Towards translation. *Nat Rev Rheumatol* 13 (2):110–122. doi:10.1038/nrrheum.2016.213.

Morita, Y., T. Sato, S. Watanabe, and Y. Ju. 2015. Determination of precise optimal cyclic strain for tenogenic differentiation of mesenchymal stem cells using a non-uniform deformation field. *Exp Mech* 55 (3):635–640. doi:10.1007/s11340-014-9965-0.

Morita, Y., S. Watanabe, Y. Ju, and B. Xu. 2013. Determination of optimal cyclic uniaxial stretches for stem cell-to-tenocyte differentiation under a wide range of mechanical stretch conditions by evaluating gene expression and protein synthesis levels. *Acta Bioeng Biomech* 15 (3):71–79.

Morita, Y., S. Suzuki, Y. Ju, and N. Kawase. 2014. Differences between protein expression and extracellular matrix state on uniaxial stretching for tenogenic differentiation. *J Mech Med Biol* 14 (02):1450025.

Morita, Y., T. Sato, S. Watanabe, and Y. Ju. 2017. Evaluation of precise optimal cyclic strain for tenogenic differentiation of MSCs. In *Mechanics of Biological Systems and Materials,* Vol. 6, pp. 149–155. Conference Proceedings of the Society for Experimental Mechanics Series. Springer, Cham, Switzerland.

Nelson, F. R., R. Zvirbulis, and A. A. Pilla. 2013. Non-invasive electromagnetic field therapy produces rapid and substantial pain reduction in early knee osteoarthritis: A randomized double-blind pilot study. *Rheumatol Int* 33 (8):2169–2173. doi:10.1007/s00296-012-2366-8.

Pneumaticos, S. G., P. C. N. Phd, W. C. McGarvey, D. R. Mody, and S. G. Trevino. 2000. The effects of early mobilization in the healing of achilles tendon repair. *Foot Ankle Int* 21 (7):551–557.

Praemer, A., S. Furner, D. P. Rice, and J. L. Kelsey. 1992. Musculoskeletal conditions in the United States. Chicago, IL: American Academy of Orthopedic Surgeons.

Pryce, B. A., S. S. Watson, N. D. Murchison, J. A. Staverosky, N. Dunker, and R. Schweitzer. 2009. Recruitment and maintenance of tendon progenitors by TGFbeta signaling are essential for tendon formation. *Development* 136 (8):1351–1361. doi:10.1242/dev.027342.

Randelli, P., A. Menon, V. Ragone, P. Creo, U. Alfieri Montrasio, C. Perucca Orfei, G. Banfi, P. Cabitza, G. Tettamanti, and L. Anastasia. 2016. Effects of the pulsed electromagnetic field PST(R) on human tendon stem cells: A controlled laboratory study. *BMC Compl Altern Med* 16:293. doi:10.1186/s12906-016-1261-3.

Rees, J. D., A. M. Wilson, and R. L. Wolman. 2006. Current concepts in the management of tendon disorders. *Rheumatol (Oxford)* 45 (5):508–521. doi:10.1093/rheumatology/kel046.

Riley, G. 2008. Tendinopathy—from basic science to treatment. *Nat Clin Pract Rheumatol* 4 (2):82–89. doi:10.1038/ncprheum0700.

Riley, G. P., R. L. Harrall, C. R. Constant, M. D. Chard, T. E. Cawston, and B. L. Hazleman. 1994. Tendon degeneration and chronic shoulder pain: changes in the collagen composition of the human rotator cuff tendons in rotator cuff tendinitis. *Ann Rheum Dis* 53 (6):359–66.

Rotherham, M., and A. J. El Haj. 2015. Remote activation of the Wnt/beta-catenin signalling pathway using functionalised magnetic particles. *PLoS One* 10 (3):e0121761. doi:10.1371/journal.pone.0121761.

Sant, S., D. F. Coutinho, A. K. Gaharwar, N. M. Neves, R. L. Reis, M. E. Gomes, and A. Khademhosseini. 2016. Self-assembled hydrogel fiber bundles from oppositely charged polyelectrolytes mimic micro-/nanoscale hierarchy of collagen. *Adv Funct Mater* 1606273. doi:10.1002/adfm.201606273.

Santos, L. J., R. L. Reis, and M. E. Gomes. 2015. Harnessing magnetic-mechano actuation in regenerative medicine and tissue engineering. *Trends Biotechnol* 33 (8):471–479. doi:10.1016/j.tibtech.2015.06.006.

Santos, L., M. Silva, A. I. Goncalves, T. Pesqueira, M. T. Rodrigues, and M. E. Gomes. 2016. In vitro and in vivo assessment of magnetically actuated biomaterials and prospects in tendon healing. *Nanomed (Lond)* 11 (9):1107–1122. doi:10.2217/nnm-2015-0014.

Sapir, Y., B. Polyak, and S. Cohen. 2014. Cardiac tissue engineering in magnetically actuated scaffolds. *Nanotechnol* 25 (1):014009. doi:10.1088/0957-4484/25/1/014009.

Sapir-Lekhovitser, Y., M. Y. Rotenberg, J. Jopp, G. Friedman, B. Polyak, and S. Cohen. 2016. Magnetically actuated tissue engineered scaffold: Insights into mechanism of physical stimulation. *Nanoscale* 8 (6):3386–3399. doi:10.1039/c5nr05500h.

Schweitzer, R., J. H. Chyung, L. C. Murtaugh, A. E. Brent, V. Rosen, E. N. Olson, A. Lassar, and C. J. Tabin. 2001. Analysis of the tendon cell fate using Scleraxis, a specific marker for tendons and ligaments. *Development* 128 (19):3855–3866.

Screen, H. R. C., S. R. Byers, A. D. Lynn, V. Nguyen, D. Patel, and S. J. Bryant. 2010. Characterization of a novel fiber composite material for mechanotransduction research of fibrous connective tissues. *Adv Funct Mat* 20 (5):738–747. doi:10.1002/adfm.200901711.

Screen, H. R. C., J. C. Shelton, D. L. Bader, and D. A. Lee. 2005. Cyclic tensile strain upregulates collagen synthesis in isolated tendon fascicles. *Biochem Biophys Res Commun* 336:424–429.

Seeliger, C., K. Falldorf, J. Sachtleben, and M. van Griensven. 2014. Low-frequency pulsed electromagnetic fields significantly improve time of closure and proliferation of human tendon fibroblasts. *Eur J Med Res* 19:37. doi:10.1186/2047-783x-19-37.

Sensini, A., C. Gualandi, L. Cristofolini, G. Tozzi, M. Dicarlo, G. Teti, M. Mattioli-Belmonte, and M. Letizia Focarete. 2017. Biofabrication of bundles of poly(lactic acid)-collagen blends mimicking the fascicles of the human Achille tendon. *Biofabrication* 9 (1):015025. doi:10.1088/1758-5090/aa6204.

Shilpa, S., D. F. Coutinho, A. K. Gaharwar, N. M. Neves, R. L. Reis, M. E. Gomes, A. Khademhosseini. 2017. Self-assembled hydrogel fiber bundles from oppositely charged polyelectrolytes mimic micro-/nanoscale hierarchy of collagen. *Adv Funct Mat*.

Shuakat, M. N., and T. Lin. 2014. Recent developments in electrospinning of nanofiber yarns. *J Nanosci Nanotechnol* 14 (2):1389–1408.

Shukunami, C., A. Takimoto, M. Oro, and Y. Hiraki. 2006. Scleraxis positively regulates the expression of tenomodulin, a differentiation marker of tenocytes. *Dev Biol* 298 (1):234–247. doi:10.1016/j.ydbio.2006.06.036.

Silver, F. H., J. W. Freeman, and G. P. Seehra. 2003. Collagen self-assembly and the development of tendon mechanical properties. *J Biomechan* 36:1529–1553.

Sorrenti, S. J. 2006. Achilles tendon rupture: Effect of early mobilization in rehabilitation after surgical repair. *Foot Ankle Int* 27 (6):407–410. doi:10.1177/107110070602700603.

Soslowsky, L. J., S. Thomopoulos, S. Tun, C. L. Flanagan, C. C. Keefer, J. Mastaw, and J. E. Carpenter. 2000. Neer Award 1999. Overuse activity injures the supraspinatus tendon in an animal model: A histologic and biomechanical study. *J Shoulder Elbow Surg* 9 (2):79–84.

Spiesz, E. M., C. T. Thorpe, S. Chaudhry, G. P. Riley, H. L. Birch, P. D. Clegg, and H. R. Screen. 2015. Tendon extracellular matrix damage, degradation and inflammation in response to in vitro overload exercise. *J Orthop Res* 33 (6):889–897. doi:10.1002/jor.22879.

Strauch, B., M. K. Patel, D. J. Rosen, S. Mahadevia, N. Brindzei, and A. A. Pilla. 2006. Pulsed magnetic field therapy increases tensile strength in a rat Achilles' tendon repair model. *J Hand Surg Am* 31 (7):1131–1135. doi:10.1016/j.jhsa.2006.03.024.

Subramanian, A., and T. F. Schilling. 2015. Tendon development and musculoskeletal assembly: Emerging roles for the extracellular matrix. *Development* 142 (24):4191–4204. doi:10.1242/dev.114777.

Sun, L., L. Qu, R. Zhu, H. Li, Y. Xue, X. Liu, J. Fan, and H. Fan. 2016. Effects of mechanical stretch on cell proliferation and matrix formation of mesenchymal stem cell and anterior cruciate ligament fibroblast. *Stem Cells Int* 2016:9842075. doi:10.1155/2016/9842075.

Teh, T. K., S. L. Toh, and J. C. Goh. 2013. Aligned fibrous scaffolds for enhanced mechanoresponse and tenogenesis of mesenchymal stem cells. *Tissue Eng Part A* 19 (11–12):1360–1372. doi:10.1089/ten. TEA.2012.0279.

Thomopoulos, S., G. R. Williams, J. A. Gimbel, M. Favata, and L. J. Soslowsky. 2003. Variation of biomechanical, structural, and compositional properties along the tendon to bone insertion site. *J Orthop Res* 21 (3):413–419. doi:10.1016/S0736-0266(03)00057-3.

Thorpe, C. T., S. Chaudhry, Lei, II, A. Varone, G. P. Riley, H. L. Birch, P. D. Clegg, and H. R. Screen. 2015. Tendon overload results in alterations in cell shape and increased markers of inflammation and matrix degradation. *Scand J Med Sci Sports* 25 (4):e381–e391. doi:10.1111/sms.12333.

Tucker, J. J., J. M. Cirone, T. R. Morris, C. A. Nuss, J. Huegel, E. I. Waldorff, N. Zhang, J. T. Ryaby, and L. J. Soslowsky. 2017. Pulsed electromagnetic field therapy improves tendon-to-bone healing in a rat rotator cuff repair model. *J Orthop Res* 35 (4):902–909. doi:10.1002/jor.23333.

Vogel, K. G., and D. Heinegård. 1985. Characterization of proteoglycans from adult bovine tendon. *J Biol Chem* 260 (16):9298–9306.

Voleti, P. B., M. R. Buckley, and L. J. Soslowsly. 2012. Tendon healing: Repair and regeneration. *Ann Rev Biomed Eng* 14:47–71.

Wada, A., H. Kubota, K. Miyanishi, H. Hatanaka, H. Miura, and Y. Iwamoto. 2001. Comparison of postoperative early active mobilization and immobilization in vivo utilising a four-strand flexor tendon repair. *J Hand Surg Br* 26 (4):301–306. doi:10.1054/jhsb.2000.0547.

Wang, J. H. 2006. Mechanobiology of tendon. *J Biomech* 39 (9):1563–1582. doi:10.1016/j.jbiomech.2005.05.011.

Wang, L., Y. Wu, B. Guo, and P. X. Ma. 2015. Nanofiber yarn/hydrogel core-shell scaffolds mimicking native skeletal muscle tissue for guiding 3D myoblast alignment, elongation, and differentiation. *ACS Nano* 9 (9):9167–9179. doi:10.1021/acsnano.5b03644.

Wang, T., Z. Lin, M. Ni, C. Thien, R. E. Day, B. Gardiner, J. Rubenson et al. 2015. Cyclic mechanical stimulation rescues achilles tendon from degeneration in a bioreactor system. *J Orthop Res* 33 (12):1888–1896. doi:10.1002/jor.22960.

Wang, T., Z. Lin, R. E. Day, B. Gardiner, E. Landao-Bassonga, J. Rubenson, T. B. Kirk et al. 2013. Programmable mechanical stimulation influences tendon homeostasis in a bioreactor system. *Biotechnol Bioeng* 110:1495–1507.

Wang, W., J. He, B. Feng, Z. Zhang, W. Zhang, G. Zhou, Y. Cao, W. Fu, and W. Liu. 2016. Aligned nanofibers direct human dermal fibroblasts to tenogenic phenotype in vitro and enhance tendon regeneration in vivo. *Nanomedicine (Lond)* 11 (9):1055–1072. doi:10.2217/nnm.16.24.

Wenstrup, R. J., J. B. Florer, E. W. Brunskill, S. M. Bell, I. Chervoneva, and D. E. Birk. 2004. Type V collagen controls the initiation of collagen fibril assembly. *J Biol Chem* 279 (51):53331–53337. doi:10.1074/jbc.M409622200.

Xu, W., Y. Wang, E. Liu, Y. Sun, Z. Luo, Z. Xu, W. Liu et al. 2013. Human iPSC-derived neural crest stem cells promote tendon repair in a rat patellar tendon window defect model. *Tissue Eng Part A* 19 (21–22):2439–2451. doi:10.1089/ten.TEA.2012.0453.

Yoon, J. H., and J. Halper. 2005. Tendon proteoglycans: Biochemistry and function. *J Musculoskelet Neuron Interact* 5 (1):22–34.

Younesi, M., A. Islam, V. Kishore, J. M. Anderson, and O. Akkus. 2014. Tenogenic induction of human MSCs by anisotropically aligned collagen biotextiles. *Adv Funct Mater* 24 (36):5762–5770. doi:10.1002/adfm.201400828.

Young, B. B., G. Zhang, M. Koch, and D. E. Birk. 2002. The roles of types XII and XIV collagen in fibrillogenesis and matrix assembly in the developing cornea. *J Cell Biochem* 87 (2):208–220. doi:10.1002/jcb.10290.

Youngstrom, D. W., I. Rajpar, D. L. Kaplan, and J. G. Barrett. 2015. A bioreactor system for in vitro tendon differentiation and tendon tissue engineering. *J Orthop Res* 33 (6):911–918. doi:10.1002/jor.22848.

Zhang, C., H. Yuan, H. Liu, X. Chen, P. Lu, T. Zhu, L. Yang et al. 2015. Well-aligned chitosan-based ultrafine fibers committed teno-lineage differentiation of human induced pluripotent stem cells for Achilles tendon regeneration. *Biomaterials* 53:716–730. doi:10.1016/j.biomaterials.2015.02.051.

Zhang, G. Y., P. S. Robinson, L. J. Soslowsky, R. V. Iozzo, and D. E. Birk. 2004. Development of tendon structure and function in the decorin-deficient mouse: Relationship between decorin and biglycan expression. *Faseb J* 18 (5):A788–A789.

Zhang, J., and J. H. Wang. 2010a. Mechanobiological response of tendon stem cells: Implications of tendon homeostasis and pathogenesis of tendinopathy. *J Orthop Res* 28 (5):639–643. doi:10.1002/jor.21046.

Zhang, J., and J. H.-C. Wang. 2010b. Production of PGE2 increases in tendons subjected to repetitive mechanical loading and induces differentiation of tendon stem cells into non-tenocytes. *J Orthopaedic Res* 28 (2):198–203.

11 Liver Tissue Engineering

Sara Morini, Natalia Sánchez-Romero, Iris Plá Palacín, Pilar Sainz Arnal, Manuel Almeida, Laurens Verscheijden, Joana I. Almeida, Alberto Lue, Sara Llorente, Helen Almeida, Pablo Royo Dachary, Agustín García Gil, Trinidad Serrano-Aulló, and Pedro M. Baptista

CONTENTS

11.1   INTRODUCTION

The human liver is the largest solid organ in the body, accounting for 2%–5% of body weight, and carries out a complex variety of functions, including metabolic homeostasis of nutrients and several hormones in the systemic circulation, the regulation of tolerance and inflammation, and the synthesis of mainly all plasma proteins. The liver also metabolizes xenobiotics, drugs, endogenous hormones and waste compounds, and produces bile to help digestion and as a preferred way of excretion of liver waste products. The liver possesses a unique regenerative capacity, with the potential for restoration of liver mass and function even in response to massive damage, when less than one-third of the cells remain functional.

Liver disease and the loss of liver function are currently the twelfth most frequent cause of death in the United States and the fourth most frequent for middle-aged adults.[1] Liver transplantation is the only definitive treatment for liver failure. Numerous attempts have been made to increase the supply of available livers for transplant, including split livers and living-related partial donor procedures.[2] Despite these surgical improvements, organ shortage remains critical. To palliate this growing problem, several research groups have made a significant effort towards the development of liver technologies that could provide temporary support for patients with liver failure. These intend to provide enough time for the regeneration of the native liver tissue or act as a temporary bridge to transplantation.

Liver tissue engineering approaches held promise as an alternative to organ transplantation. In this chapter, we will illustrate the actual strategies of liver tissue engineering, with a particular focus on the most representative liver models currently used in liver bioengineering. These models include bidimensional (2D) and tridimensional (3D) culture models and decellularization/recellularization of whole-liver scaffolds. We will also describe some of the necessary enabling technologies, such as spinner flasks or stirred-tank bioreactors, and the need for a universal formulation of culture medium with the addition of oxygen carriers that allows stable and efficient growth of the different cell types used. Finally, we will finish with the regulatory and technological challenges in the organ engineering field.

11.2   2D CULTURE OF HEPATIC CELLS

Primary human liver cells are considered the gold standard for constructing *in vitro* relevant liver culture models for a multitude of applications. Due to their origin in the human liver, they are the real surrogates of the native functionality of the organ *in vivo*. Hence, they can provide highly predictive results in toxicological and pharmacological *in vitro* research. On the one hand, as each cell preparation is generated from a different donor, this can offer the chance to evaluate several genetic polymorphisms using individual cell batches. On the other hand, the isolation procedures and the differences between the donors (i.e., age, sex, body mass index, liver fat content, liver damage) can introduce some variations in experimental results, which make the standardization of models difficult. Various cell types have been extensively studied and utilized for liver tissue engineering. These include primary human and porcine hepatocytes, immortalized hepatocytes, human hepatic cells lines, and stem cells.

11.2.1 PRIMARY PORCINE HEPATOCYTES

Taking advantage of the physiological similarity between pigs and humans, porcine hepatocytes are used to bypass the limited availability of human hepatocytes. The main limitations in using xenogenic porcine hepatocytes are the transfer of zoonotic diseases, protein-protein incompatibility, and the possible immune responses generated during treatment. Therefore, alternative cell sources are needed.

11.2.2 PRIMARY HUMAN HEPATOCYTES

Cultures of primary human hepatocytes (PHH) are considered to be the gold standard for the investigation of hepatic metabolism and toxicity of xenobiotics. They can maintain functional activity for 24–72 hours, can be used for enzyme induction and inhibition studies, allow for medium-throughput screening of compounds, and are useful for examining interspecies and inter-individual differences in metabolism.[3,4] However, the availability of PHH is limited. PHH are usually isolated from resected liver tissue or whole livers not suitable for transplantation. The success of cell isolation is then influenced by several factors, such as donor characteristics, intraoperative factors, cell isolation conditions, tissue collection, and processing.

Typically, PHH are isolated from the tissue by enzymatic digestion with collagenase. Then, the cells are plated on collagen-coated plastic dishes that allow cell attachment to form a confluent monolayer. Under these conditions, primary hepatocytes change their morphology, polarization, gene expression, and liver-specific functions, in a process also known as de-differentiation. This phenomenon represents a major stumbling block to these *in vitro* systems. As the isolated cells have lost the structure of their native niche, intercellular and extracellular matrix (ECM) interactions, and cell membrane structures, response to drugs or chemical agents can be different when compared with those occurring *in vivo*. These experimental outcomes will also affect cell functionality over time. Liver-specific functions, such as albumin production and cytochrome P450 expression, decrease rapidly over the first 24–48 hours of culture as the cells begin to lose their differentiated status.[5,6] Numerous studies tried to improve PHH survival and liver-specific functions *in vitro* through modifications in microenvironmental signals including soluble factors (medium formulation), cell-ECM interactions and heterotypic cell-cell interactions with non-parenchymal cells. The adjustment of cell media with hormones, corticosteroids, growth factors, vitamins, amino acids or trace elements, or non-physiological molecules (i.e., phenobarbital and dimethyl sulfoxide), and oxygen tension have been shown to modulate hepatocyte function.[7–9]

In contrast from the culture on a monolayer of ECM molecules (i.e., collagen I), hepatocytes have also been sandwiched between two layers of type I collagen gel. Under these conditions, hepatocytes exhibit desirable morphology with polarized bile canaliculi, as well as stable functions for several weeks.[10] Nonetheless, phase I/II detoxification processes have been shown to become imbalanced over time in this format.[11] Moreover, the presence of an overlapping layer of ECM may impair the diffusion of molecular stimuli (e.g., drug candidates).

11.2.3 HUMAN HEPATOCYTE CELL LINES

To enhance cell availability and overcome the limited *in vitro* proliferation of PHH, several attempts have been made to produce new or use existing human hepatocyte cell lines. Transfection of simian virus 40 T antigen (SV40 T antigen) has been used to develop many immortalized cell lines.[12] Human hepatocytes have also been immortalized through lipofectamine-mediated co-transfection of albumin-promoter-regulated antisense constructs and gene coding for the cellular transcription factor E2F and D1 cyclin.[13] However, the potential tumorigenic effects of these cells have not been adequately assessed in these studies. Common immortalized liver-derived cell lines in use are Fa2N-4, HepG2, Hep3B, PLc/PRFs Huh7, HBG, and HepaRG.[14,15] The HepG2 line expresses many liver-specific genes. However, the expression profile of genes involved in phase I and phase II metabolism has been shown to vary between passages, and as a result, data can be difficult to interpret across laboratories and with continuous cell expansion.[15] HepaRG, a human hepatoma-derived cell line, preserves the expression of several liver-specific functions as well as many cytochrome P450s, nuclear receptors, membrane transporters, and phase II enzymes.[16] HepaRG cells have a stable karyotype, can differentiate into either hepatocyte or biliary lineages, have a high proliferative capacity, and have shown to produce data that is both reproducible and consistent between various experiments. However, the expression of liver-specific functions in HepaRG cells is still lower than that of primary hepatocytes, and they represent a phenotype from a single donor, which reduce their predictive value for the human population.[14,16,17]

11.2.4 CO-CULTURE OF PHH WITH NON-PARENCHYMAL CELLS

The non-parenchymal cell (NPC) fraction contains several cell types of various origin, such as Kupffer cells (KC) (i.e., liver macrophages), liver endothelial cells (LEC), and the hepatic stellate cells (i.e., pericytes) (HSC). It is believed that these cells play an essential role in physiological liver function as well as in acute liver damage, in acute inflammation, and in chronic liver diseases. The crosstalk of hepatocytes with NPC through soluble factors and cell-cell connections plays a fundamental role in liver physiology and the maintenance of hepatocyte differentiation.[18] Various attempts have been made to realize co-culture models mimicking the *in vivo* liver environment. Nguyen TV and colleagues created co-culture models of PHH and KC to evaluate hepatocyte reactions in a pro-inflammatory environment.[19] Co-culture of PHH with LEC was shown to support hepatocytes in the maintenance of their specific functions, in addition to the formation of capillary-like structures by co-cultured endothelial cells.[20] Hepatic co-cultures have mainly been employed to investigate different physiologic and pathophysiologic processes, including the acute phase response, mutagenesis, xenobiotic toxicity, oxidative stress, and lipid and drug metabolism.[18]

11.2.5 STEM CELLS

Theoretically, stem cells represent an inexhaustible cell source and potentially provide large numbers of cells that can be stored for later use. The historically named "oval cells" are the putative stem cells of the adult liver that widely proliferate during prolonged liver injury and can differentiate into hepatocytes and cholangiocytes. Unfortunately, the low number of "oval cells" residing in the liver limits their isolation and purification. Moreover, there may be an increased risk of tumorigenesis in processes in which hepatocytes are derived from "oval cells." The increased risk suggests that "oval cells" could be targets for carcinogens and generate hepatocytes with genomic damage. Alternatively, excessive rounds of replication needed for the oval cell/hepatocyte transition may increase the risk for genetic abnormalities in newly formed hepatocytes. Contrarily to adult hepatocytes, fetal hepatocytes can actively grow *in vitro*. These are progenitor cells that can be isolated from human fetal livers at early stages of gestation.[21] These cells can proliferate extensively *in vitro* and can differentiate to hepatocyte and cholangiocytes *in vivo*. However, their limited availability, ethical issues, their potential tumorigenicity, and incomplete differentiation limits their use in clinical applications.[22] Nevertheless, they have been already used successfully in clinical trials of chronic liver disease.[23]

The pluripotency, availability, and unlimited proliferation *in vitro* make the embryonic stem cells (ESC) very attractive for regenerative therapy. Many laboratories have developed protocols to isolate human ESC and induce them to form hepatocyte-like cells.[24,25] Although, the clinical use of ESC for cell therapies still have some significant challenges, such as ethical concerns, immunocompatibility, and possible teratoma formation from residual non-differentiated cells. Moreover, all differentiation protocols result in a highly variable functionality within the cell population. Finally, cells begin to lose hepatic characteristics after a few days much like standard culture conditions.[26] Induced pluripotent stem cells (iPSC) can be used to encompass the ethical and immune concerns. Some efforts aimed at differentiating iPSC to hepatocytes are promising.[27,28] Nonetheless, expression levels of xenobiotic metabolism genes in iPSC are still not identical to those found in whole liver or freshly isolated primary hepatocytes. Concomitantly, enzyme levels decrease rapidly during the culture period (similar to 2D hepatocyte culture), the possible teratoma formation by non-differentiated cells, the concerns of using viral vectors for reprogramming, and changes in cell cycle regulators, are still limiting the application of these cells for liver regenerative therapy.[29,30]

Hepatocyte-like cells can also be generated from adult cells by over-expressing hepatocyte lineage-specific transcription factors.[31] These cells are usually described as induced hepatocyte-like cells (iHep). They can express hepatic genes and functions and engraft and reconstitute hepatic tissue in murine models. More studies are required to improve the development of this technology, but this method appears to be very promising for liver tissue engineering, as it can bypass all the complicated steps involved in generating iPSC and their differentiation to functional hepatocytes.[32] All these different strategies are displayed in Figure 11.1.

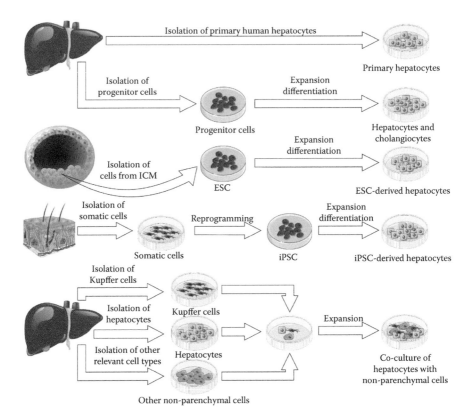

FIGURE 11.1 Currently available strategies for the isolation and generation of hepatic cells and tissues.

11.3 3D MODELS OF HEPATIC TISSUES

Due to the shortage of liver donors, tissue engineering has become a hopeful strategy to replace existing treatments for different liver diseases. The primary goal of regenerative medicine is to stimulate the diseased liver to regenerate itself or substitute it with an *ex vivo* bioengineered liver capable of performing the usual hepatic functions.[33] Hence, it is essential to develop appropriate supporting biomaterials that recreate the liver ECM and display some hepatic properties to recreate the *in vivo* microenvironment in an *ex vivo* setting. Consequently, to mimic the natural liver microenvironment, it is crucial to identify the most appropriate cells and a suitable supporting scaffold. Substances commonly employed for this purpose include natural polymeric materials, synthetic polymers or ceramics, biodegradable polymers with or without adsorbed proteins or immobilized functional groups, or, more recently, the product of the decellularization of an organ. Subsequently, the cells are seeded in the scaffold and cultured in a bioreactor until a functional tissue is generated. Thus, the original goal of this field is to recreate liver constructs that can be transplanted into patients to repair damaged hepatic tissues. One of the main shortcomings is the generation of functional vascularization. Therefore, the use of endothelial cells and stellate cells in the formation of hepatocyte spheroids has risen as a potential alternative.[34,35]

A significant effort has been required in the search of biomaterials capable of mimicking the ECM, as well as trying to replicate the benefits offered regarding cell adhesion, growth, viability, and maintenance of the differentiation state and metabolic function. To this aim, distinct biomaterials have been used, where alginate scaffolds have been established as one of the most widely employed in this field, owing to its porosity, hydrophilic properties, and biocompatibility. Alginate scaffolds in combinations with hepatocytes or mesenchymal stem/stromal cells (MSC) have demonstrated to improve the survival of animal models with 70–80% partial hepatectomy.[36,37] Chitosan is also used as a scaffold due to its low cytotoxicity, and high biodegradability and biocompatibility. Shang and co-workers generated a hybrid sponge made of hyaluronic acid and galactosylated chitosan to reproduce the liver environment and seeded hepatocytes and endothelial cells.[38] Due to the abundance of collagen in the hepatic ECM, its ubiquity, and biocompatibility, collagen I is a commonly used molecule in this field. Thus, 3D collagen scaffolds have been employed to culture a large variety of stem cells for several tissue engineering applications.[39,40] Nevertheless, the weak mechanical properties of collagen scaffolds limit their applications to some extent. To solve this problem, collagen scaffolds are cross-linked by chemical or physical methods or modified with natural/synthetic polymers or inorganic materials.[41] Additionally, these tissue-engineered liver constructs are also excellent biological surrogates for many pharmaceutical and biomedical applications. Hence, hepatic tissue bioengineering provides new alternatives and possibilities to many diseases and some hope for patients with untreatable diseases. Extracorporeal liver support devices emerged as a possible solution to reduce the number of patients waiting for liver transplantation.

11.3.1 Artificial and Bioartificial Liver Devices

Up to date, there are two types of liver support devices: artificial liver (AL) and bioartificial liver (BAL) devices, which differ in the system used for detoxification. Since the very beginning, the function of these devices was to detoxify the blood of injured livers; however, with a series of improvements, liver support devices let patients recover from an acute injury avoiding, in many cases, transplantation.[42] AL devices remove toxins from blood and plasma through different techniques such as hemofiltration, hemodialysis, hemodiafiltration, plasmapheresis, hemadsorption, plasma fractionation, or albumin dialysis.[42–44] MARS (molecular adsorbent recirculating system) and Prometheus are the two central albumin dialysis systems. With MARS, there is an albumin-impermeable membrane, which separates the high-flux albumin coated dialyzer from albumin-filled dialysate.[45,46] Whereas in Prometheus devices, albumin and protein-bound toxins pass across an albumin permeable membrane returning toxin-free albumin to the patient.[44,45,47] Several clinical studies have reported that MARS and Prometheus devices showed hepatic function enhancements in patients with acute liver failure (ALF) and acute-on-chronic liver failure.[44] On the other hand, BAL devices were designed to supply metabolic and hepatic functions, which are not available from AL devices.[48] It is possible for BAL devices to supply metabolic and hepatic functions due to the combination of hollow fibers, or porous matrix membranes, with primary hepatocytes and hepatoma cell lines.[48,49] Although an ideal source, the use of primary human hepatocytes is limited by their availability and reduced proliferative capacity *in vitro*, making them an unreliable source to use in BAL devices.[47,49] Hence, the most common BAL devices used so far in clinical trials are the HepatAssist and ELAD (extracorporeal liver assist device).[46,50] ELAD devices use a human immortalized cell line from the human hepatoma cell line HepG2, while the others use porcine hepatocytes.[51] In Europe, Asia, and US, several studies are carried out with BAL devices to assess efficacy and safety in the treatment of different end-stage liver diseases.

The use of human hepatic cells within the ELAD system has been granted the Orphan Drug Designation for ALF by the European Medicines Agency (EMA) and the US Food and Drug Administration (FDA). The clinical assessment of the efficacy and safety of ELAD system in subjects with Alcohol-Induced Liver Failure has concluded and the commercial introduction in the US is expected sometime in 2018.[52] Hep-Art Medical Devices B.V has developed a robust bio-artificial liver system, the Academic Medical Center-BAL (AMC-BAL). This support device has been developed to assist ALF patients to liver recovery or to the moment of donor-liver availability. The core of the system is a bioreactor loaded with hepatocytes. When based on porcine liver cells, the system appeared safe during a clinical phase 1 study with ALF patients.

To comply with the European Union (EU) standards, the company developed the human liver cell line cBAL 111 to replace the porcine cells and generated a protocol to mature the human HepaRG liver cell line in the AMC-BAL. The company is one of the participants of a large EU consortium (named BALANCE) and currently they are preparing the first clinical trial with the AMC-BAL, using HepRG liver cells, in patients with ALF.[53]

Although liver support devices are, in some cases, a promising solution for patients waiting for a liver transplantation, it is essential to address several challenges first. On the one side, AL systems are not capable of replacing necessary synthetic and metabolic functions of the liver. On the other, BAL systems can perform these functions, but they are currently too expensive to produce and operate, there are no optimal cell sources, and the bioreactor technologies still need further optimization.

11.3.2   3D BIOPRINTING

3D bioprinting is a novel technology, which consists in the biofabrication of artificial constructs through a layer-by-layer methodological approach. In this process, tiny quantities of biomaterials and cells are placed very carefully allowing the production of constructs with the most accurate detail.[54] Medical applications and uses for 3D bioprinting are multiplying including the creation of customized implants, prosthetics, drug discovery, and drug delivery or dosage forms.[55] Thus, the current medical uses of this technique can be classified into two significant categories: tissue and organ fabrication, and pharmaceutical research. Also, the employment of custom-made implants, equipment, and surgical tools have a positive repercussion in the patient recovery time, and thus, in the success of the treatment. Hence, 3D bioprinting presents some potential advantages beyond tissue engineering and regenerative medicine such as highly specific cell placing/positioning and concentration.[56] Compared to traditional methods, 3D bioprinting has the potential to make the fabrication of complex organs like the liver, an organ with high cell density, a lot easier.[57] In this sense, Organovo™ has generated 3D vascularized liver constructs with endothelial, hepatic stellate, and hepatocytes achieving high cell viability and robust drug metabolization. Furthermore, the use of liver spheroids, instead of single cells in bioprinting, protect the hepatocytes from the shear stress generated during this procedure.[58] Also, the employment of this technique for screening purposes could determine if a specific drug would be useful for a particular person.[59] Snyder and co-workers designed a microfluidic bioprinting model to study liver damage from radiation.[60] Other researchers carried out a drug-testing platform with immortalized hepatocytes encapsulated in alginate.[61] Finally, in October 2016, Organovo™ announced a plan to develop 3D-bioprinted human liver tissues for direct transplantation to patients.[62] Their preclinical studies in animal models have shown decent vascularization, engraftment, and function. Thus, the next step is to try to do it in humans with a therapeutic tissue program. In this program, Organovo will be focused on developing clinical solutions for acute-on-chronic liver failure and pediatric metabolic liver diseases.

11.3.3 Liver Organoids and Spheroids

Liver organoids are more complex models that try to mimic 3D cell-cell and cell-ECM interactions in more appropriate physiological conditions. Thus, liver organoids are 3D *in vitro* cultures that can better replicate the microenvironment of *in vivo* tissue.[63,64] Hence, it is essential to keep in mind that primary hepatocytes can form spheroids called hepatospheres when creating these structures. These assemblies are spontaneous, non-adherent aggregates where the functionality of the cells is preserved due to the majority of attachments among cells. Integrin-ECM binding followed by cadherin-cadherin interactions originate the spheroids.[65] The size in these configurations is very critical because it has been demonstrated that cell viability decreases with increasing spheroid size.[66,67] Hence, spheroids exemplify the most energy efficient organization. In the 1980s, Laundry and colleagues started to use the term *spheroid* to describe 3D rat hepatocyte aggregates.[68] Until then, the term *aggregation pattern* had been employed to explain the capacity of cells to produce aggregates in 24 hours. In the of the liver, the hepatic spheroids must be capable of reproducing hepatic tissue. Several approaches can be used, such as: (1) non-adherent dishes under static conditions, (2) agitation cultures, or (3) hanging drops. Culturing hepatocytes in a low-adherent well is the simplest way to obtain spheroids in static conditions. After an initial attachment, these cells form a monolayer, which gradually separates from the plate giving rise to a spheroid. Uncoated plates with albumin, a positive charge, or the elimination of serum have shown to be beneficial in spheroid formation,[69] while laminin, collagens, or fibronectin prevent it. On the other hand, it has been confirmed that agitation conditions, including rocked and rotary cultures, improve spheroid formation. Sakay and colleagues tried to obtain a large number of porcine hepatocyte spheroids to be used in a BAL through a rotational culture in a spinner flask fitted with a silicon tubing apparatus for oxygen supply. They concluded that this suspension culture method could be one of the most promising module types.[70] Brophy and colleagues demonstrated that hepatocytes formed spheroids when suspended in rocked (oscillatory) motion.[71] Rocking promoted mixing, oxygenation, and increased frequency of collisions between freshly isolated hepatocytes, which in turn accelerated their aggregation into clusters and the formation of spheroids. Their results showed that the rocked technique is a relatively simple yet rapid and highly efficient method of forming 3D, suspension-stable, hepatic tissue constructs.[72] Nevertheless, all the techniques described earlier have the same problem when obtaining aggregates with regular geometry. In 2003, Kelm and co-workers published a universal protocol to produce hepatospheres with a highly reproducible size with less than 10% of variations.[73]

Hence, liver organoids and spheroids are a powerful tool to better evaluate the cellular changes that lead to tumorigenesis and cancer progression.[74,75] Also, in recent years, many organoid system models have been developed for drug screening and testing. Drug diffusion kinetics change radically in 3D culture, probably due to these structures reproduce much better the interactions between cells and matrix. This theory could explain why drugs that are effective in 2D models are often ineffective when tested in patients.[74,75] Thus, several commercial 3D platforms have been developed for drug screening and drug studies. These platforms include the 3D InSight™ Human Liver Microtissues of Insphero, the "Hepatopac" platform,[76] the HepaChip® in vitro microfluidic system,[77] and the Hμrel® microliver platforms.[78] Hence, this field seems to be progressing to a more personalized medicine owing to the use of host-tissue based organoids in patients with liver malignancies to screen pharmacologic agents for activity against tumors and toxicity in the normal tissues.

11.4   WHOLE-LIVER BIOENGINEERING

Tissue decellularization consists on the perfusion of several detergent solutions or other reagents through the whole organ vascular network, obtaining an acellular bioscaffold.[79,80] This procedure results in a complete decellularized organ that preserves the native ECM microarchitecture and many essential bioactive signals that are difficult to replicate in vitro and are vital for cell viability, attachment, and differentiation.[81] Furthermore, the vascular network and the bile duct system can also be well preserved.[82]

In the last years, decellularized scaffolds have been used for a variety of regenerative medicine strategies for tissue and organ replacement. The different techniques used for the decellularization process can be classified into the categories of chemical agents,[83,84] biologic agents,[83,85] and physical agents.[86,87] Usually, these methods are combined to maximize the effectiveness of the decellularization. The use of decellularized thin tissues already has some clinical applications in various areas, such as orthopedic and dental, plastic/reconstructive surgery, and the cardiovascular field. ECM-based materials are currently used for the treatment of tendon/ligament damage (i.e., *GraftJacket®* regenerative tissue matrix—Wright Medical Inc., USA), in oral and maxillofacial surgery (i.e., *AlloDerm®* - LifeCell Corp., USA), and in cardiovascular surgery (i.e., *CardioCel®* - Admedus HIS Inc., Australia).[88] Nevertheless, solid organ replacement is still unavailable because organs are much larger than tissues, they possess a complex 3D architecture and need a sizeable vascular network to become functional. As a result, the different agents used for decellularization (mostly detergents) are introduced into the organ through its vascular system, allowing an efficient removal of all native cells and residual DNA while preserving the structural integrity of the ECM. Regarding the liver, several research groups have used different perfusion protocols to generate decellularized 3D liver scaffolds. For example, Uygun and colleagues demonstrated a decellularization method for ischemic rat livers based on portal vein perfusion using different concentrations of Sodium Dodecyl Sulphate (SDS): 0.01% SDS for 24h followed by 0.1% SDS for 24h, and then 1% SDS for 24h.[89] After 88 hours, the authors obtained a translucent acellular scaffold, which retained the gross shape of the liver. Baptista and co-workers used a different perfusion protocol depending on liver size and structure.[79] Several animal models were used, namely mice, rats, ferrets, rabbits, and pigs. The protocol proposed by the authors involved perfusing deionized water through the portal vein (40 times the volume of the liver) followed by 1% Triton X-100 with 0.1% ammonium hydroxide. After decellularization, the hepatic parenchyma was transparent, and the vascular tree was visible. De Kock and collaborators optimized a method where the liver was first perfused with 1% Triton-X 100 solution for 30 min followed by perfusion with 1% SDS solution for 30 minutes. In this study, researchers achieved whole rat liver decellularization within 60 minutes by portal perfusion with mild detergent solutions, resulting in an acellular, translucent, naturally-derived liver scaffold capable of withstanding fluid flows and maintaining the *in vivo* 3D liver architecture.[90] After the decellularization process, scaffolds are sterilized and mounted in a bioreactor system for their recellularization with different types of *in vitro* cultured cells. One of the most challenging aspects of the recellularization process is to generate the appropriate number and ratio of different cells that represent the hepatic tissue, according to their native distribution.[91]

There are also different techniques for scaffold recellularization. Soto-Gutiérrez and co-workers compared three different seeding conditions: direct parenchymal injection, continuous perfusion, and multistep infusion.[92] The direct parenchymal injection works by directly injecting the cells into the hepatic lobes of the acellular scaffold. The continuous perfusion method involves perfusing the cells as a suspension in the culture medium, at a controlled flow rate through the liver scaffold in a bioreactor system. As for multistep infusion, it involves the delivery of the cells as smaller multiple batches separated by fixed time intervals. The authors concluded that the best results were obtained using the multistep infusion system, which resulted in a hepatocyte engraftment rate of 90% with cells able to produce albumin, metabolize ammonia and express CYP1A/2 activity during seven days of experiment. The different decellularization/recellularization methods described earlier have helped us to further understand the field of liver tissue engineering and the future direction of this technology.

11.5 ENABLING TECHNOLOGIES

Bioreactors for tissue engineering and cell therapy are used for different purposes, such as large-scale cell culture, promotion of tissue formation, or acellular scaffold seeding. They represent a robust alternative to traditional 2D cultures. In these, a designed fluid flow provides molecular transport and mechanical stimulation, which mimic the living microenvironment, providing fundamental requirements for the correct cell development and function[93] (like oxygen, pH, nutrients, metabolite transport, temperature), maintaining the 3D structure. These systems produce significant amounts of cells and the speeding up of the cell culture process. Commercially available microcarriers with different size and composition can also be used for large-scale cell culture for many different anchorage-dependent cell types, including fibroblasts, osteoblasts, chondrocytes or hepatocytes, among others.[94] Perfusion bioreactors can also be used for acellular scaffold seeding.[79,89,95–97]

11.5.1 SPINNER FLASKS AND STIRRED-TANK BIOREACTORS FOR LIVER TISSUE ENGINEERING

The ability of hepatocytes to spontaneously aggregate when seeded on low-adhesive surfaces, and for aggregates to maintain liver-specific functions to a higher degree than cells in monolayer culture, has been reported in various studies.[98,99] Formation and maintenance of spheroids in traditional static culture systems are often tricky, as cell density, culture medium depth, and other multiple factors compete to create local microenvironments that are difficult to control (especially oxygen concentration) and reproduce between labs.[100] Various groups have developed bioreactors based on stirred systems to enhance stem cell cultures and differentiation systems. Yin CH. and colleagues created a cultivation system to differentiate mouse ESC-derived embryoid bodies into hepatocytes using a spinner bioreactor.[101] Wu and co-workers generated an efficient method of forming spheroids in spinner vessels, using freshly harvested primary rat hepatocytes cultivated as spheroids. They observed that within 24 hours after inoculation, more than 80% of inoculated cells formed spheroids and the efficiency was significantly higher than in stationary Petri dishes. Moreover, microvilli-lined bile canaliculus-like channels were observed in the interior of spheroids. The authors suggested that this method could be useful for studying xenobiotic drug metabolism.[102] Spheroids can experience a high shear force. To bypass this problem, Nelson LJ. and collaborators generated a new rotational cell culture system. In their work, the authors showed that in the absence of exogenous biomatrix scaffolding, primary porcine hepatocytes cultured in chemically defined medium under microgravity conditions rapidly form macroscopic 3D spheroid structures, which exhibit ultrastructural, morphological and functional features of differentiated, polarized hepatic tissue.[103]

Stirred-tank bioreactors have mainly been used in the large-scale culture of mammalian cells to create a suitable microenvironment with tightly controlled variables, such as pH and oxygen levels. With appropriate design of impellers, the fluid mechanic microenvironment in these systems can provide relatively uniform low-shear mixing, and the commercial availability of reactor systems makes them accessible for general use. Recently, Tostões and colleagues established and tested a perfused bioreactor system for the long-term maintenance of primary cultures of human hepatocyte spheroids. They found that using this method, the generated hepatocyte spheroids reproducibly recapitulated *in vivo* hepatic functions and structure, despite inter-donor variability. They hypothesized that these reproducible time-course profiles were made possible because of the tight control of critical environmental variables at physiological values. Moreover, they observed that the spheroid's inner structure resembled the liver architecture, with functional bile canaliculi-like structures and liver-specific markers (such as CYP450 expression and phase II and III drug-metabolizing enzyme gene expression and transport activity). This system constitutes an ideal long-term culture platform for analyzing hepatic function for drug development tests.[104]

11.5.2 Universal Formulation of Culture Medium

So far, there is no "universal" culture medium available when designing a bioreactor-based process since it depends on the cell type used in each bioreactor system. If only one cell type is involved, the culture medium used for the bioreactor system will be the same as used in the expansion under standard conditions (i.e., static). When using two or more cell types, more often than not, an optimal medium is not possible, so an appropriate culture medium should be formulated so that the minimum requirements are met for each cell type. This strategy usually allows for the best possible average growth and viability/function maintenance across all cell types. Nevertheless, there are currently several groups exploring the possibilities of designing a "universal" culture medium that functions as a blood substitute for applications ranging from organ bioengineering to microfluidic platforms of body-on-a-chip.

11.5.3 Oxygen Carriers

Oxygen is a fundamental component of the hepatic microenvironment, mediating cellular survival, differentiation, growth, metabolism, and function. Suitable oxygen supply is essential for the function of highly metabolic tissues such as muscle or liver. Oxygen is provided to hepatocytes *in vivo* by a mixture of arterial and venous blood at an estimated rate of ~1.2 nmol/s/10^6 cells.[105] In contrast, hepatocyte consumption has been measured *in vitro* to be 0.9 nmol/s/10^6 cells during the first 48 hours after seeding and 0.4 nmol/s/10^6 cells during stable long-term cultures.[106,107] During standard *in vitro* culture, oxygen is supplied by diffusion from the air-liquid interface. Frequently, hepatocytes are cultured below 1–1.5 mm of a stagnant medium layer under ambient oxygen partial pressure. Here, oxygen partial pressure on the cell surface is expected to reach 12 mmHg,[108] resulting in an oxygen uptake/supply rate of 0.35 nmol/s/10^6 cells in culture, ~3-fold less than the oxygen demand of freshly seeded hepatocytes.[109] This oxygen diffusion limits the density of hepatocytes that can be used in a bioreactor or seeded in culture. Even though the increase of oxygen tension to 500 mmHg or above would meet the oxygen demand of the cells, various studies have indicated that supraphysiological oxygen levels can promote the formation of free radicals that compromise cellular viability, especially during long-term cultures.[110,111]

The addition of an oxygen carrier to the culture medium can represent an ideal method to increase the solubility of oxygen in the liquid, enhancing oxygen supply without raising oxygen tension. There are mainly two classes of oxygen carriers that have been used with hepatic cells and tissue: perfluorocarbon-based (PFC) and hemoglobin-based (HBOC). PFC can be chemically synthesized, are biologically inert, highly stable, not subject to oxidation or free radical reactions, cannot be metabolized, and do not present any potential risk of virus infection.[112,113] However, their oxygen solubility is, at most, around 20 times higher than that of blood,[114] and they only increase the oxygen concentration within one order of magnitude, far below blood oxygen levels.[115,116] PFC is used as a suspension of its emulsions, and the maximum level of PFC is around 15% (v/v). Hence, the total oxygen solubility of such PFC-containing physiological saline or culture medium is, at most, four times higher than those of culture medium. This oxygenation capacity is also not sufficient to meet the need of substantial liver tissue equivalents.[117] One exciting application of PFC-based oxygen carriers in liver tissue engineering is shown by the study of Nahmias and colleagues.[115] They reported incorporation of PFC emulsions into the upper collagen gel layer in sandwich-collagen gel culture of rat hepatocytes. The cells showed a significant increase in viability and functions (i.e., long-term albumin secretion and long-term urea secretion), compared to usual polystyrene dish- or plate-based cultures of hepatocytes, where the oxygen supply limits the hepatocyte functions. Thus, the authors concluded that culturing hepatocytes on oxygenated matrix mimics the oxygen-rich environment of the liver and provides a simple method for enhanced long-term function. HBOC are usually obtained from expired human blood units and potentially infected by viruses. Nonetheless, they have a much higher potential to transport oxygen as 1g of hemoglobin can bind approximately 1.37 ml of oxygen at $22°C$,[118] much higher solubility comparable to that of blood and release oxygen very efficiently at low oxygen tension according to the specific oxygen association-dissociation curve.[115] Despite that, free hemoglobin molecules, when in contact with living tissues, can induce vasoconstriction and renal toxicity in mammals and can also act as a trigger in the production of radical oxygen species (ROS). To circumnavigate these stumbling blocks, encapsulation or crosslinking are usually employed. Purified hemoglobin can be encapsulated in a phospholipid bilayer that resembles the membrane of red blood cells, resulting in a relatively stable liposome structure that prevents the leakage of free hemoglobin and its direct contact with cells.[119] Polyethyleneglycol (PEG)-decorated liposome-encapsulated hemoglobin (LEH) is one of the promising designs as a red blood cell substitute for infusion to humans.[120] LEH has a high capacity to bind and release oxygen physiologically, it carries no blood type antigens or pathogens, and can be stored for long periods (~ two years). *In vivo*, LEH has been shown not to cause damage to organs, including the liver. However, the half-life of LEH in circulation is still under red blood cells.[121] Sakai and collaborators checked its possible toxicity and efficacy in the short-term (24 hours) in mature and fetal rat hepatocytes cultured in conventional monolayers and a 2D thin flat-plate bioreactor. They observed remarkably enhanced viability and functionality of mature rat hepatocytes in the presence of LEH. Unfortunately, substantial toxicity was explicitly observed for hepatic progenitors. Because hemoglobin-free liposomes had no toxicity, the observed toxicity was attributed to the hemoglobin molecules.[122] Montagne and colleagues used the same PEG-decorated LEH to investigate the effects of longer exposures (up to 14 days) in adult and rat fetal liver cells, with the purpose of engineering sizeable artificial liver equivalents in the future. Primary fetal and adult rat liver cells were directly exposed to LEH for 6–14 days in static culture or a perfused flat plate bioreactor. The authors showed that the functions and viability of adult rat hepatocytes exposed to LEH were not adversely affected by static monolayer culture and were improved in the bioreactor system. However, some toxicity of LEH was observed with fetal liver cells after four days of culture. Albeit LEH could be considered a suitable oxygen carrier for long-term culture of the mature hepatocyte, currently it is not ideal for perfusing fetal hepatocyte cultures in direct contact with the liposomes. Consequently, LEH will have to be made less toxic, or a more complex bioreactor, which avoids the direct contact between hepatocytes and perfusates, must be designed if fetal liver cells are planned to be used for liver tissue engineering.[123]

11.6 ETHICAL, REGULATORY AND TECHNOLOGICAL CHALLENGES

11.6.1 REGULATORY LANDSCAPE

Development of a tissue-engineered product for clinical use can be challenging. Because of the novelty, complexity, and technical specificity, it is essential to understand the regulations that guarantee the quality and safety of these novel products. For this goal, we will be focused on two regulatory agencies: FDA and EMA. Both organizations have similar objectives, but their systems of operation are different, and the approval of one of them does not imply the endorsement by the other.

11.6.1.1 Food and Drug Administration

The FDA's Center for Biologics Evaluation and Research is responsible for ensuring the safety, purity, potency, and effectiveness of many biologically derived products. The term *tissue engineered medical products* (TEMP) has been defined in a standard document of the American Society for Testing and Materials, and this terminology has been included in the FDA-recognized consensus standards database.[124]

TEMP can consist of a variety of different constituents, such as cells, scaffolds, device, and others (or any combination of these). Because there is no legal definition for TEMP in FDA regulations and because of many parts—each with different regulatory pathway—make up the tissue-engineered product, the FDA classifies these products as combination products. Congress recognized the existence of combination products when it enacted the Safe Medical Device Act of 1990, and it was defined in the 21 Code of Federal Regulation 1270/1271 Part C 210/211/820.[125] It is necessary to determine the Primary Mode of Action (PMOA) of the overall product[123] to determine which center will be responsible for a particular combination product.[126] The PMOA is defined as "the single mode of action of a combination product that provides the most important therapeutic effect of the combination product" [https://www.fda.gov/downloads/CombinationProducts/AboutCombinationProducts/UCM571726.pdf].

11.6.1.2 European Medicines Agency

In the European Union (EU), cells with a scaffold are regulated under the advanced therapy medicinal products (ATMPs) directives as viable cells with a scaffold. EMA and the Committee for Advanced Therapies (CAT) are responsible for reviewing applications for marketing authorization for Advanced Therapy Medicinal Products (ATMPs). In 2007, the European Parliament and Council of the European Union (EU) issued an amendment to Directive 2001/83/EC and Regulation (EC) No. 776/2004 to include regulatory provisions for ATMPs defined in Regulation EC No 1394/2007.[127–129]

According to this regulation, when a product contains viable cells or tissues, the pharmacological, immunological, or metabolic action of those cells or tissues shall be considered as the principal mode of action of the product. Therefore, biomaterials might not be the only actor in these fields. However, the biomaterial biocompatibility is still an essential requisite, and the new products will be subjected to the same regulatory rules as the others biomedical devices (Regulation EC No 1394/2007).

In addition to the requirements laid down in Article 6 of Regulation No 726/2004, the application for the authorization of an ATMP containing medical devices, biomaterials, scaffolds, or matrices shall include a description of the physical characteristics and performance of the product. It should also include the description of the product design method, by the Annex 1 to Directive 2001/83/EC.

11.6.2 BENCH TO BEDSIDE AND CLINICAL TRIALS

In the last few years, almost every tissue and organ in the body has been bioengineered *in vitro* with success. Currently, there is already a significant number of clinical products available on the market composed of decellularized scaffolds and seeded with allogeneic or xenogeneic cell sources. Some of the most relevant, available tissues include skin, small intestine, esophagus, bladder, bone, carotid arteries, amongst others.[130–132] Familiar to them all, the lack of requirement of an extensive vascular network to become functional.

One of the first clinical applications of tissue engineered products were skin grafts constructed from decellularized human skin, such as Alloderm®, which have been immensely successful and in use for two decades now.[133] The facilitated access makes it easy to assess engraftment and viability and to replace the graft if necessary.[134] Since then, the tissue engineering sector has grown exponentially with breakthroughs reached in this area in the last years, as well as the relevance of the economic activity, with increasing numbers of products entering the marketplace and into clinical trials. Currently, the availability of tissue engineered-based products is a fact, and although the investigations have shown promising results, significant challenges still need to be overcome in the area of whole organ engineering. These technological challenges are the main limitation to transfer this new technology into clinical trials. These include identifying appropriate species for providing decellularized tissues, selecting ideal cell sources, achieving robust vascularization, optimizing bioreactor perfusion technology along with scalability, and preventing graft immunological rejection.[135–137]

11.6.3 GMP Production

Control of clinical product manufacturing in both EU and the US is exerted by using Good Manufacturing Practice (GMP) regulations and guidelines, to protect the patient from receiving poor quality, unsafe, and products that vary from their specifications. GMP regulation includes Good Practice for Tissue and Cells[138] and Good Engineering Practice (GEP). GMP facilities follow GMP guidelines promulged by each regulatory agency, and they have specialized facility designs and highly trained personnel to produce the first clinical prototype faithfully in a controlled and reproducible fashion. The EU regulates this field by the publication of GMP directives and GMP Guidelines, which are prepared and published in one volume by the European Commission under the auspices of Directorate General Enterprise.[139] The US control procedures are comparable to EU's practices, whereby the GMP Regulations are published in the Code of Federal Regulations by various executive departments and agencies of Federal Government.[140] Recently, FDA and EMA are making significant progress toward mutually recognizing each other's GMP inspections. In March 2011, both agencies launched a joint pilot program. This program aimed to foster greater international collaboration and information sharing to help to distribute inspection capacity better, allowing more sites to be monitored and reducing unnecessary duplication through the implementation of the International Council for Harmonisation (ICH) and relevant regulatory requirements.[141]

11.6.4 Stem Cells

The discovery and manipulation of human embryonic stem cells (ESC) have been described as one of the most significant breakthroughs in biomedicine of the century.[142] In fact, it offers a great promise for understanding the underlying mechanisms of human development and differentiation, and it represents the hope for new cellular therapies against human diseases. However, human ESC research also raises sharp ethical and political controversies.[143] The central moral question about human ESC research concerns the respect for human life. Since the usage of these cells involves the destruction of human embryos, the question of when life begins has been highly controversial and closely linked to debates over abortion. Another dilemma about human ESC research is the availability of the biological material. Although the donation of embryos and gametes is based on very well-established guidelines,[144–146] a recent scandal involved inappropriate payments to gametes donors, severe deficiencies in the informed consent process, and a serious encouragement on scientists in serving as donors.[147,148] Poor medical assistance along medical procedures has been identified as well. On the other hand, the usage of surplus human embryos available from *in vitro* fertilization (IVF) used for stem cell research is also subject to debate since it does not protect the reproductive interests of donors in the infertility treatments. Apart from that, IVF embryos may carry a high rate of genetic errors (such as mitochondrial DNA defects) and therefore would be unsuitable for human ESC research.[149] Another hazard that human ESC faces is their potential uncontrollable growth (i.e., development of teratomas) and the possibility for them to be rejected by the host immune system. Much remains to be discovered about stem cells, but some of the ethical and technological dilemmas presented earlier need to be discussed with the scientific community to ensure that the potential of human ESC research is carried out in an ethically appropriate manner for everyone involved.

About a decade ago, iPSCs were engineered for the first time by transduction of only four factors in human and mouse fibroblasts, resulting in a cell type sharing many characteristics with ESC.[150] This discovery led to enthusiasm in the field, as ethical difficulties associated with the use of human ESC could easily be circumvented. Moreover, the derivation of iPSC introduced the possibility of autologous cell transplantation for virtually every human tissue and enabled a more elegant way to personalize disease models and toxicity screenings.[151,152] Indeed, recently early clinical phase I and II trials have been started using cells derived from iPSC (and ESC). However, no trial has yet been considered for the transplantation of iPSC-derived hepatocyte-like cells.[153] Despite the enormous potential, rapid progress and considerable investments made into the field of iPSC tissue engineering, several (technological) hurdles remain to hinder iPSC cell transplantation to become a clinical reality. First, viral vectors are often used as the most efficient system to deliver reprogramming factors to the recipient genome. This is accompanied with a risk of randomly inserting DNA into the genome making iPSC-derived cells less suitable for transplantation due to the risk of tumorigenicity of undifferentiated cells. These problems, however, potentially be avoided by using non-integrating reprogramming techniques. Kaji and co-workers used a non-viral transfection system which reduces genome modification in iPSC and enables complete elimination of exogenous reprogramming factors.[154] Kim and colleagues generated stable iPSC from human fibroblasts through a DNA vector-free, direct protein transduction system which abolish the weaknesses of viral and other DNA-based reprogramming methods.[155]

Also, it has been proposed that non-intentional aberrations in genomic integrity occur, which has proven to correlate to donor cell type, a method of reprogramming and prolonged culture of cells.[156] This is also confirmed by the teratoma-forming propensity of different iPSC lines indicating that optimization of the used technologies would benefit safety in clinical trials.[157] There are also concerns regarding genetic stability during reprogramming and differentiation. Differentiation protocols for iPSC mostly lead to immature phenotypes, which also turns out to be a problem for the production of iPSC-derived hepatocytes. Cells express immature/fetal markers such as alpha-fetoprotein and only show low expression of mature markers such as CYP450 and other phase II biotransformation-enzymes.[158,159] Moreover, differentiation often leads to heterogeneous cultures having both hepatocyte-like cells, but also stem cell-like impurities, which further questions the suitability of cells for transplantation and requires optimization of currently used differentiation procedures before translation to the clinic will be feasible.[160] In summary, before clinical-grade iPSC and ESC can be produced, robust, safe and efficacious cells need to be generated, which also will be required by FDA and EMA. To demonstrate the clinical feasibility of iPSC and ESC, more preclinical and clinical studies must be conducted.[161]

As these studies are required for every stem cell-line used, autologous transplantation currently remains out of reach (for iPSC). However, for allogeneic transplantation, cautions and critical attitude can help to avoid an early disappointment of the enormous potential of the use of pluripotent stem cells in tissue engineering.[162]

11.7 CONCLUSIONS AND FINAL PERSPECTIVES

This chapter has focused on recent strategies of liver tissue engineering with potential for developing innovative treatments and efficient models for the study of the liver. The knowledge of regenerative biology and medicine has exponentially increased in the last years, giving new hopes to the development of effective treatments for liver diseases. Much work is still ahead to obtain final therapies or models that accurately represent the liver to its fullest. First of all, standardization of cellular removal protocols, cell isolation methods, and medium formulation are needed to achieve a high grade of reproducibility of the results between the distinct laboratories all around the world. Moreover, further work is required to guarantee that decellularized scaffolds are not recognized by the host immune system. Despite the obstacles, some 2D liver culture models, especially co-cultures with hepatocytes and non-parenchymal cells (which better represent the *in vivo* microenvironment of the liver), can be considered useful models to study the acute phase response, mutagenesis, the xenobiotic toxicity, lipid, and drug metabolism in the liver. Currently, liver organoids are also considered a possible tool to assess cell changes that lead to tumorigenesis and cancer progression and can also be useful for drug screening and testing. The improvement and the standardization of the protocols used are needed to enhance the production rate, and to ameliorate the features of the obtained organoids.

One of the biggest challenges still open in whole-organ liver bioengineering is represented by the identification and selection of the most suitable cell sources for the most effective seeding of acellular scaffolds. An optimal recellularization is crucial to produce a clinically functional organ. Nonetheless, complete recellularization using all the necessary liver cell types, including Kupffer, sinusoidal endothelial, and stellate cells, and the generation of a fully functional liver have not been accomplished yet. To date, several efforts have been made to isolate liver endothelial cells or hepatocytes and to maintain them growing in culture in the long term. Nevertheless, all these attempts are denied because these cells lose the ability to proliferate once outside of their natural environment. The iPSC source could be considered a very promising source for liver regeneration. This technology may provide a potential source of cells which could be used for whole-liver bioengineering. However, bioengineering a fully functional organ with a comparable size with the human one has yet to be achieved by using iPSC technology.

Also, to get the optimal appropriate cell numbers to bioengineer a fully functional liver of human size remains a huge issue in the field. Therefore, the development of a scalable and robust platform for the large-scale expansion of various cell types will be strictly essential for the recellularization of whole-liver scaffolds.

There is also a requirement for an optimization of the cell seeding techniques, which play a critical role in the generation of a bioengineered whole-organs. A successful seeding of the decellularized scaffold vascular network is essential to the survival of the bioengineered organ. An optimal cell seeding, oxygen and nutrient supply to cells, as well as the control of biochemical concentrations and gradients in large scaffolds is complex and highly depend on cell metabolism and bioreactor configurations. Therefore, the identification of the most suitable cell perfusion route, as well as the optimization of the perfusion flow rate and the mechanical environment in which the whole-organ scaffold is placed is extremely important.

Regarding the bioreactor systems used in liver tissue engineering, the clinical assessment of the efficacy and safety of BAL systems in the treatment of different end-stage liver diseases is near completion (second half of 2018). Then, the field can focus on reducing the costs, find more suitable cell sources and optimize bioreactor technologies, to ensure feasibility, with a special regard when considering potential availabilities within government health systems.

Finally, a major challenge still exists about the creation and the development of large animal models for bioengineered organ transplantation. This will help enormously in standardization of the implantation techniques, limiting eventual side effects. Moreover, long-term follow-up studies will be required to translate the usage of bioengineered organs into clinical practice, with the final objective to obtain organs for human clinical transplantation.

Altogether, the strategies for liver tissue engineering showed in this chapter may have an impact on innovative personalized tissue engineering-based treatments as alternative and effective therapies for different liver diseases or pathologies.

ACKNOWLEDGMENTS

This work was supported by Instituto de Salud Carlos III, through a predoctoral fellowship i-PFIS IFI15/00158 (I. P-P), Gobierno de Aragón and Fondo Social Europeo through a predoctoral fellowship DGA C066/2014 (P. S-A), and by two predoctoral fellowships from Fundação para Ciência e Tecnologia, Portugal (PD/BD/114057/2015 and SFRH/BD/116780/2016) (S.M. and J.I. A., respectively). N. S-R was supported by a POCTEFA/RefBio II research grant and FGJ Gobierno de Aragón. P.M.B. was supported by an H2020-MSCA-IF-2014 from the European Research Agency and by the PI15/00563 Research Project from Instituto de Salud Carlos III, Madrid, Spain.

REFERENCES

1. Mann, R.E. et al. The epidemiology of alcoholic liver disease. *Alcohol Res Health* **27**, 209–219 (2003).
2. Bosch, F.X. et al. Primary liver cancer: Worldwide incidence and trends. *Gastroenterology* **127**, S5–S16 (2004).
3. LeCluyse, E.L. Human hepatocyte culture systems for the in vitro evaluation of cytochrome P450 expression and regulation. *Eur J Pharm Sci* **13**, 343–368 (2001).
4. Sivaraman, A. et al. A microscale in vitro physiological model of the liver: Predictive screens for drug metabolism and enzyme induction. *Curr Drug Metab* **6**, 569–591 (2005).

5. LeCluyse, E.L. et al. Strategies for restoration and maintenance of normal hepatic structure and function in long-term cultures of rat hepatocytes. *Adv Drug Deliver Rev* **22**, 133–186 (1996).

6. Nelson, K.F. et al. Long-term maintenance and induction of cytochrome P-450 in primary cultures of rat hepatocytes. *Biochem Pharmacol* **31**, 2211–2214 (1982).

7. Guillouzo, A. Liver cell models in in vitro toxicology. *Environ Health Perspect* **106 Suppl 2**, 511–532 (1998).

8. Miyazaki, M. et al. Long-term survival of functional hepatocytes from adult rat in the presence of phenobarbital in primary culture. *Exp Cell Res* **159**, 176–190 (1985).

9. Kidambi, S. et al. Oxygen-mediated enhancement of primary hepatocyte metabolism, functional polarization, gene expression, and drug clearance. *Proc Nat Acad Sci USA* **106**, 15714–15719 (2009).

10. LeCluyse, E.L. et al. Formation of extensive canalicular networks by rat hepatocytes cultured in collagen-sandwich configuration. *Am J Physiol* **266**, C1764–C1774 (1994).

11. Richert, L. et al. Evaluation of the effect of culture configuration on morphology, survival time, antioxidant status and metabolic capacities of cultured rat hepatocytes. *Toxicol In Vitro* **16**, 89–99 (2002).

12. Li, J. et al. Establishment of highly differentiated immortalized human hepatocyte line with simian virus 40 large tumor antigen for liver based cell therapy. *ASAIO J* **51**, 262–268 (2005).

13. Werner, A. et al. Cultivation and characterization of a new immortalized human hepatocyte cell line, HepZ, for use in an artificial liver support system. *Ann Ny Acad Sci* **875**, 364–368 (1999).

14. Guguen-Guillouzo, C. et al. Stem cell-derived hepatocytes and their use in toxicology. *Toxicology* **270**, 3–9 (2010).

15. Guguen-Guillouzo, C. & Guillouzo, A. General review on in vitro hepatocyte models and their applications. *Methods Mol Biol* **640**, 1–40 (2010).

16. Aninat, C. et al. Expression of cytochromes P450, conjugating enzymes and nuclear receptors in human hepatoma HepaRG cells. *Drug Metab Dispos* **34**, 75–83 (2006).

17. Marion, M.J., Hantz, O. & Durantel, D. The HepaRG cell line: Biological properties and relevance as a tool for cell biology, drug metabolism, and virology studies. *Methods Mol Biol* **640**, 261–272 (2010).

18. Bhatia, S.N. et al. Effect of cell-cell interactions in preservation of cellular phenotype: Cocultivation of hepatocytes and nonparenchymal cells. *FASEB J* **13**, 1883–1900 (1999).

19. Nguyen, T.V. et al. Establishment of a hepatocyte-kupffer cell coculture model for assessment of proinflammatory cytokine effects on metabolizing enzymes and drug transporters. *Drug Metab Dispos* **43**, 774–785 (2015).

20. Salerno, S. et al. Human hepatocytes and endothelial cells in organotypic membrane systems. *Biomaterials* **32**, 8848–8859 (2011).

21. Mahieu-Caputo, D. et al. Repopulation of athymic mouse liver by cryopreserved early human fetal hepatoblasts. *Hum Gene Ther* **15**, 1219–1228 (2004).

22. Diekmann, S. et al. Present and future developments in hepatic tissue engineering for liver support systems–State of the art and future developments of hepatic cell culture techniques for the use in liver support systems. *Cytotechnology* **50**, 163–179 (2006).

23. Cardinale, V. et al. Transplantation of human fetal biliary tree stem/progenitor cells into two patients with advanced liver cirrhosis. *BMC Gastroenterol* **14**, 204 (2014).

24. Hamazaki, T. et al. Hepatic maturation in differentiating embryonic stem cells in vitro. *FEBS Lett* **497**, 15–19 (2001).

25. Schwartz, R.E. et al. Defined conditions for development of functional hepatic cells from human embryonic stem cells. *Stem Cells Dev* **14**, 643–655 (2005).
26. Guillouzo, A. & Guguen-Guillouzo, C. Evolving concepts in liver tissue modeling and implications for in vitro toxicology. *Expert Opin Drug Metab Toxicol* **4**, 1279–1294 (2008).
27. Sullivan, G.J. et al. Generation of functional human hepatic endoderm from human induced pluripotent stem cells. *Hepatology* **51**, 329–335 (2010).
28. Liu, H. et al. Generation of endoderm-derived human induced pluripotent stem cells from primary hepatocytes. *Hepatology* **51**, 1810–1819 (2010).
29. Soldatow, V.Y. et al. In vitro models for liver toxicity testing. *Toxicol Res (Camb)* **2**, 23–39 (2013).
30. Palakkan, A.A. et al. Liver tissue engineering and cell sources: Issues and challenges. *Liver Int* **33**, 666–676 (2013).
31. Huang, P. et al. Induction of functional hepatocyte-like cells from mouse fibroblasts by defined factors. *Nature* **475**, 386–389 (2011).
32. Sekiya, S. & Suzuki, A. Direct conversion of mouse fibroblasts to hepatocyte-like cells by defined factors. *Nature* **475**, 390–393 (2011).
33. Mason, C. & Dunnill, P. A brief definition of regenerative medicine. *Regen Med* **3**, 1–5 (2008).
34. Inamori, M. et al. An approach for formation of vascularized liver tissue by endothelial cell-covered hepatocyte spheroid integration. *Tissue Eng Part A* **15**, 2029–2037 (2009).
35. Abu-Absi, S.F. et al. Three-dimensional co-culture of hepatocytes and stellate cells. *Cytotechnology* **45**, 125–140 (2004).
36. Lin, J. et al. Use an alginate scaffold-bone marrow stromal cell (BMSC) complex for the treatment of acute liver failure in rats. *Int J Clin Exp Med* **8**, 12593–12600 (2015).
37. Shteyer, E. et al. Reduced liver cell death using an alginate scaffold bandage: A novel approach for liver reconstruction after extended partial hepatectomy. *Acta Biomat* **10**, 3209–3216 (2014).
38. Shang, Y. et al. Hybrid sponge comprised of galactosylated chitosan and hyaluronic acid mediates the co-culture of hepatocytes and endothelial cells. *J Biosci Bioeng* **117**, 99–106 (2014).
39. Rad, A.T. et al. Conducting scaffolds for liver tissue engineering. *J Biomed Mater Res A* **102**, 4169–4181 (2014).
40. Glowacki, J. & Mizuno, S. Collagen scaffolds for tissue engineering. *Biopolymers* **89**, 338–344 (2008).
41. Chevallay, B. & Herbage, D. Collagen-based biomaterials as 3D scaffold for cell cultures: Applications for tissue engineering and gene therapy. *Med Biol Eng Comp* **38**, 211–218 (2000).
42. Carpentier, B. et al. Artificial and bioartificial liver devices: Present and future. *Gut* **58**, 1690–1702 (2009).
43. Phua, J. & Lee, K.H. Liver support devices. *Curr Opin Crit Care* **14**, 208–215 (2008).
44. Banares, R. et al. Liver support systems: Will they ever reach prime time? *Curr Gastroenterol Rep* **15**, 312.
45. Rifai, K. Extracorporeal albumin dialysis. *Hepatol Res* **38**, S41–S45 (2008).
46. Brophy, C.M. & Nyberg, S.L. Extracorporeal treatment of acute liver failure. *Hepatol Res* **38**, S34–S40 (2008).
47. Pless, G. Artificial and bioartificial liver support. *Organogenesis* **3**, 20–24 (2007).
48. Park, J.K. & Lee, D.H. Bioartificial liver systems: Current status and future perspective. *J Biosci Bioeng* **99**, 311–319 (2005).
49. Cao, S. et al. New approaches to supporting the failing liver. *Annu Rev Med* **49**, 85–94 (1998).

50. McKenzie, T.J. et al. Artificial and bioartificial liver support. *Semin Liver Dis* **28**, 210–217 (2008).
51. Adham, M. Extracorporeal liver support: Waiting for the deciding vote. *ASAIO J* **49**, 621–632 (2003).
52. Assess Safety and Efficacy of ELAD (Extracorporeal Liver Assist System) in Subjects With Alcohol-Induced Liver Failure. (2011).
53. Hep-Art, Bio-artificial Liver Support.
54. Murphy, S.V. & Atala, A. 3D bioprinting of tissues and organs. *Nat Biotechnol* **32**, 773–785 (2014).
55. Klein, G.T. et al. 3D printing and neurosurgery–ready for prime time? *World Neurosurg* **80**, 233–235 (2013).
56. Cui, X. et al. Thermal inkjet printing in tissue engineering and regenerative medicine. *Recent Pat Drug Deliv Formul* **6**, 149–155 (2012).
57. Zhang, Y.S. et al. 3D Bioprinting for tissue and organ fabrication. *Ann Biomed Eng* **45**, 148–163 (2017).
58. Bhise, N.S. et al. A liver-on-a-chip platform with bioprinted hepatic spheroids. *Biofabrication* **8**, 014101 (2016).
59. Banks, J. Adding value in additive manufacturing: Researchers in the United Kingdom and Europe look to 3D printing for customization. *IEEE Pulse* **4**, 22–26 (2013).
60. Snyder, J.E. et al. Bioprinting cell-laden matrigel for radioprotection study of liver by pro-drug conversion in a dual-tissue microfluidic chip. *Biofabrication* **3**, 034112 (2011).
61. Chang, R. et al. Biofabrication of a three-dimensional liver micro-organ as an in vitro drug metabolism model. *Biofabrication* **2**, 045004 (2010).
62. Organovo Introduces 3D Bioprinted Human Liver as Leading Therapeutic Tissue in Preclinical Development. (2016).
63. Huch, M. et al. Long-term culture of genome-stable bipotent stem cells from adult human liver. *Cell* (2014).
64. Huch, M. et al. In vitro expansion of single Lgr5+ liver stem cells induced by Wnt-driven regeneration. *Nature* **494**, 247–250 (2013).
65. Godoy, P. et al. Recent advances in 2D and 3D in vitro systems using primary hepato-cytes, alternative hepatocyte sources and non-parenchymal liver cells and their use in investigating mechanisms of hepatotoxicity, cell signaling and ADME. *Arch Toxicol* **87**, 1315–1530 (2013).
66. Lin, R.Z. & Chang, H.Y. Recent advances in three-dimensional multicellular spheroid culture for biomedical research. *Biotechnol J* **3**, 1172–1184 (2008).
67. Glicklis, R. et al. Modeling mass transfer in hepatocyte spheroids via cell viability, spheroid size, and hepatocellular functions. *Biotechnol Bioeng* **86**, 672–680 (2004).
68. Landry, J. et al. Spheroidal aggregate culture of rat liver cells: Histotypic reorganiza-tion, biomatrix deposition, and maintenance of functional activities. *J Cell Biol* **101**, 914–923 (1985).
69. Koide, N. et al. Formation of multicellular spheroids composed of adult rat hepatocytes in dishes with positively charged surfaces and under other nonadherent environments. *Exp Cell Res* **186**, 227–235 (1990).
70. Sakai, Y., Naruse, K., Nagashima, I., Muto, T. & Suzuki, M. Large-scale preparation and function of porcine hepatocyte spheroids. *Int J Artif Organs* **19**, 294–301 (1996).
71. Nyberg, S.L. et al. Rapid, large-scale formation of porcine hepatocyte spheroids in a novel spheroid reservoir bioartificial liver. *Liver Transpl* **11**, 901–910 (2005).
72. Brophy, C.M. et al. Rat hepatocyte spheroids formed by rocked technique maintain dif-ferentiated hepatocyte gene expression and function. *Hepatology* **49**, 578–586 (2009).
73. Kelm, J.M., Timmins, N.E., Brown, C.J., Fussenegger, M. & Nielsen, L.K. Method for generation of homogeneous multicellular tumor spheroids applicable to a wide variety of cell types. *Biotechnol Bioeng* **83**, 173–180 (2003).

74. Saito, M. et al. Reconstruction of liver organoid using a bioreactor. *World J Gastroenterol* **12**, 1881–1888 (2006).
75. Dedhia, P.H., Bertaux-Skeirik, N., Zavros, Y. & Spence, J.R. Organoid Models of human gastrointestinal development and disease. *Gastroenterology* **150**, 1098–1112 (2016).
76. Chan, T.S. et al. Meeting the challenge of predicting hepatic clearance of compounds slowly metabolized by cytochrome P450 using a novel hepatocyte model, HepatoPac. *Drug Metabol Disposition: The Biol Fate Chem* **41**, 2024–2032 (2013).
77. Schütte, J. et al. A method for patterned in situ biofunctionalization in injection-molded microfluidic devices. *Lab Chip* **10**, 2551–2558 (2010).
78. Baxter, G.T. Hurel—An in vivo-surrogate assay platform for cell-based studies. *Alter Lab Animals: ATLA* **37 Suppl 1**, 11–18 (2009).
79. Baptista, P.M. et al. The use of whole organ decellularization for the generation of a vascularized liver organoid. *Hepatology* **53**, 604–617 (2011).
80. Ott, H.C. et al. Perfusion-decellularized matrix: Using nature's platform to engineer a bioartificial heart. *Nat Med* **14**, 213–221 (2008).
81. Kim, B.S. et al. Biomaterials for tissue engineering. *World J Urol* **18**, 2–9 (2000).
82. Fukumitsu, K. et al. Bioengineering in organ transplantation: Targeting the liver. *Transplant Proc* **43**, 2137–2138 (2011).
83. Brown, A.L. et al. 22 week assessment of bladder acellular matrix as a bladder augmentation material in a porcine model. *Biomaterials* **23**, 2179–2190 (2002).
84. Xu, C.C. et al. A biodegradable, acellular xenogeneic scaffold for regeneration of the vocal fold lamina propria. *Tissue Eng* **13**, 551–566 (2007).
85. Reddy, P.P. et al. Regeneration of functional bladder substitutes using large segment acellular matrix allografts in a porcine model. *J Urol* **164**, 936–941 (2000).
86. Lee, R.C. & Kolodney, M.S. Electrical injury mechanisms: Electrical breakdown of cell membranes. *Plast Reconstr Surg* **80**, 672–679 (1987).
87. Phillips, M. et al. Nonthermal irreversible electroporation for tissue decellularization. *J Biomech Eng* **132**, 091003 (2010).
88. Parmaksiz, M. et al. Clinical applications of decellularized extracellular matrices for tissue engineering and regenerative medicine. *Biomed Mat* **11**, 022003 (2016).
89. Uygun, B.E. et al. Organ reengineering through development of a transplantable recellularized liver graft using decellularized liver matrix. *Nat Med* **16**, 814–820 (2010).
90. De Kock, J. et al. Simple and quick method for whole-liver decellularization: A novel in vitro three-dimensional bioengineering tool? *Arch Toxicol* **85**, 607–612 (2011).
91. Faulk, D.M. et al. Decellularization and cell seeding of whole liver biologic scaffolds composed of extracellular matrix. *J Clin Exp Hepatol* **5**, 69–80 (2015).
92. Soto-Gutierrez, A. et al. A whole-organ regenerative medicine approach for liver replacement. *Tissue Eng Part C Meth.* **17**, 677–686 (2011).
93. Zhang, L. et al. Research progress in liver tissue engineering. *Biomed Mater Eng* **28**, S113–S119 (2017).
94. Li, C. et al. Chemically crosslinked alginate porous microcarriers modified with bioactive molecule for expansion of human hepatocellular carcinoma cells. *J Biomed Mater Res B Appl Biomater* **102**, 1648–1658 (2014).
95. Watanabe, F.D. et al. Clinical experience with a bioartificial liver in the treatment of severe liver failure. A phase I clinical trial. *Ann Surg* **225**, 484–491; discussion 491–484 (1997).
96. Gan, J.H. et al. Hybrid artificial liver support system for treatment of severe liver failure. *World J Gastroenterol* **11**, 890–894 (2005).
97. Catapano, G. et al. Transport advances in disposable bioreactors for liver tissue engineering. *Adv Biochem Eng Biotechnol* **115**, 117–143 (2009).

98. Powers, M.J. et al. Cell-substratum adhesion strength as a determinant of hepatocyte aggregate morphology. *Biotechnol Bioeng* **53**, 415–426 (1997).
99. Lillegard, J.B. et al. Normal atmospheric oxygen tension and the use of antioxidants improve hepatocyte spheroid viability and function. *J Cell Physiol* **226**, 2987–2996 (2011).
100. Ebrahimkhani, M.R. et al. Bioreactor technologies to support liver function in vitro. *Adv Drug Deliv Rev* **69–70**, 132–157 (2014).
101. Yin, C.H. et al. Production of mouse embryoid bodies with hepatic differentiation potential by stirred tank bioreactor. *Biosci Biotechnol Biochem* **71**, 728–734 (2007).
102. Wu, F.J. et al. Efficient assembly of rat hepatocyte spheroids for tissue engineering applications. *Biotechnol Bioeng* **50**, 404–415 (1996).
103. Nelson, L.J. et al. Low-shear modelled microgravity environment maintains morphology and differentiated functionality of primary porcine hepatocyte cultures. *Cells Tissues Organs* **192**, 125–140 (2010).
104. Tostoes, R.M. et al. Human liver cell spheroids in extended perfusion bioreactor culture for repeated-dose drug testing. *Hepatology* **55**, 1227–1236 (2012).
105. Lemasters, J.J. Hypoxic, ischemic, and reperfusion injury to liver. in *The Liver: Biology and Pathobiology* (eds. Arias, I.M., Boyer, J. M., Chisani, F. V., Fausto, N., Schachter, D., Shafritz, D. A.), pp. 257–279 (Lippincott Williams & Wilkins, Philadeplhia, PA, 2001).
106. Balis, U.J. et al. Oxygen consumption characteristics of porcine hepatocytes. *Metab Eng* **1**, 49–62 (1999).
107. Foy, B.D. et al. A device to measure the oxygen uptake rate of attached cells: Importance in bioartificial organ design. *Cell Transplant* **3**, 515–527 (1994).
108. Yarmush, M.L. et al. Hepatic tissue engineering. Development of critical technologies. *Ann N Y Acad Sci* **665**, 238–252 (1992).
109. Stevens, K.M. Oxygen requirements for liver cells in vitro. *Nature* **206**, 199 (1965).
110. Fariss, M.W. Oxygen toxicity: Unique cytoprotective properties of vitamin E succinate in hepatocytes. *Free Radic Biol Med* **9**, 333–343 (1990).
111. Martin, H. et al. Morphological and biochemical integrity of human liver slices in long-term culture: Effects of oxygen tension. *Cell Biol Toxicol* **18**, 73–85 (2002).
112. Lowe, K.C. Engineering blood: Synthetic substitutes from fluorinated compounds. *Tissue Eng* **9**, 389–399 (2003).
113. Riess, J.G. Oxygen carriers ("blood substitutes")—Raison d'Etre, chemistry, and some physiology. *Chem Rev* **101**, 2797–2919 (2001).
114. Lowe, K.C. et al. Perfluorochemicals: Their applications and benefits to cell culture. *Trends Biotechnol* **16**, 272–277 (1998).
115. Nahmias, Y. et al. A novel formulation of oxygen-carrying matrix enhances liver-specific function of cultured hepatocytes. *FASEB J* **20**, 2531–2533 (2006).
116. Radisic, M. et al. Biomimetic approach to cardiac tissue engineering: Oxygen carriers and channeled scaffolds. *Tissue Eng* **12**, 2077–2091 (2006).
117. Sakai, Y. et al. Toward engineering of vascularized three-dimensional liver tissue equivalents possessing a clinically significant mass. *Biochem Eng J* **48**, 348–361 (2010).
118. Dijkhuizen, P. et al. The oxygen binding capacity of human haemoglobin. Hufner's factor redetermined. *Pflugers Arch* **369**, 223–231 (1977).
119. Takeoka, S. et al. Effect of Hb-encapsulation with vesicles on H_2O_2 reaction and lipid peroxidation. *Bioconjug Chem* **13**, 1302–1308 (2002).
120. Takahashi, A. Characterization of neo red-cells (Nrcs), their function and safety *in-vivo* tests. *Artif Cell Blood Sub* **23**, 347–354 (1995).
121. Sakai, H. et al. Haemoglobin-vesicles as artificial oxygen carriers: Present situation and future visions. *J Intern Med* **263**, 4–15 (2008).

122. Naruto, H. et al. Feasibility of direct oxygenation of primary-cultured rat hepatocytes using polyethylene glycol-decorated liposome-encapsulated hemoglobin (LEH). *J Biosci Bioeng* **104**, 343–346 (2007).
123. Montagne, K. et al. Use of liposome encapsulated hemoglobin as an oxygen carrier for fetal and adult rat liver cell culture. *J Biosci Bioeng* **112**, 485–490 (2011).
124. Mao, S. et al. Imitation of drug metabolism in human liver and cytotoxicity assay using a microfluidic device coupled to mass spectrometric detection. *Lab on a chip* **12**, 219–226 (2011).
125. FDA, U.S. Regulation of Human Cells, Tissues, and Cellular and Tissue-Based Products (HCT/Ps). 21 CFR, Parts 1270 and 1271.
126. Food and Drug Administration, HHS. Definition of primary mode of action of a combinant product. Final rule. *Fed Regist* **70**(164), 49848–49862 (August 25, 2005).
127. Clerico, A. et al. Thirty years of the heart as an endocrine organ: Physiological role and clinical utility of cardiac natriuretic hormones. *Am J Physiol Heart Circ Physiol* **301**, H12–H20 (2011).
128. Stastna, M. et al. Identification and functionality of proteomes secreted by rat cardiac stem cells and neonatal cardiomyocytes. *Proteomics* **10**, 245–253 (2010).
129. Shah, A. et al. Oxidative stress augments the secretion of atrial natriuretic peptide in isolated rat atria. *Peptides* **32**, 1172–1178 (2011).
130. Yannas, I.V. et al. Synthesis and characterization of a model extracellular matrix that induces partial regeneration of adult mammalian skin. *Proc Natl Acad Sci USA* **86**, 933–937 (1989).
131. Atala, A. et al. Tissue-engineered autologous bladders for patients needing cystoplasty. *Lancet* **367**, 1241–1246 (2006).
132. Warnke, P.H. et al. Growth and transplantation of a custom vascularised bone graft in a man. *Lancet* **364**, 766–770 (2004).
133. van der Veen, V.C. et al. Biological background of dermal substitutes. *Burns: J Int Soc Burn Inj* **36**, 305–321 (2010).
134. Bottcher-Haberzeth, S. et al. Tissue engineering of skin. *Burns: J Int Soc Burn Inj* **36**, 450–460 (2010).
135. Baptista, P.M. et al. Whole organ decellularization—A tool for bioscaffold fabrication and organ bioengineering. *Conference proceedings:... Annual International Conference of the IEEE Engineering in Medicine and Biology Society. IEEE Engineering in Medicine and Biology Society. Annual Conference* **2009**, 6526–6529 (2009).
136. Bayrak, A. et al. Human immune responses to porcine xenogeneic matrices and their extracellular matrix constituents in vitro. *Biomaterials* **31**, 3793–3803 (2010).
137. Bastian, F. et al. IgG deposition and activation of the classical complement pathway involvement in the activation of human granulocytes by decellularized porcine heart valve tissue. *Biomaterials* **29**, 1824–1832 (2008).
138. Gardenbroek, T.J. et al. Erratum to: The ACCURE-trial: The effect of appendectomy on the clinical course of ulcerative colitis, a randomised international multicenter trial (NTR2883) and the ACCURE-UK trial: A randomised external pilot trial (ISRCTN56523019). *BMC Surg* **16**, 1 (2016).
139. Berry, M.N. and Friend, D.S. High-yield preparation of isolated rat liver parenchymal cells: A biochemical and fine structural study. *J Cell Biol* **43**, 506–520 (1969).
140. Grazal, J.S. and Earl, D.S. EU and FDA GMP regulations: Overview and comparison. *Qual Assur J* **2**, 55–60 (1997).
141. Report from the EMA-FDA QbD pilot program. (European Medicines Agency, EMA/213746/2017, 19/04/2017). http://www.ema.europa.eu/docs/en_GB/document_library/Other/2017/04/WC500225533.pdf.
142. Hoey, D.A. et al. The mechanics of the primary cilium: An intricate structure with complex function. *J Biomech* **45**, 17–26 (2012).

143. Lo, B. and Parham, L. Ethical issues in stem cell research. *Endocr Rev* **30**, 204–213 (2009).

144. Guidelines for human embryonic stem cell research/Board on Life Sciences, National Research Council, Board on Health Sciences Policy, Institute of Medicine. (500 Fifth Street, NW Washington, DC 200001, 2005).

145. Lomax, G.P. et al. Responsible oversight of human stem cell research: The California Institute for Regenerative Medicine's medical and ethical standards. *PLoS Med* **4**, e114 (2007).

146. Carson, S.A. et al. Proposed oocyte donation guidelines for stem cell research. *Fertil Steril* **94**, 2503–2506 (2010).

147. Holden, C. Korean stem cell scandal. Schatten: Pitt panel finds "misbehavior" but not misconduct. *Science* **311**, 928 (2006).

148. Chong, S. and Normile, D. Stem cells. How young Korean researchers helped unearth a scandal. *Science* **311**, 22–25 (2006).

149. Bavister, B.D. et al. Challenges of primate embryonic stem cell research. *Cloning Stem Cells* **7**, 82–94 (2005).

150. Takahashi, K. and Yamanaka, S. Induction of pluripotent stem cells from mouse embryonic and adult fibroblast cultures by defined factors. *Cell* **126**, 663–676 (2006).

151. Sayed, N. et al. Translation of human-induced pluripotent stem cells: From clinical trial in a dish to precision medicine. *J Am Col Cardiol* **67**, 2161–2176 (2016).

152. Cramer, A.O. and MacLaren, R.E. Translating induced pluripotent stem cells from bench to bedside: Application to retinal diseases. *Cur Gene Therapy* **13**, 139–151 (2013).

153. Trounson, A. and DeWitt, N.D. Pluripotent stem cells progressing to the clinic. *Nat Rev Mol Cell Biol* **17**, 194–200 (2016).

154. Kaji, K. et al. Virus-free induction of pluripotency and subsequent excision of reprogramming factors. *Nature* **458**, 771–775 (2009).

155. Kim, D. et al. Generation of human induced pluripotent stem cells by direct delivery of reprogramming proteins. *Cell Stem Cell* **4**, 472–476 (2009).

156. Tapia, N. and Scholer, H.R. Molecular obstacles to clinical translation of iPSCs. *Cell Stem Cell* **19**, 298–309 (2016).

157. Miura, K. et al. Variation in the safety of induced pluripotent stem cell lines. *Nat Biotechnol* **27**, 743–745 (2009).

158. Schwartz, R.E. et al. Pluripotent stem cell-derived hepatocyte-like cells. *Biotechnol Adv* **32**, 504–513 (2014).

159. Gomez-Lechon, M.J. and Tolosa, L. Human hepatocytes derived from pluripotent stem cells: A promising cell model for drug hepatotoxicity screening. *Arch Toxicol* **90**, 2049–2061 (2016).

160. Serra, M. et al. Process engineering of human pluripotent stem cells for clinical application. *Trends Biotechnol* **30**, 350–359 (2012).

161. Neofytou, E. et al. Hurdles to clinical translation of human induced pluripotent stem cells. *J Clin Invest* **125**, 2551–2557 (2015).

162. ISSCR. Guidelines for Stem Cell Science and Clinical Translation. http://www.isscr.org/home/publications/2015-guidelines-draft (2015).

Index

Note: Page numbers in italic and bold refer to figures and tables respectively.